Unmanned Aerial Vehicles in Civilian Logistics and Supply Chain Management

Tarryn Kille
University of Southern Queensland, Australia

Paul R. Bates
University of Southern Queensland, Australia

Seung Yong Lee
University of Southern Queensland, Australia

A volume in the Advances in Logistics, Operations, and Management Science (ALOMS) Book Series

IGI Global
DISSEMINATOR OF KNOWLEDGE

Published in the United States of America by
IGI Global
Business Science Reference (an imprint of IGI Global)
701 E. Chocolate Avenue
Hershey PA, USA 17033
Tel: 717-533-8845
Fax: 717-533-8661
E-mail: cust@igi-global.com
Web site: http://www.igi-global.com

Library of Congress Cataloging-in-Publication Data

Names: Kille, Tarryn, 1979- editor. | Bates, Paul R., 1959- editor. | Lee, Seung Yong, 1971- editor.
Title: Unmanned aerial vehicles in civilian logistics and supply chain management / Tarryn Kille, Paul R. Bates, and Seung Yong Lee, editors.
Description: Hershey PA : Business Science Reference, [2019] | Includes bibliographical references.
Identifiers: LCCN 2018037397| ISBN 9781522579007 (hardcover) | ISBN 9781522579014 (ebook)
Subjects: LCSH: Business logistics--Technological innovations. | Drone aircraft--Economic aspects.
Classification: LCC HD38.5 .U554 2019 | DDC 658.7028/4--dc23 LC record available at https://lccn.loc.gov/2018037397

This book is published in the IGI Global book series Advances in Logistics, Operations, and Management Science (ALOMS) (ISSN: 2327-350X; eISSN: 2327-3518)

British Cataloguing in Publication Data
A Cataloguing in Publication record for this book is available from the British Library.

All work contributed to this book is new, previously-unpublished material.
The views expressed in this book are those of the authors, but not necessarily of the publisher.

For electronic access to this publication, please contact: eresources@igi-global.com.

Advances in Logistics, Operations, and Management Science (ALOMS) Book Series

ISSN:2327-350X
EISSN:2327-3518

Editor-in-Chief: John Wang, Montclair State University, USA

MISSION

Operations research and management science continue to influence business processes, administration, and management information systems, particularly in covering the application methods for decision-making processes. New case studies and applications on management science, operations management, social sciences, and other behavioral sciences have been incorporated into business and organizations real-world objectives.

The **Advances in Logistics, Operations, and Management Science** (ALOMS) Book Series provides a collection of reference publications on the current trends, applications, theories, and practices in the management science field. Providing relevant and current research, this series and its individual publications would be useful for academics, researchers, scholars, and practitioners interested in improving decision making models and business functions.

COVERAGE

- Computing and information technologies
- Political science
- Operations Management
- Marketing engineering
- Decision analysis and decision support
- Networks
- Information Management
- Finance
- Production management
- Organizational Behavior

The Advances in Logistics, Operations, and Management Science (ALOMS) Book Series (ISSN 2327-350X) is published by IGI Global, 701 E. Chocolate Avenue, Hershey, PA 17033-1240, USA, www.igi-global.com. This series is composed of titles available for purchase individually; each title is edited to be contextually exclusive from any other title within the series. For pricing and ordering information please visit http://www.igi-global.com/book-series/advances-logistics-operations-management-science/37170. Postmaster: Send all address changes to above address.

Titles in this Series

For a list of additional titles in this series, please visit:
https://www.igi-global.com/book-series/advances-logistics-operations-management-science/37170

Global Supply Chains and Multimodal Logistics Emerging Research and Opportunities
Deepankar Sinha (Indian Institute of Foreign Trade, India)
Business Science Reference • © 2019 • 220pp • H/C (ISBN: 9781522582984) • US $165.00

Agile Approaches for Successfully Managing and Executing Projects...
Hür Bersam Bolat (Istanbul Technical University, Turkey) and Gül Tekin Temur (Bahçeşehir University, Turkey)
Business Science Reference • © 2019 • 424pp • H/C (ISBN: 9781522578659) • US $235.00

Technology Optimization and Change Management for Successful Digital Supply Chains
Ehap Sabri (KPMG LLP, USA & University of Texas at Dallas, USA)
Business Science Reference • © 2019 • 323pp • H/C (ISBN: 9781522577003) • US $210.00

Strategy and Superior Performance of Micro and Small Businesses in Volatile Economies
João Conrado de Amorim Carvalho (Unidade de Ensino Superior Dom Bosco, Brazil) and Emmanuel M.C.B. Sabino (FORUM (Centro de Formação, Estudos e Pesquisas), Brazil)
Business Science Reference • © 2019 • 389pp • H/C (ISBN: 9781522578888) • US $225.00

Logistics and Transport Modeling in Urban Goods Movement
Jesus Gonzalez-Feliu (Ecole des Mines de Saint-Etienne, France)
Business Science Reference • © 2019 • 273pp • H/C (ISBN: 9781522582922) • US $185.00

Hierarchical Planning and Information Sharing Techniques in Supply Chain Management
Atour Taghipour (Normandy University, France)
Business Science Reference • © 2019 • 287pp • H/C (ISBN: 9781522572992) • US $185.00

Global Perspectives on Human Capital-Intensive Firms
Cécile Cézanne (Université Côte d'Azur, France) and Laurence Saglietto (Université Côte d'Azur, France)
Business Science Reference • © 2019 • 267pp • H/C (ISBN: 9781522574262) • US $185.00

For an entire list of titles in this series, please visit:
https://www.igi-global.com/book-series/advances-logistics-operations-management-science/37170

IGI Global
DISSEMINATOR OF KNOWLEDGE

701 East Chocolate Avenue, Hershey, PA 17033, USA
Tel: 717-533-8845 x100 • Fax: 717-533-8661
E-Mail: cust@igi-global.com • www.igi-global.com

Editorial Advisory Board

Table of Contents

Detailed Table of Contents

The drone industry is rapidly developing around the world, and the numbers of drones are increasing. In order to maintain safety and secure stability of drone flights, regulations and laws related to drone operations are established in each country. This chapter reviews the rules and laws of drones established by the International Civil Aviation Organization, the United States, China, Japan, Australia, India, and Korea. In order to protect victims and develop the drone industry, the author proposes that it is necessary and desirable for the legislation of a unified and global "Draft Convention for the Unification of Certain Rules Relating to Drone Operations and Transport."

In this chapter, the authors present a problem of a performance of unmanned aerial vehicles (UAVs) group flights for a solution of different tasks using criteria of efficiency (safety, regularity, efficiency, economy) and criteria of reliability (connectivity, structural redundancy, survivability, and compactness of connections; the relative distance between UAVs; centrality and periphery of UAVs in the group; the level of system centralizing; etc.). It used graph theory for quantitative estimation of effectiveness of UAVs group flight. It presented all types of UAVs connections in the group (a star, ring, tree, with a common tire, mixed, cellular, etc.). The algorithm for finding central drone repeater (CDR) in a group of the UAVs for sending a control signal to other UAVs in the group was obtained. Examples of for determining the central drone and of the optimal topology in a group of the UAVs in flight are presented.

The notion of using drones for commercial purposes has evolved in the past 5 years from the initial "boom" of excitement around this, somewhat of a novelty and curiosity, to more calculated and sophisticated use of unmanned aircraft systems (UAS), or drones. In the hands of true professionals, drones can offer highly efficient

Preface

Previously limited to military operations, Unmanned Aerial Vehicles (UAVs) appear to be gaining much traction and adoption in the commercial sector (Anand, 2007). One such application is to utilize these aircraft, also referred to as 'drones', for delivery and cargo monitoring activities in logistics and supply chain management (Brar, Rabbat, Raithatha, Runcie, & Yu, 2015). Global media attention has recently illuminated the opportunities and challenges posed by UAVs (Sheets, Rowling, & Jones, 2015). They have sparked international public interest for their potential impact in logistics (Hail & Coyne, 2014). For example, the government of the United Arab Emirates, in 2014, announced their plan to use drones to distribute official government documents (Dorling, Heinrichs, Messier, & Magierowski, 2017) . In Africa, drones are being used to assist in aid work (Sandvik, 2015). A pilot study initiated by the Stony Brook University's Global Health Institute and Vayu, a drone start-up, successfully delivered medicine and laboratory samples to remote places in Madagascar in 2016 (Wurbel, 2017).

As an emerging technology, UAVs have been trialed in logistics operations by a number of high profile organizations. Operators of postal services such as Australia Post (Cuthbertson, 2016), Singapore Post and Swiss Post (Mileham, 2016) as well as online retailers such as Amazon (Burzichelli, 2016) and Rakuten (Aimi, 2016) have all reported testing of UAVs in logistics. Google and Wal-Mart have also tested the use of UAVs (Weissbach & Tebbe, 2016). At the same time, global logistics giants have declared that they are studying the use of UAVs. DHL reported on a successful trial with the use of 'parcelcopter' within a Bavarian community (Erceg, Erceg, & Vasilj, 2017). Previously, DHL had trialed the use of UAV delivery to the German island of Juist (Agatz, Bouman, & Schmidt, 2015). Now, the success of the combined trials allows delivery services to remote areas that often take additional time for road transport to reach. For this reason, the 'parcelcopter' is being integrated within DHL's supply chain (Agatz et al., 2015).

The operation of UAVs in logistics has been extended to development aid work. The UPS Foundation recently invested in the collaboration between Zipline (a robotics company), and Gavi (a Vaccine Alliance), to deliver vaccines and blood in Rwanda (Tilley, 2016). Such use of UAV technology has the potential to make a significant impact on the everyday lives of the public.

In the modern world, emerging technologies can present both opportunities and challenges. Publicity and promotion of this technology can increase chances of the new technology receiving subsidies and access to grants if considered of significant importance to the national and/or local economy. However, such publicity, without appropriate research and investigation can lead to the creation of false expectations, poor implementation, or inefficient operations.

The difference between military and non-military operations of UAVs is important. In the past, media coverage focused on the military use of UAVs where the term 'drone' was coined (Jarnot, 2012). The media coverage of military UAVs created a negative attitude of the emerging technology and generated a sense of fear and concern in public opinion (Bartzen Culver, 2014).

Now the term 'drone' continues to be used interchangeably with UAVs in non-military operations. However, as the use of 'drone' technology increasingly becomes part of our everyday civilian lives, public anxiety and fear of the impact of UAV operations in logistics also increases (Elias, 2012; Freeman & Freeland, 2014). While testing of UAVs in logistics and supply chain management is apparent, rigorous and published case studies are illusive in academic literature.

Such research is needed to ensure industry representatives; governments and business leaders are able to make informed decisions regarding the operation of drone activity in logistics and supply chain management. This book intends to make a significant contribution to the literature in the field by offering a platform for rigorous and high quality research.

Current research in the field is focused on technological development (to improve the functionality and endurance of drone delivery in logistics) and regulatory challenges posed by such operations. However, there has been limited attention applied to operational and integration challenges associated with this emerging technology in logistics and supply chain management.

As such, this book presents a collection of research and case studies, which specifically investigate the opportunities, and challenges of the use of UAVs in logistics and supply chain management. More-over, the book aims to provide industry representatives, governments, business leaders, and students with recommendations which aim to improve understanding, and address these challenges associated with this emerging technology.

THE CHALLENGES

The use of UAVs in commercial logistics and supply chain applications has the potential to drastically alter many industries, and, in the process alter our attitudes and behaviors regarding their impact on our daily lives. However, the proliferation of drones challenges our traditional notions of safety, security, privacy, ownership, liability and regulation.

The global political and regulatory environment encompassing UAV operations is problematic (Kreps, 2014). User and airspace safety are likely to continue to be important considerations in regulation and public policy regarding drones. For example, in the United States, drones that weigh more than 250 grams and less than 55 pounds (approximately 25 kilogram) currently have to be registered with the Federal Aviation Administration before they can be operated outdoors (Federal Aviation Administration, 2016). There are further restrictions to operations concerning UAVs with a weight in the range between 250 grams and 25 kilograms (Federal Aviation Administration, 2016). Further work is needed on regulation that supports the inevitable integration of commercial UAVs into controlled airspace (Elias, 2012).

Some argue that the technological development of UAVs is often hindered by a go-slow approach to regulation (Rao, Gopi, & Maione, 2016), suggesting that the law has struggled to keep up with rapid growth in technology (Jenkins, 2013). In some respects, the rapid enhancements in UAV technology create both opportunities and challenges.

In terms of technology advances, the civilian UAV industry will benefit from the technologies developed in other industries. Likewise, other industries will benefit from drone technology. For example, the miniaturization of electronics and the enhancements in battery technology and other innovative design features of electric propulsion systems have allowed small UAVs to develop in ways that will influence manned aircraft designs over the coming years (Perritt & Sprague, 2017).

Similarly, Vertical Takeoff and Landing (VTOL) designs, as observed in military operations, are expanding to UAV operations. Some UAV operators appear to be moving to a hybrid VTOL/helicopter design (Antcliff, Moore, & Goodrich, 2016). This technology and developments is likely to impact both manned and UAV aircraft designs in the future.

The pressing matter for air traffic management is collision avoidance (Elias, 2012). There is a current mandate that all manned aircraft (with some exceptions) must be equipped with ADS-B out technology by 2020 (Zimmerman, 2013). ADS-B offers useful solutions to the challenge of collision avoidance in that the system allows aircraft to transmit position, speed, and directional data to each other while in flight. However, appropriate mandates and UAV inclusive regulation will need to be considered carefully.

This then leads to additional requirement for robust specifications and standards regarding airworthiness and certification for those UAVs entering airspace currently occupied by manned aircraft. While the introduction of UAV operations must not compromise safety of manned aircraft, the expansion of commercial UAV applications will need to be supported with appropriate and relevant regulation (Weibel & Hansman, 2005).

Autonomy will continue to be the focus of technological advancements in UAV design. This will extend to the application of coordinated flights of multiple UAVs. While there are significant mathematical and technical challenges to these enhancements, further consideration is required from the Human System Integration (HSI) perspective. The expanded use of UAVs will require transport managers, engineers and UAV operators to consider the key human factors issues to improve safety, usability, and human operator performance. For example, the ideal blend of automation and human interaction must consider the strengths of humans (e.g., flexibility, and decision making), and the strengths of machines (e.g., accuracy, and rapid computation) (Mouloua, Gilson, Daskarolis-Kring, Kring, & Hancock, 2001).

The ability of the UAV to carry sophisticated imaging equipment attached to light weight vehicles, to hover in position despite adverse weather conditions, to fly simple missions, and takeoff and land automatically, combined with their comparatively (compared to manned aircraft) lower operational costs, has allowed UAVs to support many industries where current manned aircraft and helicopters are unable to reach. As technology develops, larger UAVs will be a direct source of competition for manned aircraft. While some industries may be challenged, many will be advantaged.

For example, it is likely that unmanned air cargo flights will occur earlier than the widespread implementation of small-package delivery by microdrones (Perritt & Sprague, 2017). The segregation and management of low-level airspace will drive any further developments, particularly in largely populated urban centers already experiencing high air traffic congestion. One of the most positive applications in small-package delivery however, is related to health. In particular, the use of UAVs for the delivery of vaccines, medicines and humanitarian aid. While the opportunities in this area of logistics appear obvious, there continue to be considerable challenges to such implementation.

ORGANIZATION OF THE BOOK

The proliferation and diversity in applications of UAVs has created a paradigm shift in air transportation, and paved a pathway for a new area for research across a range of disciplines. Additionally, the dramatic growth of UAVs in civil society has uncovered significant challenges imposed on UAV operators, the commercial air transport sector, air traffic control, government policymakers and highlighted a range of socio-political risks. The significant technological and engineering discoveries have created far-reaching applications for UAVs. Their high mobility, versatility and effect of use in environments considered dangerous to humans, combined with the efficiency in reducing operating time, makes them a practical tool for a variety of tasks.

The challenge for aviation professionals and government officials is how to establish this new and emerging air transport vehicle while protecting the vitality of the industry and the freedom of the community. Business leaders, government officials, and the travelling public would prefer a 'silver bullet' to resolve the issues associated with planning and implementing UAV related applications into the current air transport regulatory regime. In reality, air transport and UAV operations are a complex and highly dynamic system, which requires the consideration of appropriate policies, strategies, tools, and processes. In addressing these challenges, the book is organized into ten chapters. A brief description of each of the chapters follows:

Chapter 1 discusses the importance and integration of appropriate Non-Technical Skills (NTS) training, particularly Situation Awareness (SA) and workload management, to further improve UAV mission effectiveness. Considering logistics applications, the chapter explores technical design and human factors challenges impacting on UAV operations. The chapter reviews historical research and applicable studies in the field and offers solutions to these human factors challenges. While technical design solutions to UAV systems and interfaces are examined, the authors contend that specific training strategies, which focus on the human UAV operator, should also be considered.

Chapter 2 presents the role, functions and prospects in civil UAV development, as well as technical and regulatory barriers to the introduction of unmanned aircraft into special cargo transportation technology. The authors argue that the degree of UAV involvement in freight traffic will continue to grow rapidly as the range of UAV flight and carrying capacity increases, and the air law is liberalized.

Chapter 3 provides an insightful explanation of the different types and potential applications of unmanned aerial vehicles as well as the potential economic impact of their development and use. The chapter analyzes the application of drones in Turkey and Croatia. Although different in terms of their size and the number of inhabitants, both countries are facing similar challenges with respect to UAV application. By understanding the development of UAV implementation in both countries, useful conclusions are drawn.

Chapter 4 discusses the use of drones in humanitarian logistics. This chapter considers a range of aspects regarding the use of drones in healthcare, by undertaking a detailed SWOT analysis. The research highlights that the strengths and opportunities outweigh the weaknesses and threats, suggesting that drones will revolutionize the way medical supplies are delivered in the coming years.

Chapter 5 considers the potential operation of long-range drones to support the logistic response to a natural disaster using a case study of Cyclone Pam that struck Tafea Province of Vanuatu in March 2015. The study reviews how the core capabilities of such drones might be employed in order to overcome the key challenge facing humanitarian logisticians responding to such disasters. The chapter also offers an investigation of the people, process and technology issues that need to be resolved in order to operationalize the concept.

Chapter 6 extends on the use of drones in health by examining the role of UAVs in the delivery of medical and emergency supplies to remote areas. The chapter discusses a number of potential considerations for operators wishing to use UAVs to deliver medical and emergency supplies to remote areas. These considerations stem fundamentally from the nature of the international regulatory framework for unmanned aircraft operations and the peculiarities of using a UAV to deliver medical and emergency supplies. Reflections of these considerations are also provided with a discussion of a UAV delivery project involving medical and emergency supplies in Northland, New Zealand.

Chapter 7 reviews the rules and laws of drones established by the International Civil Aviation Organization, the United States, China, Japan, Australia, India and Korea. The chapter considers that while drones will be able to fly over other countries in the future, there is currently no international legal guidance and regulations for regulating the jurisdiction and limited amount of compensation for the personal or property damage caused by the drone accidents of third persons on the ground in another country. In an effort to resolve these issues, the author recommends a draft Convention for the Unification of Certain Rules Relating to Drone Operations and Transport including the establishment of standards for flying in densely populated areas and near airports.

Chapter 8 considers the problem of a performance of Unmanned Aerial Vehicle (UAV) group flights for a solution of different tasks using criteria of efficiency and criteria of reliability. The research, highly relevant to future logistics applications of UAVs, presents the algorithm for finding the Central Drone Repeater (CDR) in a group of UAVs for sending a control signal to other UAVs in the group. The chapter offers some examples of determining the central drone and the optimal topology in a group of UAVs in flight.

Chapter 9 takes a philosophical orientation, highlighting that in the hands of true professionals, drones can offer highly efficient and profitable solutions for industrial, and commercial inspections and other data capturing tasks. The chapter discusses the challenges of integrating a commercial drone operation into an existing organization's safety management system. The chapter considers the human factors challenges, policy and procedural challenges, logistical challenges, financial challenges, and the challenges of managing change and expectations. The author argues that the success of the emerging UAS industry will be determined by the willingness and ability of organizations to adopt not only conventional aviation safety philosophies and disciplines but also work together to develop a specific UAS Safety philosophy, supported by Safety Management and Quality Management Systems.

Chapter 10 concludes by summarizing the key challenges, findings and solutions presented in the book. The authors address key management considerations relevant to the future integration of UAV operations in logistics and supply chain management. The chapter is organized around six themes that underlie the current discourse regarding the future of UAVs in civilian commercial operations including: policy certification; technology; training; social economic forces; air cargo; and the effect of UAVs on other sectors of the air transport industry.

IMPACT OF THIS RESEARCH

Our previous field of research has examined the impact of government policy (regulation) on the airfreight service needs of regional communities in Australia (Kille, Bates, & Murray, 2013a). Conducted in three phases, our research has employed a mixed-method approach investigating qualitative and quantitative data. Our research has revealed that for regional businesses: (1) cost is an important motivator in business logistics decision making; (2) access to multi-mode freight connectivity is critical; and (3) airfreight services must be considered in state-wide integrated freight infrastructure planning processes (Kille, Bates, & Murray, 2014, 2018). Our research has aimed to assist regional communities, policy-makers, logistics, and aviation industry representatives as they seek to overcome the challenges hindering the development of aviation and freight service delivery in regional Australia (Kille, Bates, & Murray, 2013b).

Considering the factors associated with regional /rural business identified in our earlier research, it is apparent that drone technology has the potential to respond to all three issues. However, adoption of this technology must be applied with careful attention to rigorous and comprehensive research, which addresses the multi faceted challenges of drone operations in logistics.

For this reason, this book publication aims to extend our contribution to the field by offering a comprehensive overview and potential solutions (via drone operations) which address issues of cost, access, connectivity and integrated infrastructure.

To our knowledge, no book publication currently exists that specifically provides a detailed investigation of the emerging use of UAVs in logistics and supply chain management. A number of studies have been published that look at impacts of drone operations in logistics in isolation of other important factors and/or considerations. However, technology and the application of drones in logistics is developing rapidly and a platform for research in this activity is required.

While industry representatives, governments and business leaders facing implications of this emerging technology would benefit from this publication, the book may also be used for educational purposes. Students currently studying air transport, commerce, business, law, logistics and supply chain management will benefit from the opportunity to access a comprehensive publication that considers the multi-faceted factors that impact on this sector of transport and business operations.

REFERENCES

Agatz, N., Bouman, P., & Schmidt, M. (2015). *Optimzation Approaches for the Traveling Salesman Problem with Drone, ERIM Report Series Research in Management.* Rotterdam: Erasmus Research Institute of Management.

Aimi, G. (2016). Last-mile logistics. *MHD Supply Chain Solutions*, *46*(4), 60.

Anand, S. (2007). Domestic use of unmanned aircraft systems: An evaluation of policy constraints and the role of industry consensus standards. *ASTM Standardization News*, *35*(9), 30.

Antcliff, K. R., Moore, M. D., & Goodrich, K. H. (2016). *Silcon Valley as an Early Adopter for On-Demand Civil VTOL Operations.* Paper presented at the 16th AIAA Aviation Technology, Integration, and Operations Conference.

Bartzen Culver, K. (2014). From Battlefield to Newsroom: Implications of Drone Technology in Journalism. *Journal of Mass Media Ethics*, *29*(1), 52–64. doi:10.1080/08900523.2013.829679

Brar, S., Rabbat, R., Raithatha, V., Runcie, G., & Yu, A. (2015). *Drones for Deliveries.* Berkeley, CA: Sutardja Center Entrepreneurship Technol.

Burzichelli, C. D. (2016). Delivery Drones: Will amazon air see the national airspace. *Rutgers Computer & Technology Law Journal*, *42*, 162.

Cuthbertson, A. (2016). *Australia post to launch drone delivery service.* Retrieved 15 February, 2019, from https://www.newsweek.com/australia-post-drone-delivery-service-drones-449442

Dorling, K., Heinrichs, J., Messier, G. G., & Magierowski, S. (2017). Vehicle routing problems for drone delivery. *IEEE Transactions on Systems, Man, and Cybernetics. Systems*, *47*(1), 70–85. doi:10.1109/TSMC.2016.2582745

Elias, B. (2012). *Pilotless drones: Background and considerations for congress regarding unmanned aircraft operations in the national airspace system.* Congressional Research Service.

Erceg, A., Erceg, B. C., & Vasilj, A. (2017). Unmanned aircraft systems in logistics legal regulation and worldwide examples toward use in Croatia. *Business Logistics in Modern Management*, 43-62.

Federal Aviation Administration. (2016). *Operation and Certification of Small Unmanned Aircraft Systems 14 CFR Part 107*. Retrieved 12 December, 2018, from https://www.faa.gov/uas/media/RIN_2120-AJ60_Clean_Signed.pdf

Freeman, P. K., & Freeland, R. S. (2014). Politics & Technology: US policies restricting unmanned aerial systems in agriculture. *Food Policy*, *49*(1), 302–311. doi:10.1016/j.foodpol.2014.09.008

Hail, A. R., & Coyne, C. J. (2014). The political economy of drones. *Defence and Peace Economics*, *25*(5), 445–460. doi:10.1080/10242694.2013.833369

Jarnot, G. (2012). History. In R. Barnhart, S. Hottman, D. Marshall, & E. Shappee (Eds.), *Introduction to Unmanned Aircraft Systems*. New York: CRC Press.

Jenkins, B. (2013). Watching the watchmen: Drone Privacy and the need for oversight. *Kentucky Law Journal (Lexington, Ky.)*, *102*, 161.

Kille, T., Bates, P., & Murray, P. S. (2013a). *Farm to Table: Air Transport Issues for an Island State.* Paper presented at the 2013 Australasian Transport Research Forum Conference, Queensland, Australia.

Kille, T., Bates, P., & Murray, P. S. (2013b). Measuring Success: The Search for Assessment Criteria in Determining the Impact of Deregulation in Regional Aviation. *International Journal of Aviation Management*, *2*(1/2), 4–34. doi:10.1504/IJAM.2013.053046

Kille, T., Bates, P., & Murray, P. S. (2014). A systematic review of the role of regional aviation in the airfreight industry. *Science-based Technologies*, *24*(4), 471–485. doi:10.18372/2310-5461.24.7520

Kille, T., Bates, P., & Murray, P. S. (2018). Regional Devlopment: Air Freight Service Needs for Regional Communities. In M. Khosrwo-Pour (Ed.), *Encyclopedia of Information Science and Technology* (4th ed.). Hershey, PA: IGI Global. doi:10.4018/978-1-5225-2255-3.ch684

Kreps, S. (2014). Flying under the radar: A study of public attitudes towards unmanned aerial vehicles. *Research and Politics*, *1*(1), 1–7. doi:10.1177/2053168014536533

Mileham, R. (2016). Prime movers. *Engineering & Technology, 11*(4), 70–72. doi:10.1049/et.2016.0408

Mouloua, M., Gilson, R., Daskarolis-Kring, E., Kring, J., & Hancock, P. (2001). *Ergonomics of UAV/UCAV mission success: Considerations for data link, control and display issues.* Paper presented at the Human Factors and Ergonomics Society 45th Annual Meeting, Santa Monica, CA. 10.1177/154193120104500231

Perritt, H., & Sprague, E. O. (2017). *Domesticating Drones: The technology, Law, and Economics of Unmanned Aircraft.* New York: Routledge.

Rao, B., Gopi, A. G., & Maione, R. (2016). The societal impact of commercial drones. *Technology in Society, 45*, 83–90. doi:10.1016/j.techsoc.2016.02.009

Sandvik, K. (2015). African drone stories. *BEHEMOTH: A Journal on Civilization, 8*(2), 73-96.

Sheets, P., Rowling, C. M., & Jones, T. M. (2015). The view from above (and below): A comparison of American, British and Arab news coverage of US Drones. *Media. War and Conflict, 8*(3), 289–311. doi:10.1177/1750635215593973

Tilley, A. (2016). *UPS Experiments with Drone Delivery In Partnership With Zipline*. Retrieved 20 February, 2019, from https://www.forbes.com/sites/aarontilley/2016/05/09/ups-experiments-with-drone-delivery-in-partnership-with-zipline/#3a424e7e45a4

Weibel, R. E., & Hansman, R. J. (2005). *Safety considerations for operation of unmanned aerial vehicles in the national airspace system.* Technical Report ICAT-2005-1. MIT International Center for Air Transportation.

Weissbach, D., & Tebbe, K. (2016). Drones in sight: Rapid growth through M&A's in a soaring new industry. *Strategic Direction, 32*(6), 37–39. doi:10.1108/SD-04-2016-0044

Wurbel, H. (2017). *Framework for the evaluation of cost-effectiveness of drone use for the last-mile delivery of vaccines* (Master of Global Health). University of Barcelona. Retrieved from https://logcluster.org/sites/default/files/gm_files/master_final_project_heike_wurbel_13_jun2017_003.pdf

Zimmerman, J. (2013). *ADS-B 101: what is it and why you should care.* Air Facts Journal.

Acknowledgment

With sincere thanks to my co-editors Professor Paul Bates and Dr. Seung Young Lee for their invaluable knowledge, advice, support, and friendship. A tremendous thank you to the chapter authors for their interest and commitment to their subject areas and for their dedication to the writing process. Thank you to my husband, Murray and my daughters, Emma and Ashley, for their incredible patience and continued support of my work.

Tarryn Kille
University of Southern Queensland, Australia

Chapter 1
Non-Technical Skills (NTS) Training for UAV Operators:
Situational Awareness and Workload Management

Tarryn Kille
University of Southern Queensland, Australia

Seung Yong Lee
University of Southern Queensland, Australia

Paul R. Bates
https://orcid.org/0000-0003-1073-7172
University of Southern Queensland, Australia

David Murray Kille
University of Southern Queensland, Australia

ABSTRACT

In civilian operations, the utilization of UAVs is diverse and broad, the variations of UAVs are extensive, and the application needs and performance characteristics also vary significantly. To this end, the emerging opportunities for UAV operations have generated an urgent need for trained operators to ensure these systems are used effectively and safely. This chapter discusses the importance and integration of appropriate non technical skills (NTS) training, particularly situation awareness (SA) and workload management, to further improve UAV mission effectiveness. The chapter explores technical design and human factors challenges impacting on UAV operations. By reviewing historical research and applicable studies in the field, the chapter also offers recommendations and solutions. While technical design solutions to UAV systems and interfaces are examined, the authors contend that specific training strategies, which focus on the human UAV operator, should also be considered.

DOI: 10.4018/978-1-5225-7900-7.ch001

INTRODUCTION

Technological improvements, advancements in communications and the development of battery technology has supported the emergence of small, low cost UAVs which allow civilians to work with this new air transport technology. UAVs offer good solutions as they can cover vast areas and can be used to explore and assess damage after a catastrophic event (Astuti, Longo, Melita, Muscato, & Orlando, 2008). Additionally, their high mobility, versatility and effective use in environments considered dangerous to humans (Kontitsis, Tsourveloudis, & Valavanis, 2003), combined with reduction in operating time, makes them a practical tool for a variety of tasks.

These tasks include search and rescue using high quality imaging (Rathinam et al., 2007); analysis of gas composition in volcanoes (Astuti et al., 2008); rivers, bridges and coastal areas surveillance (Rathinam et al., 2007), monitoring forest fires (Casbeer, Beard, McLain, Li, & Mehra, 2005), transportation of vaccines (Haidari et al., 2016), and last mile logistics (Brar, Rabbat, Raithatha, Runcie, & Yu, 2015). Civilian applications continue to expand in areas such as agriculture, mapping, the measurement of structures and anthropology (De la Torre, Ramallo, & Cervantes, 2016).

In civilian operations, the utilization of UAVs is diverse and broad (Gupta, Ghonge, & Jawandhiya, 2013), the variations of UAVs are extensive and the application needs and performance characteristics also vary significantly. To this end, the emerging opportunities for UAV operations have generated an urgent need for trained operators to ensure these systems are used effectively and safely.

In recent years, research associated with civilian application of UAVs has focused on the use of algorithms and specific hardware features that allow the UAV to operate autonomously, or perform more effectively (De la Torre et al., 2016). However, this research often fails to emphasize that despite the use of artificial intelligence there remains a requirement for significant human interaction in the operation of UAVs.

As the ubiquitous use of UAVs continues, transport managers, engineers and UAV operators need to consider the key human factors issues to improve safety, usability, and human operator performance. For example, the optimum blend of automation and human interaction should consider the strengths of humans (e.g., flexibility, and decision making), and strengths of machines (e.g., accuracy, and rapid computation) (Mouloua, Gilson, Daskarolis-Kring, Kring, & Hancock, 2001).

The UAV operator plays an important part of effective UAV missions, and more research in the area of human integration in the automated systems of UAVs is required. Therefore, the intent of this chapter is to highlight the importance and integration of appropriate Non Technical Skills (NTS) training, particularly Situation Awareness (SA) and workload management, to further improve UAV mission effectiveness.

BACKGROUND

Consider that within any flight, there are many layers to the complexity of the operations. For example, there are layers of technology, protocol as well as behavioural layers. Non-technical skills are considered an essential layer as they reduce the likelihood that an error or failure will lead to an accident. Non-technical skills are social, mental and personal management skills that contribute to safer and more efficient operations. Non-technical skills support the technical skills of flight crew and contribute to dependable and effective performance in the often-complex work systems that characterise the aviation industry. Non-technical skills in the aviation industry typically include: situational awareness; decision-making; communication; team work; leadership; managing stress; workload/task management; and coping with fatigue.

The concept of non-technical skills in flight crew training is not new in aviation training. Training programs with a focus on non-technical skills emerged in direct response to research highlighting the significant contribution of human factors to aviation incidents and accidents (Thomas, 2004). Crew Resource Management (CRM), which considers the effective use of all accessible resources to reach safe and efficient flight operations, has long been the focal point enabling the reduction of human error and improvement in safety outcomes (Lauber, 1987; Wiener, Kanki, & Helmreich, 1993). In flight crew training, the cornerstone of CRM was the growth of non-technical skills such as leadership, communication, conflict resolution, decision making as well as fatigue and stress management (Helmreich & Wilhelm, 1991). While the concept of CRM training has evolved through a series of paradigm shifts, it is the development of threat and error management strategies that has been the focus of attention most recently (Helmreich, Merritt, & Wilhelm, 1999).

Helmreich and Merritt (2000) assert that sound threat and error management training of aviation personnel is not necessarily the single solution to removing the human contribution to incidents and accidents. Rather, it is specifically customized training programs that will make the key contribution in reducing risk. It is widely appreciated that flight crew assume the final line of defence in the often porous and flawed socio-technical system within which aircraft operate (Reason, 1997). As such, by providing crew members with skills in threat and error management, aviation organisations can build additional barriers of defence against active and latent conditions that often lie conspicuously within the organisation or the wider operating environment.

There is much research to suggest that human error contributes significantly to accidents and incidents in commercial aviation (Wiegmann & Shappell, 2001). There are some common errors that are also applicable to UAV operators such as failing to observe alerts of a visual and auditory nature; failing to maintain sound SA; and poor decision-making.

The UAV operator interface is the primary facilitator of human-machine interaction and coordination. Thus, the design of an interface that provides the operator with an appropriate level of awareness of the tasks and activities under their control and that of the other operators involved in the mission becomes critical in the effort to mimimize information overload, distraction, coordination errors and miscommunication.

While supporting awareness for the operator needs more than simple provision of information, Drury and Scott (2008) maintain that understanding the information is a crucial prerequisite for providing appropriate awareness to the human operators within intelligent systems. However, there is limited UAV-specific guidance available to designers to identify what information needs to be provided to UAV operators to encourage and promote appropriate awareness and the required level of detail that it needs to provide (Drury & Scott, 2008). Displaying all the information available can cognitively overload an operator (Buttigieg & Sanderson, 1991), especially when the operator is experiencing time pressure and in system architectures where an operator may be expected to control multiple UAVs (Cummings & Mitchell, 2007). Furthermore, there is limited research on the importance of skills training in areas of SA and workload management, which can also improve the overall flight performance.

Literature associated with 'inhabited' aviation and air traffic control is well stocked with studies of awareness that may be useful to informing work on UAV operations (for example, Endsley (2000)). However, piloting a UAV is vastly different from piloting an inhabited aircraft. UAV pilots do not have: (1) the proprioceptive cues that pilots of manned aircraft have to feel any changes in aircraft attitude or alterations to the engine vibration signalling speed changes or engine trouble; (2) an extensive field of vision of the actual aircraft environment; and (3) the reaction pattern rigorously taught in pilot training programs.

In the literature concerning human factors in UAV operations, a significant proportion of the research is grounded in military operations (De la Torre et al., 2016). Correspondingly, State safety regulators are increasingly implementing rules on the use of UAVs for safety and security reasons. These developments further emphasise the importance of controlling the use of UAVs in airspace, and also the importance of appropriate and supportive training for future UAV operators, where challenges associated with human factors are included.

Human error is the common reason stated for an aircraft accident. However, human error is often linked to a latent condition hidden within the entire operation. Such conditions may include high (or low) workload, fatigue, and limited knowledge of the situation or inadequate training. These are just a few of the many causes that can reduce pilot performance and result in an accident (Manning, Rash, LeDuc, Noback, & McKeon, 2004). In the UAV context, further support for this is found in a study by Tvaryanas, Thompson, and Constable (2005). In this research it was found that a significant number of the UAV incidents that occurred over 10 years of U.S. Army, Air Force, and Navy/Marines UAV operations were due to human factors issues such as workload, attention, and crew coordination and communication.

Although automation is being used increasingly in the development of UAVs, it is important to note that automation can also have the effect of increasing the workload of the operator and reducing situational awareness (Ruff, Narayanan, & Draper, 2002). Furthermore, high levels of automation can also reduce the opportunity for the operator to promptly intervene when overriding automation may be necessary (McCarley & Wickens, 2005). For this reason, automation and its range of functions should not completely remove human intervention (Hopcroft, Burchat, & Vince, 2006).

In human factors research, it is apparent that greater performance success occurs when the systems and equipment are designed to account for human beings, rather than removing humans from the system entirely (Abbott, Slotte, & Stimson, 1996). The benefits and strengths of combining humans and robots to achieve cooperative tasks has become widely acknowledged (Crandall & Cummings, 2007). Manipulating the levels of autonomy of the robot to cater for human input provides a good opportunity to achieve an optimal combination in mixed human/robot teams. At the heart of this assumption is the notion that robot performance is improved with human input (Kaupp & Makerenko, 2008). Thus, if human input is critical to UAV operations, then it is necessary to understand the variables that affect human performance, specifically in tasks where an operator interacts with a UAV. This will likely lead to the best performance of the operator to complete a task in the most efficient and effective way while reducing the risk of possible accidents or failures. This will also provide greater appreciation for NTS when recruiting and training these people.

Considering the rapid pace of UAV evolution, it is critical now for airspace designers, planners, transport and safety regulators to recognize the potential for UAVs to introduce new operational challenges within the entire air transport system. The integration of UAVs into non-segregated air spaces (i.e. airspace that may include manned aircraft) raises many complex and potentially demanding technical, political and economic issues (Nisser & Westin, 2006). Some of the most urgent questions to be addressed, however, are those corresponding to human factors (Gawron, 1998).

Studies into aviation human factors can be dated back some seven decades ago. Since then, there has been a significant body of theoretical and empirical research that has been dedicated to human performance aspects of manned flight. While this body of research is useful in helping to address some of the impending challenges facing UAV operations, it is apparent that there are significant differences between the key human factors issues associated with manned flight and remotely piloted flight. This chapter is important to the UAV body of literature as it attempts to highlight these key differences. It is hoped that developing an appreciation of two key non-technical skills training requirements associated with UAV operations will be instructive and informative to future UAV policy makers, designers, regulators and training providers.

As technology and automation develop rapidly, there are many emerging human factors challenges associated with UAV operations now and into the future. The emergence of threat and error management training has highlighted the important safety link between human factors challenges and appropriate non-technical skill training. This chapter focuses on two of these non-technical skill training requirements relating to remotely piloted aircraft systems in civilian aviation operations including: situational awareness; and workload management. While the two are heavily integrated, the following sections will address each issue consecutively. Each important non-technical skill is discussed with regards to the history and importance of the skill to air transport pilots of manned aircraft, while highlighting the pressing nature of such issues in UAV operations. This will be followed by suggested solutions to address these challenges. The solutions present research addressing the improvement to UAV operator skills of SA and workload management while discussing the technical and non-technical perspectives.

MAIN FOCUS OF THE CHAPTER

Issues, Controversies, Problems

Situational Awareness (SA)

From the perspective of an aircraft pilot, SA means holding a mental picture of the existing inter-relationship of flight conditions, geographic location, aircraft energy state and configuration, as well as any other factors that may impact the aircraft's safety (e.g., terrain proximity, airspace restrictions, weather and potential obstructions). Some possible consequences of poor situational awareness may include outcomes such as controlled flight into terrain (CFIT), los of control, loss of separation, airspace violation, heavy icing, or an unexpected confrontation with wake vortex turbulence.

Endsley (2000) suggested a definition of SA as "the (1) perception [noticing] of the elements in the environment within a volume of time and space, the (2) comprehension of their meaning, and the (3) projection of their status in the near future" (p. 5). This three-stage definition can be further understood by highlighting three significant features. Firstly, SA involves cognition and working memory (the short-term information store that can be quickly forgotten if not practiced repeatedly), rather than action and response. Strong SA can bolster good selection of action, yet it is not naturally a part of that choice. Secondly, SA is associated with dynamic, evolving situations, and as such, is not the same as the more static knowledge provided by long-term memory (e.g., understanding of how systems work, or what procedures to follow). Thirdly, the content or the outcome of SA is different from the process of maintaining SA (Adams, Tenney, & Pew, 1995).

A pilot who is situationally aware is able to promptly access an accurate mental picture of the dynamic environment that is extensive and more enduring than that which is held within the limited capacity of the working memory, or "consciousness". A pilot that displays good SA may not be thinking consciously about the aircraft operating close by. However, if required to suddenly respond appropriately to this situation, can do so quickly and accurately because of an ability to quickly access specific information from memory (Kintsch & Ericsson, 1995). It is evident that the non-technical skill of good situational awareness has the effect of reinforcing a pilot's ability to respond to the unexpected (Wickens, 1999). Thus, aircraft design features that enhance routine performance may reduce SA, creating an effect, which 'tunnels' a pilots attention to focus on the expected events and tasks.

The notion of SA also has direct implications in the design and operation of UAVs. In the context of UAVs, SA is broadly defined as the operator's awareness of the status and alterations in a machine's operation (Mouloua, Gilson, Kring, & Hancock, 2001). This awareness should provide the operator with the ability to react quickly and appropriately to unexpected events (Weimer, 1995). High levels of SA will support positive UAV mission performance (Mouloua, Gilson, Kring, et al., 2001) while poor SA is often linked to operator errors (Barnes and Matz (1998).

Endsley (2000) claimed that it is "goals" that are the cornerstone to the development of SA. In the goal-driven process, the operator is seeking information that assists in achieving the goal, the operator chooses mental models to complement the goal, the operator then combines the goals and their related mental picture to explain, incorporate and filter information.

A study conducted by Wickens and Dixon (2006) investigated how mission goals can impact on the operator in simulated UAV flights. In this study, operators were asked to track the UAV to waypoints while also required to report on "command targets," seek out camouflaged "targets of opportunity" and identify and respond to system failures onboard. In one experiment, the operators were exposed to two potential automation failures (i.e., autopilot failure and auto-alert system failure). The tests helped to uncover that unreliability in both types of failures has the effect of degrading operator performance. However, it was also found, that this degrade performance was less significant for unreliability in autopilot systems. Wickens and

Dixon (2006) propose that the difference was due to the operator's priorities in that environment. In these experiments, the operator considered that the navigation and performance of the UAV was of greater importance to the success of the mission than health monitoring of the UAV.

If SA needs can be identified for safe UAV operations, and we can dissect SA into parts that an operator includes in his assessment of the current UAV state, we can vastly enhance human-UAV communication and control and create more appropriate interfaces for UAV operators. For example, Mouloula, Gilson, Kring, and Hancock (2001) assert that displays that allow the system structure and processes to be more visible to the operator will help in highlighting options for action, that can assist the operator to maintain SA.

The research attempting to resolve SA issues in UAV operations stems from discussions with cognitive psychologists regarding the 'shared fate' phenomenon. This phenomenon suggests that because the ground operators do not share the same fate as the UAV, they may maneuver the UAV in a more aggressive manner that may increase the likelihood of accidents.

Workload Management

The design of aircraft cockpits has evolved rapidly over the past three decades. The increasing use of automation in the cockpit has resulted in an important systems management dimension to stick-and-rudder flying. Consequently, research has seen and increased focus on the mental workload of flight crew and air traffic controllers (Funk, Suroteguh, Wilson, & Lyall, 1998; Schutte & Trujillo, 1996; Wickens, 2003).

Workload has been an important focus in research as errors can be prompted when mental task demands exceed human operator capabilities. The outcome of such errors can severely impact safety. In an effort to respond to increased automation, the mental demands imposed by a human-machine interface need careful consideration in conjunction with the physical task-load that automation can also create.

Workload can be defined as the composite of the demand for labor and the human response to this demand (Mouloua, Gilson, Daskarolis-Kring, et al., 2001). Measurement and assessment of workload has become a significant theme in research. Subsequently, the research has promoted the development of systems for human-machine communication in an effort to enhance safety, health, efficiency, and operator success in the long term (Rubio, Diaz, Martin, & Puente, 2004). Within air transport operations, workload levels can vary between long periods of low workload and short periods of high workload. Work design that is effective often aims to circumvent such extremes of either high or low demand, which can be a threat to the continued maintenance and effectiveness of skills (Sauer, Wastell, & Hockey, 1996). Additionally, extended periods of high workload can lead to reduced attention, higher levels of stress and fatigue, reduced flexibility, and information processing short-comings (Connors, Harrison, & Akins, 1985).

A range of tools is available which attempt to assess and forecast mental workload. These methods are commonly separated into the three categories: (1) performance based measures; (2) subjective measures, and (3) physiology based measures (Meshkati, Hancock, & Rahimi, 1992). The subjective methods have demonstrated dominance in measuring the workload of the operator. The rationale behind the frequent use of subjective methods is founded in the practical advantages (simplicity of implementation) and access to ongoing data. These elements can bolster the ability of subjective methods to offer perceptive measures of the operator's mental workload (Rubio et al., 2004).

Wiegmann and Shappell (2001) suggest that human error is a considerable contributing factor to accidents in commercial aviation. Such errors are also common to UAV pilots including: failure to commence appropriate maneuvers; failure to recognize auditory or visual alerts; failure to maintain sound SA; and poor decision-making.

In the field of human factors, workload is considered the total of the task demands that are imposed on an operator, as well as the subjective response of the operator to those specific demands. Factors such as expertise, noise, stress, time pressures, and distraction can shape how the human performs a given task. For example, one specific task may represent an appropriate level of workload for an experienced operator. However, the same task may fatigue or exhaust a novice. Generally, there is an important distinction made between task-load (the objective demands of a task) and workload (the subjective demand experienced while performing the task).

In the context of UAV operations, one workload concern is associated with the possibility that automated Communication, Navigation and Surveillance (CNS) systems may interfere with operator workload. Past experiences have allowed the aviation community to appreciate that automation can sometimes have the unintended consequence of increasing workload. Sarter, Woods, and Billings (1999) suggested that automation has not resulted in reducing workload and errors. Rather, automation has simply changed the nature of workload and error or shifted these problems in time. This is evident in the early glass cockpit aircraft designs, such as the Airbus A320, and the Boeing B757 where the Flight Management Systems (FMSs) presented the unintended effect of reducing workload during the low demand phases of flight (e.g., during cruise), but increasing already high workload during high demand phases of flight (such as late arrival runway changes, which require FMS reprogramming.

Many of the automated systems provide a support to pilots during low workload phases of the flight, but often hinder the pilot at times when automation is required most. In air transport, high workload flows are often experienced during descent and approach to an airfield, working on checklists, briefing crew and communication with Air Traffic Control (ATC). Directions and clearances/requests for ATC can often require additional changes to the planned flight path. This flight management combined with the close proximity to the ground forces the pilot to be more conscious and aware of errors, while being more cautious when making decisions (Sarter et al., 1999).

It is ironic that the workload often recedes just prior to high workload phases, which often creates a considerable transition challenge for the pilot. Low (2004) found that reduced workload and improved system performance could be attained with the application of intermediate levels of automation, which require continued human involvement, instead of highly or fully automated systems that remove the operator from an active role. Mouloula et al. (2001) suggest that because UAV control is likely to be highly automated and the operator's tasks are essentially focused on supervisory monitoring, UAV systems need to be designed to remove both over-load and under-load. One plausible solution is to merge manual and automated control, assigning the operator with higher order tasks.

Humans have a limited capacity for information (Broadbent, 1958; Klingberg, 2000). As soon as the information flow is greater than the human operator's ability to mentally process the information, any additional information can no longer be attended to, or may displace other information and tasks already being processed. When this occurs, there are only two options available: shedding less important tasks, or performing all of the tasks at a less than optimal level.

Effective workload management ensures that the essential operations are completed by planning and logically sequencing them to reduce any opportunity for work overload. As a pilot increases their level of experience, they learn to identify and recognize future workload requirements and subsequently preparing for high workload periods during periods of low workload. While automation has the potential for reducing operator workload, in the case of UAV operations, the challenge for safety-efficient operations will be to determine if automation actually reduces workload and what level and degree automation will be integrated into such operations.

Early UAV research assumed that UAV operators would have some degree of responsibility for actually flying the UAVs in addition to other tasks such as navigation and management of payload. Studies such as Wikens, Dixon, and Chang (2003) and Ruff et al. (2002) reveal that while UAV operators are assigned various control responsibilities such as navigation, and execution of mission, they are limited in their ability to control multiple vehicles, even when supported by automation. However, enhancements in technology and technical design is giving rise to a future where UAV operators may be operating more than one or two UAVs (Cummings, Bruni, Mercier, & Mitchell, 2007). Effective solutions to workload management challenges will be required.

SOLUTIONS AND RECOMMENDATIONS

Developing UAV Operator Skills in Situation Awareness

Maintaining SA is critical to aviation safety. For the operation of highly autonomous UAVs, there are a number of factors that can collectively be extremely challenging to maintain SA. These factors include:

1. Display design that may not be ideal for maintaining SA;
2. The removal of the pilot from the aircraft, resulting in sensory isolation;
3. Delays in data links, low-grade quality of images from onboard sensors; and
4. Lengthy periods of monitoring highly automated systems, leading to the operator feeling 'out of the loop'.

These factors are discussed in the following sections, while presenting current research proposing solutions to these factors.

Principles of UAV Display

Australian mission control element (MCE) operators for the deployment of the Global Hawk in 2001 rated status displays and controls in the MCE as consistently unacceptable (Hopcroft et al., 2006). Within the report, there were a number of areas identified as troublesome including the physical location and arrangement of the displays spaced too far apart, and the difficulty in reading displays due to fonts and colours applied). While the Global Hawk is a highly automated UAV in terms of automated decision making and flight control and is predominantly used for surveillance activities (Stacy, Craig, Staromlynska, & Smith, 2002), the case study serves to illustrate the key design failures that impact on UAV operation and a pilot's ability to maintain SA.

The concerns highlighted by the Global Hawk operators suggest that there are some standard design guidelines that UAV designers may have failed to observe. Seven crucial principles of aviation display design have been defined by Wickens and Hollands (2000).:

1. **Principle of Information Need:** When information is required more frequently, it should be displayed in a location that is easily accessible by the operator;
2. **Principle of Legibility:** For displays to be of use to the operator, they should be understandable. As such, displays should be of the appropriate size with adequate contrast, brightness, illumination and volume (for auditory displays), to provide relevant information in a timely and comprehensible fashion;

3. **Principle of Display Integration/Proximity Compatibility Principle:** When operators are required to compare and integrate more than one source of information, those sources should be located in close proximity within the display;
4. **Principle of Pictorial Realism:** The display should pictorially represent the information and associated trends;
5. **Principle of the Moving Part:** The moving element on a display should reflect the moving element as perceived in the mental model of the operator;
6. **Principle of Predictive Aiding:** Predictive information of the aircraft state is valuable only when it is accurate, reliable and easy to comprehend by the operator; and
7. **Principle of Discriminality:** A display element in the context of one situation should not be represented in a similar way (look, sound or feel) to another element that may occur in the same display situation.

Unfortunately, the MCE displays used in the Global Hawk Australian demonstration of 2001 appear to have ignored many of these display design principles. For example, operators commented on the difficulty of reading displays. This stems from the importance of the principle of legibility. There were also complaints on the spread of the displays, which appears to highlight the violation of the principle of proximity. These examples illustrate that there are fundamental design issues associated with the presentation that need careful consideration when investigating the human machine integration component of UAV operation. As such, in resolving SA issues in UAV operations, it is important to consider the principles outlined by Wickens and Hollands (2000).

The Removal of the Pilot From the Aircraft Resulting in Sensory Isolation

The second impact on situational awareness for UAV operators relates to the consequence of the operator being separated from the aircraft. Such a separation means that the operator is then deprived of many of the sensory cures that are normally available to the pilot of a manned aircraft. Unlike an inhabited aircraft, where the pilot receives direct sensory information from the environment in which his/her vehicle is operating, the UAV operator can only receive sensory information from onboard sensors via data link. For many UAV operators, this generally includes visual imagery, which often covers a limited field-of-view. The sensory cues that are lost for UAV operators include ambient visual information, vestibular input, and sound. When compared to the pilot of a manned aircraft, the UAV operator is required to perform in what is considered "sensory isolation" from the vehicle in his/her control. Research is critical to pinpoint specific ways in which sensory isolation impacts on operator performance in the many tasks and various stages of UAV flight. More importantly, research is required to investigate advanced display designs, which may provide compensation for the lack of direct sensory input from the physical environment in which the aircraft is operating.

Researchers such as Ruff et al. (2002), Calhoun, Draper, and Ruff (2002), and Dixon, Wickens, and Chang (2003) looked closely at the benefits of multimodal displays to UAV operators. Ruff et al. (2002) studied the benefits of haptic displays in warning UAV operators to the onset of turbulence. For UAV operators with a conventional display, turbulence is realised entirely by disturbances in the camera image that comes from UAV onboard sensors. Ruff et al. (2002) revealed that haptic information transmitted through the joystick control column enhanced the UAV operator's SA in a simulated UAV approach and landing task although these improvements were experienced in limited circumstances (i.e., only when turbulence occurred far from the runway). Nevertheless, such results suggest that there is merit in multi-modal displays in an effort to compensate for lost sensory information found in UAV operations. More research is required to understand the costs and benefits of multimodal displays in resolving UAV operator's sensory isolation issues, and to determine optimal design for these displays.

Considering the lack of multisensory cues available to UAV operators, Wickens and Hollands (2000) argue that the installation of multisensory interfaces would be of great benefit as multi-sensory interfaces may help to reduce the high workload that can sometimes be experienced by a UAV operator, particularly when one sensory mode (e.g., visual) is overwhelmed with information sources (Calhoun et al., 2002). This suggestion in founded in the Multiple Resource Theory, which proposes that different sensory modalities utilise different attentional resources (Calhoun et al., 2002).

We have established that pilots of manned aircraft have access to much multisensory information, which helps to understand the status of their aircraft within its environment (Draper, Ruff, Repperger, & Lu, 2000). This information includes the visual input, vestibular and auditory information that provides pilots with the cues regarding speed of travel, angle of bank, the geographic elements within the vicinity, weather conditions, and the health of the aircraft. Hopcroft et al. (2006) suggest that UAV operators can also access much of this information via their instruments and displays. Unfortunately, monitoring these displays and their instruments may be considered more laborious and not suited to the building of SA in the way pilots of manned aircraft do.

Draper et al. (2000) raised the possibility of augmented reality and synthetic vision to better understand the benefits of supplemental sensor input. Similarly, Van Erp and Van Breda (1999) have suggested that such augmented reality displays can offer significant improvements in accuracy and lower the cognitive demands of target tracking with a payload sensor, leading to improved UAV flight control. More recently Ruano, Cuevas, Gallego, and Garcia (2017) presented an Augemented Reality (AR) tool for UAV operators. This tool has two functionalities for medium-altitude long-endurance UAVs: route orientation; and target identification. Route orientation allows the operator to identify upcoming waypoints and the route that the UAV will follow. Target identification provides a fast target localization, even in the presence of obstructions. The experiments demonstrate that such an AR tool significantly improves situational awareness of the UAV operator (Ruano et al., 2017).

Although these initial studies provide some evidence that multi-sensory interfaces may improve the performance of UAV operators, more investigation is required. The problems that result from sensory isolation for control of highly autonomus vehicles as opposed to manually controlled vehicles is still not clear, and any interactions of interface type with the experience level of the UAV operator is not clearly understood. Questions that future studies may answer include:

1. Will the provision of multisensory cues be as effective in situations where the UAV operator has less control over the aircraft (i.e., highly autonomous UAVs)?;
2. Would a licensed aircraft pilot benefit more from multisensory cues than a non-pilot due to previous experience in manned aircraft? (Tvaryanas et al., 2005); and
3. Are imitating cues normally present in a manned aircraft (e.g., ambient noise) beneficial to UAV operators, or are they more beneficial simply to make more effective use of attentional resources? (Draper et al., 2000)

Data Link Delays and Sensor Imagery

The problems associated with delays in data links and dropouts are of great importance to UAV flights due to the ability to interfere with the aircraft control (Mouloula et al., 2001). The UAV operation involving highly autonomous UAVs with pre-programmed flight plans may be less of a concern. However, there may still be some risk in disrupting the operation where the UAV operator is required to intervene with the autonomous flight such as navigating to waypoints. The bandwidth and time interval (between stimulation and response) limitations of the data links that transmit the images from the sensors onboard the UAV to the image quality control workstation means that timeliness and the quality of the images received can be poor (McCarley & Wickens, 2005). Environmental conditions and visual displays that are highly cluttered can also make images difficult to comprehend (Calhoun, Draper, Abernathy, Patzek, & Delgado, 2005).

Van Erp (2000) suggests that problems resulting from the limitations of data links may be resolved with the transmission of task critical information only. Van Erp (2000) also comments on image parameters that can be critical in the control of unmanned ground vehicles such as magnification factors, field size, colour application, update rates, spatial resolution, monoscopic/stereoscopic viewing, viewing direction, and aiming. By extension, it would seem appropriate to identify what information is critical for control of UAVs and to examine and determine how the data link capacity may be used in the most efficient way.

Calhoun et al. (2005) revealed the SmartCam3D System (SDS) and considered a more advanced system for the augmentation of camera imagery. The SDS draws together the imagery from the camera and a 3D computer-generated representation of what the camera should be capturing. This system has a number of benefits including picture-in-picture presentation, which provides video footage as a picture surrounded by the synthetic graphics, to enhance the field of vision.

Although there is a dearth of specific UAV information relating to augmented reality displays, there is a significant body of research on augmented reality more generally (Green, Billinghurst, & XiaoQi, 2008). There is evidence of the use of augmented displays in teleoperation activities such as robot operations on space shuttles, as well as for scientific and medical visualization (Kipper & Rampolla, 2013). Yet, the study of the various human factors issues relating to the application and use of augmented reality displays relating to UAV ground stations is still required.

The appropriate level of augmentation for optimal interpretation of UAV imagery is yet to be determined. In addition, the associated risks of UAV operators placing too much confidence in the augmented imagery also need to be determined (McCarley & Wickens, 2005). Cognitive tunnelling, that may result from excessive focus on a certain element of synthetic vision can also lead to the deterioration of the sensor images and this also requires further investigation (Yeh & Wickens, 2001).

Lengthy Periods of Monitoring Highly Automated Systems Leading to the Operator Feeling 'Out of the Loop'

There is a broad spectrum in UAV operations regarding the degree to which flight control is automated. In some instances, the aircraft is manually guided using stick and rudder controls, where the UAV operator receives visual imagery from a front facing camera mounted on the vehicle. In other instances, the control is automated only partially, where the operator may select the necessary parameters through an interface in the ground control station. Again, in other instances, control of the UAV may be fully automated, where an autopilot retains flight control accessing pre-populated flight coordinates. The manner of flight control used during takeoff and landing, can also differ from the manner of control used en route. The associated benefits of each variation of flight control may be different depending on the function of the time delays in communication between operator and UAV and the quality of visual imagery as well as other sensory information provided to the operator from the UAV.

From the perspective of improving SA, more investigation is needed to understand the interaction of human operators and the automation in UAV flights. Dixon et al. (2003) revealed that when the flight control is allocated to an autopilot, this may free some attentional resources and improve the operator's fault detection abilities. This improvement in abilities was obtained even where the autopilot was not entirely reliable.

For example, in a simulated UAV supervisory monitoring task, Ruff, et al (2002) identified that there were different benefits for automation that is managed by consent (i.e., automation that suggests an appropriate course of action but does not act until the operator gives approval). There were also identified benefits where automation is managed by exception (i.e., automation which performs a recommended course of action unless it is otherwise commanded by the operator). Further research is still required to identify which of UAV operator tasks (e.g., traffic detection, flight control, system failure detection) should be automated and what levels of automation are optimal.

Training Situation Awareness in UAV Operators

We have previously established that designing an interface that offers an operator an appropriate level of awareness of the activities of the UAV under the operator's control and of other operator's involved in the mission is crucial in reducing information overload, distraction, mis-communication and coordination breakdowns. However, supporting operator SA involves more than simply providing information.

Improved UAV designs and better flight training are important processes that help increase SA capability of pilots. Bolstad, Endsley, Costello, and Howell (2010) investigated the effectiveness of six modules of training for developing and maintaining SA by using the general aviation version of the Situation Awareness Global Assessment Technique (SAGAT) installed on a computer located next to the simulator to measure SA. The study found that the training modules improved participants' performance on these targeted skills. Results also provided promising support for the effect of the training modules in improving situation awareness.

Sorenson, Stanton, and Banks (2011) compared three theoretical frameworks covering psychology, engineering and systems ergonomics for further understanding and improving pilot SA. While engineering and psychology offer considerable knowledge in our understanding of SA, both disciplines rarely consider the interaction between the individual, the artefact and the context in which they operate. However, the systems ergonomics perspective offers a more holistic framework to investigate SA by exploring the complex interaction between the individual operator, the artefact and their environment. Matthews, Eid, Johnsen, and Boe (2011) also explored SA assessment utilizing an observer and self-rating methods under highly stressful and challenging training conditions. The findings revealed that subjective SA measures would not likely produce defensible estimates of SA in extreme conditions. Another study found that following a malfunction in a flight simulator, the eye movements of an experienced pilot significantly differed from a novice pilot, suggesting a different disruption to SA (van de Merwe, van Dijk, & Zon, 2012).

Throughout a mission, a pilot commonly alters their levels of supervisory control between a full auto pilot and other modes. This is certainly common in a UAV operating environment, and depends greatly on the level of autonomy and supervision required in various stages of flight. These transitions will involve some risks and potential reduction in flight performance or an unacceptable change in workload or SA (Nguyen, Lim, Duy Nguyen, Gordon-Brown, & Nahavandi, 2018).

Hainley, Duda, Oman, and Natapoff (2013) measured a pilot's performance, SA and work load over a number of automation mode transitions, in an effort to construct objective measures of gracefulness during mode transition. The experiments demonstrated that mental workload increases, and SA decrease in a monotonic fashion, with relation to the number of manual control loops the pilot is required to close as a result of the flight mode transition. The research also highlighted the reduced attention of the pilot to fuel status, terrain and altitude during times of high workload due to the attentional demands of manual control tasks.

Cuevas and Aguiar (2017) evaluated a behavioural measure to assess SA and understand how specific operator characteristics (knowledge, skills, and abilities (KSAs)) impact on the success of the mission in UAV operations. The results revealed that participants with greater manned flight experience performed better with respect to SA elements because, as expected, pilots of manned aircraft generally receive rigorous human factors and/or CRM training throughout their flight training (Cuevas & Aguiar, 2017). The study also demonstrated a statistically significant positive correlation between gaming experience with First-Person Shooter (FPS) games and indicators for spatial orientation (Cuevas & Aguiar, 2017). The researchers explain that this result is most likely due to the requirement for spatial awareness in these kinds of games where the player is an avatar in a virtual world. For the player to succeed, they must possess the skill to receive and comprehend all the available information to correctly assess their situation (Cuevas & Aguiar, 2017).

While the majority of these studies investigated manned flight operations, there are very limited studies that investigate effective assessment and training programs to enhance SA in UAV operations. Yet, many of the findings are relevant to UAV operations and provide a solid platform from which appropriate UAV operator training programs to enhance the non-technical skill of SA may be constructed. Perhaps the most striking findings, that best inform improvements to SA training for UAV operators are summarised below:

- UAV operators with greater manned flight experience performed better with respect to SA elements (Cuevas & Aguiar, 2017).
- UAV operators with gaming experience associated with FPS games showed improved performance in spatial orientation (Cuevas & Aguiar, 2017).
- UAV operators trained to effectively manage workload in highly stressful flight situations are likely to demonstrate positive skills of social cognition (social SA).
- UAV operator training needs to include pilot appreciation of the increased risk associated with transitions and changes in automation modes depending on the stages of flight. Such transitions will involve some risks and potential reduction in flight performance or an unacceptable change in workload or SA (Nguyen et al., 2018). As such, UAV operators need to be provided with guidance and procedures to assist in SA management during transitions.

Developing UAV Operator Skills in Workload Management

One of the most significant views on human information processing and workload is the concept that attention is a limited and multi-dimensional resource. There are a number of persuasive arguments to the commonly termed 'Multiple Resource Theory' (MRT) (Wickens, 2002), which provides a sound explanation on what and how humans share different tasks. Furthermore, this theory helps describe how two (or potentially more) tasks can be executed in parallel without any decrease in performance, when being time-shared. This particular theory has been acknowledged in highly dynamic multi-task environments such as aviation flight decks, Air Traffic Management (ATM), and hospital emergency departments.

Wickens (2002) proposed the following diagram that depicts the four categorical dimensions of mental resources involving, the processing stage, perceptual modalities (e.g., input and output of visual and auditory information) followed by processing codes.

Within the Visual Modality, there are also visual channels, which encompass the variety of processing between focal and ambient vision. The visual channels help to explain how the resources can be accessed when required. Workload and coordination costs increase and it is likely that performance will be degraded when the two tasks are competing for the same resources. On the dimension regarded as the processing stage, the resource channels used for central processing activities are distinct from the channels used for selection and execution of responses. An example

Figure 1. Diagrammatic representation of Wickens' Multiple Resource Theory

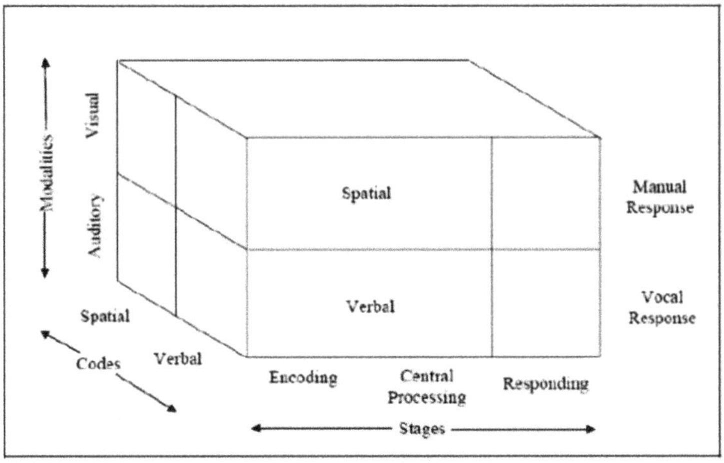

of how the resources within this dimension can be mixed successfully is when an air traffic controller can use voice to acknowledge an alteration in an aircraft flight path (a response demand) while concurrently maintaining an unhindered SA of the airspace (a perceptual-cognitive demand) (Wickens, 2002). This example explains that tasks use different processing stages and different resources. An example of a less effective mix is when trying to comprehend a speech transmission of another person, while concurrently trying to also respond in a verbal activity. In this example, two common stage-defined resources are forced to compete, leading to an increased workload, a strain on working memory, and the potential for error.

The term 'perceptual modalities' is often used to describe the various sensory inputs, most commonly the auditory versus the visual inputs. Wickens (2002) comments that dividing attention between a single visual and single auditory channel (cross-modal time sharing) is often better than trying to divide attention between two visual channels (intra-modal time sharing). A subsequent study (Wickens & Dixon, 2006) involving experiments alleviating high "triple task" workload showed that auditory interfaces in UAV systems, compared to the normal visual displays, markedly reduced workload across multiple resources.

Yet, it is not completely understood why two similar tasks restrict processing, whether it may be a result of resource competition or other peripheral factors. Nevertheless, it is considerably easier to process dual visual inputs, compared to two auditory inputs. Wickens (2002) claims that the basis of these realizations may hinge on the concept that the two different visual channels, focal and ambient vision, use separate resources. The focal vision provides humans with the ability to recognize patterns, while ambient vision (which is peripheral) is used for sensing orientation and movement.

Processing codes establish how and why individual resource channels are employed depending on the activity type being performed. Cognitive processes such as spatial and verbal processing resources, may be unclear as to which resource channel they use (Low, 2004). For example, a UAV operator, in an effort to avoid poor weather conditions, may change the vehicle course using a mathematical calculation (verbal strategy), or visualizing the vectors of the UAV, either by the observed camera fed airspace or on the map display (spatial visualization strategy).

The utilization of the MRT model allows us to forecast the different levels of interruptions and interference between tasks, and gives guidance to UAV designers in developing interfaces and displays that aim to reduce workload to an appropriate level and prohibit any potential for cognitive overload. We can use this work to assist decision making about when it may be more appropriate to use voice control rather than manual control, joystick inputs instead of predefined options, or special graphic (e.g., maps, symbols or graphs) rather than verbal material (e.g., tables or text).

In the case of UAV operations, workload will be altered in the future as we integrate data from multiple sources, perform more mental (compared to physical) tasks, and ensure that the tasks can be combined (with reference to MRT), and any additional or new tasks do not have an overwhelming effect on the computational abilities of the operator.

Training Workload Management in UAV Operators

For aviation professionals to perform in high-risk environments, they must be proficient in non-technical skills of SA and workload management. In earlier sections, some suggestions to enhance UAV operator training programs with regard to SA were provided. In this section, strategies to enhance UAV operator training specific to workload management are discussed.

Earlier, we established that perfect automation may offer benefits when workload is high, because either the task being automated is difficult (Maltz & Meyer, 2003) or other multi-task responsibilities are in direct competition for the operators' limited attentional resources (Wickens & Dixon, 2006).

In the case of UAV operations, workload levels are constantly changing between extended periods of low workload and intense periods of high workload (De la Torre et al., 2016). We have discussed the negative impacts on flight performance when cognitive demands overwhelm the capabilities of the operator, and during long periods of low cognitive demand. As such, effective work design or work scheduling generally aims to avoid these extremes of high or low demand.

The variations in workload raises an important human factor issue concerning the human weakness in maintaining vigilance during long periods of relatively low task demand. For this reason, attention and cognitive arousal must remain at sufficient level to stimulate the correct response to some (unlikely) occurrence, such as a system breakdown or critical event. Unfortunately, humans have been found to be particularly bad at sustaining alertness for lengthy periods of time.

When automation is reliable and working as designed, human attention and monitoring effectiveness naturally decline. Even though a human (UAV operator or pilot) may be highly motivated, it is unfeasible for them to retain effective visual attention with regard to an unchanging source for more than thirty minutes (Bainbridge, 1987).

Finally, De la Torre et al. (2016) explains that there are numerous advantages in the use of simulators for training future UAV pilots. One of the most significant advantages in the use of simulators is the ability to train a UAV pilot at reduced cost, in real-time. However, an important element related to enhancing the non-technical skill of workload management is that simulator training allows the UAV pilot to practice problem solving and management methods facing new and often complex scenarios.

Of all the UAV operator tasks De la Torre et al. (2016) studied in the simulator, those tasks that related to landing indicated higher error rates. Landing was also considered the task with higher mental demand. Hart and Hauser (1987) noted the mental burden on civil aviation pilots during routine flights, documenting high peak workload during takeoff and landing. Data and outcomes discovered in simulator training is often similar to that observed in real flight situations (Staffan, 2002). This emphasizes not only the benefits and effectiveness of simulator training, but also the important research and learning that can be achieved from simulator work in terms of how humans perform when interacting with UAVs.

Considering the limitations of humans in cognitive processing, the practicalities of UAV operations and varied levels of automation, the isolation from the aircraft and the challenges UAV operators face with regards to extremes in workload levels, the above-mentioned research provides some interesting insights. These insights help to provide evidence of the importance of workload management training for future UAV operators. Perhaps the most striking findings, which may best inform improvements to workload management training for UAV operators can be summarised below:

- Workload and schedule planning are an important activity in removing extreme periods of workload (during both high and low levels of cognitive demand).

- UAV operators need to be aware of attention degradation during periods of low workload and must allow for task changes as well as considering procedural activities that stimulate changes in cognitive processing.
- The use of simulators is evidently beneficial in UAV operator training with the opportunity to build skills in managing complex situations.

FUTURE RESEARCH DIRECTIONS

Earlier in the chapter, we discussed the value of multi-modal displays as a method for compensating for sensory information that UAV operators lose with conventional displays. These displays may impose significant costs for the business. Future research is required to investigate the costs and benefits of multi-modal displays in resolving the UAV operator's sensory isolation as well as the effective management of workload, and to uncover the optimal design of such displays.

We also discussed the challenges of data link bandwidths, which limit the transfer of accurate and timely information from the aircraft's on board sensors back to the UAV operator. This delayed feedback can hinder the UAV operator's performance, response to challenging scenarios and ability to resolve emergency situations. Additional research is required to identify the effect of reduced spatial and /or temporal resolution of restricted field of view on other details of UAV and payload sensor control. Also, related to this challenge is the need for further research, which investigates the optimal use of available bandwidth by prioritizing information transfer, based on the stage of flight.

More research is required to examine the effect of augmented reality in an effort to assist in improving sensor input as this technology has the opportunity to enhance UAV flight control. Van Erp and Van Breda (1999) discovered that augmented reality displays could enhance the accuracy and reduce the cognitive demands of target tracking with a payload sensor. The performance improvements provided by augmented reality were later also supported by Ruano et al. (2017). By extension, this technology provides the opportunity to enhance UAV flight control.

More research is required to provide guidance to curriculum designers and trainers to assist future UAV pilots in managing the complex tasks and demands of their roles. Dramatic improvements in technology will mean a greater reliance on automation for many UAV pilots. However, effective human interaction with the UAV will continue to be a challenge. Nevertheless constant attention to research and update of training programs will ensure safe and effective UAV operations.

CONCLUSION

Civilian applications for UAV operations are expanding rapidly. Examples of their use include (but are not limited to): providing disaster relief; transporting vaccinations; surveying landscapes; capturing geographical and agricultural data; monitoring sea, land and road operations; and delivering packages. The emergence and growth of UAV operations has highlighted the urgent need for trained operators to ensure these systems are used effectively and safely. In this chapter, we have emphasized the importance and integration of appropriate Non Technical Skills (NTS) training, particularly Situation Awareness (SA) and workload management, to further improve UAV mission effectiveness.

We have examined the technical design and human factors challenges impacting on UAV operations. Technological solutions to improving SA for UAV operators include consideration of: the principles of UAV display; the removal of the pilot from the aircraft resulting in sensory isolation; data links and sensor imagery; lengthy periods of monitoring highly automated systems which lead to the operator feeling 'out of the loop'. However, the chapter also highlighted important considerations to inform improvements to training SA in UAV operators including: UAV operators with greater manned flight experience perform better with respect to SA elements; UAV operators with gaming experience show improvements in spatial orientation; procedures and the design of systems should consider supporting crew to better manage workload during stressful events; and UAV operator training should include pilot appreciation of increased risk associated with transitions in automation levels.

In this chapter, we also discovered that SA skills are heavily integrated with workload management skills. Technological solutions to improving workload management requires aircraft systems, display and interface designers to have a deep understanding of the Wickens' (2002) Multi-Resource Theory. This understanding of human cognition, available resources, processing stages and channels, will better inform aircraft designers in the strengths and limitations of human processing within the various levels of automation provided by UAV operations. However, the chapter also highlighted important considerations to inform improvements to the training of workload management skills for UAV operators. This includes consideration of: the importance of workload and schedule planning to remove extreme (high and low) periods of workload; providing UAV operators with greater awareness of attention degradation during periods of low workload and providing guidance on strategies to address this issue; and the benefits of simulator training in allowing pilots to build skills in managing complex situations.

REFERENCES

Abbott, K. A., Slotte, S. M., & Stimson, D. K. (1996). *Federal Aviation Administration Human Factors Team Report on: The Interfaces Between Flightcrews and Modern Flight Deck Systems.* Technical Report. Retrieved August, 2018, from http://www.tc.faa.gov/its/worldpac/techrpt/hffaces.pdf

Adams, M. J., Tenney, Y. J., & Pew, R. W. (1995). Situation awareness and the cognitive management of complex systems. *Human Factors*, *37*(1), 85–104. doi:10.1518/001872095779049462

Astuti, G., Longo, D., Melita, D., Muscato, G., & Orlando, A. (2008). HIL tuning of UAV for exploration of risky environments. *International Journal of Advanced Robotic Systems*, *5*(4), 419–424. doi:10.5772/6227

Bainbridge, L. (1987). Ironies of automation. In J. Rasmussen, K. Duncan, & J. Leplat (Eds.), *New technology and human error* (pp. 271–283). New York: Wiley.

Barnes, M. J., & Matz, M. F. (1998). *Crew simulations for Unmanned Aerial Vehicles (UAV) applications: Sustained effects, shift factors, interface issues, and crew size.* Paper presented at the Human Factors and Ergonomics Society Annual Meeting, Santa Monica, CA. 10.1177/154193129804200132

Bolstad, C. A., Endsley, M. R., Costello, A. M., & Howell, C. D. (2010). Evaluation of computer-based situation awareness training for general aviation pilots. *The International Journal of Aviation Psychology*, *20*(3), 269–294. doi:10.1080/10508414.2010.487013

Brar, S., Rabbat, R., Raithatha, V., Runcie, G., & Yu, A. (2015). *Drones for Deliveries.* Berkeley, CA: Sutardja Center Entrepreneurship Technol.

Broadbent, D. (1958). *Perception and Communication.* London: Pergamon Press. doi:10.1037/10037-000

Buttigieg, M., & Sanderson, P. (1991). Emergent features in visual display design for two types of failure detection tasks. *Human Factors*, *33*(6), 631–351. doi:10.1177/001872089103300603 PMID:1800293

Calhoun, G. L., Draper, M. H., Abernathy, M. F., Patzek, M., & Delgado, F. (2005). Synthetic vision systems for improving unmanned aerial vehicle operator situation awareness. *Proceedings of SPIE: Enhanced and Synthetic Vision.* 10.1117/12.603421

Calhoun, G. L., Draper, M. H., & Ruff, H. A. (2002). Utility of a tactile display for cue faults. *Proceedings of the Human Factors and Ergonomics Society 46th Annual Meeting.*

Casbeer, D., Beard, R., McLain, T., Li, S., & Mehra, R. (2005). *Forest fire monitoring with multiple small UAVs*. Paper presented at the American control conference. 10.1109/ACC.2005.1470520

Connors, M., Harrison, A., & Akins, F. (Producer). (1985). *Living aloft: Human requirements for extended spaceflight*. Retrieved from https://ntrs.nasa.gov/search.jsp?R=19850024459

Crandall, J. W., & Cummings, M. L. (2007). *Developing performance metrics for the supervisory control of multiple robots*. Paper presented at the ACM/IEEE international conference on human-robot interaction. 10.1145/1228716.1228722

Cuevas, H. M., & Aguiar, M. (2017). Assessing Situation Awareness in Unmanned Aircraft Systems Operations. *International Journal of Aviation, Aeronautics, and Aerospace*, *4*(4).

Cummings, M. L., Bruni, S., Mercier, S., & Mitchell, P. J. (2007). Automation Architecture for Single Operator, Multiple UAV Command and Control. *The International C2 Journal*, *1*(2), 1-24.

Cummings, M. L., & Mitchell, P. J. (2007). Operator scheduling strategies in supervisory control of multiple UAVs. *Aerospace Science and Technology*, *4*(11), 339–348. doi:10.1016/j.ast.2006.10.007

De la Torre, G., Ramallo, M., & Cervantes, E. (2016). Workload perception in drone flight training simulators. *Computers in Human Behavior*, *64*, 449–454. doi:10.1016/j.chb.2016.07.040

Dixon, S. R., Wickens, C. D., & Chang, D. (2003). Comparing quantitative model predictions to experimental data in multiple-UAV flight control. *Proceedings of the Human Factors and Ergonomics Society 47th Annual Meeting*. 10.1177/154193120304700122

Draper, M. H., Ruff, H. A., Repperger, D. W., & Lu, L. G. (2000). Multi-sensory interface concepts supporing turbulence detection by UAV controllers. *Proceedings of the First Human Performance, Situational Awareness and Automation Conference*.

Drury, J. L., & Scott, S. D. (Producer). (2008). *Awareness in UAV Operations*. Retrieved from http://www.cs.uml.edu/~holly/91.550/papers/Drury_Scott_UAV-Awareness.pdf

Endsley, M. R. (2000). Theoretical Underpinnings of Situation Awareness: a Critical Review. In M. R. Endsley & D. J. Garland (Eds.), *Situation Awareness Analysis and Measurement*. Mahwah, NJ: Lawrence Erlbaum Associates.

Funk, K., Suroteguh, C., Wilson, J., & Lyall, B. (1998). *Flight deck automation and task management.* Paper presented at the IEEE 1998 International Conference on Systems, Man, and Cybernetics, New York, NY. 10.1109/ICSMC.1998.725523

Gawron, V. J. (1998). *Human factors issues in the development, evaluation, and operation of uninhabited aerial vehicles.* Paper presented at the Associations of Unmanned Vehicles Systems International.

Green, S., Billinghurst, M., Chen, X. Q., & Chase, J. G. (2008). Human-Robot Collaboration: A Literature Review and Augmented Reality Approach in Design. *International Journal of Advanced Robotic Systems, 5*(1), 1–18. doi:10.5772/5664

Gupta, S. G., Ghonge, M. M., & Jawandhiya, P. M. (2013). Review of unmanned aircraft system (UAS). *Technology, 2*(4).

Haidari, L., Brown, S., Ferguson, M., Bancroft, E., Spiker, M., Wilcox, A., & Lee, B. (2016). The economic and operational value of using drones to transport vaccines. *Vaccine, 34*(34), 4062–4067. doi:10.1016/j.vaccine.2016.06.022 PMID:27340098

Hainley, J. C. J. Jr, Duda, K. R., Oman, M., & Natapoff, A. (2013). Pilot performance, workload, and situation awareness during lunar landing mode transitions. *Journal of Spacecraft and Rockets, 50*(4), 793–801. doi:10.2514/1.A32267

Hart, S. G., & Hauser, J. R. (1987). Inflight application of three pilot workload measurement techniques. *Aviation, Space, and Environmental Medicine.* PMID:3593142

Helmreich, R. L., & Merritt, A. C. (2000). Safety and error management: The role of crew resource management. In B. J. Hayward & A. C. Lowe (Eds.), *Aviation resource management* (pp. 107–119). Aldershot, UK: Ashgate.

Helmreich, R. L., Merritt, A. C., & Wilhelm, J. A. (1999). The evolution of crew resouce management. *The International Journal of Aviation Psychology, 9*, 19–32. doi:10.120715327108ijap0901_2 PMID:11541445

Helmreich, R. L., & Wilhelm, J. A. (1991). Outcomes of crew resource management training. *The International Journal of Aviation Psychology, 1*(4), 287–300. doi:10.120715327108ijap0104_3 PMID:11537899

Hopcroft, R., Burchat, E., & Vince, J. (2006). Unmanned aerial vehicles for maritime patrol: Human factors issues (No. DSTO-GD-0463). Air operations division: Defence sicenice and technology organisation Edinbrgh (Australia).

Kaupp, T., & Makerenko, A. (2008). *Measuing human-robot team effectiveness to determine an appropriate autonomy level.* Paper presented at the Robotics and automation, 2008, ICRA 2008, IEEE international conference.

Kintsch, W., & Ericsson, K. A. (1995). Long term working memory. *Psychological Review, 102*(2), 211–245. doi:10.1037/0033-295X.102.2.211 PMID:7740089

Kipper, G., & Rampolla, J. (2013). *Augmented Reality: An emerging technologies guide to AR*. Waltham, MA: Elsevier.

Klingberg, T. (2000). Limitations in information processing in the human brain: Neuroimaging of dual task performance and working memory tasks. *Progress in Brain Research, 126*, 95–102. doi:10.1016/S0079-6123(00)26009-3 PMID:11105642

Kontitsis, P., Tsourveloudis, G., & Valavanis, P. (2003). *A UAV based automated airborne survellance system.* Paper presented at the 11th Mediterranean conference on control and automation, Rhodes, Greece.

Lauber, J. K. (1987). Cockpit resource management: Background and overview. In H. W. Orlady & H. C. Foushee (Eds.), *Cockpit resource management training: Proceedings of the NASA/MAC workshop* (pp. 5-14). Moffett Field, CA: NASA-Ames Research Center.

Low, I. (2004). A Tool for the Assessment of the Impact of change in Automated ATM Systems on Mental Wokload (Report: HRS/HSP-005-REP-03). Brussels: European Air Traffic Management (EATM).

Maltz, M., & Meyer, J. (2003). Use of warnings in an attentionally demanding detection task. *Human Factors, 43*, 217–226.

Manning, S. D., Rash, C. E., LeDuc, P. A., Noback, R. K., & McKeon, J. (2004). *The Role of human causal factors in US army unmanned aerial vehicle accdients (No. USAARL-2004-11).* Fort Rucker, AL: Army Aeromedical Research Lab. doi:10.21236/ADA421592

Matthews, M. D., Eid, J., Johnsen, B. H., & Boe, O. C. (2011). A comparison of expert ratings and self-assessments of situation awareness during a combat fatigue course. *Military Psychology, 23*(2), 125–136. doi:10.1080/08995605.2011.550222

McCarley, J. S., & Wickens, C. D. (2005). *Human factors implications of UAVs in the national airspace.* Aviation Human Factors Division: University of Illinois at Urbana-Champaign.

Meshkati, N., Hancock, P., & Rahimi, M. (1992). Techniques in mental workload assessment. In J. Wilson & E. Corlett (Eds.), *Evaluation of human work. A practical ergonomics methodology* (pp. 605–627). London: Taylor & Francis.

Mouloua, M., Gilson, R., Daskarolis-Kring, E., Kring, J., & Hancock, P. (2001). *Ergonomics of UAV/UCAV mission success: Considerations for data link, control and display issues.* Paper presented at the Human Factors and Ergonomics Society 45th Annual Meeting, Santa Monica, CA. 10.1177/154193120104500231

Mouloua, M., Gilson, R., Kring, J., & Hancock, P. (2001). *Workload, Situation Awareness, and Teaming Issues for UAV/UCAV Operations.* Paper presented at the Human Factors and Ergonomics Society 45th Annual Meeting. 10.1177/154193120104500235

Nguyen, T., Lim, C. P., Duy Nguyen, N., Gordon-Brown, L., & Nahavandi, S. (2018). A Review of Situation Awareness Assessment Approaches in Aviation Environments. *Eprint ArXiv 1803 (08067)* Retrieved 29 December, 2018, from https://arxiv.org/ftp/arxiv/papers/1803/1803.08067.pdf

Nisser, T., & Westin, C. (2006). *Human Factors Challenges in Unmanned Aerial Vehicles (UAVs): A Literature Review*. Lund University: School of Aviation.

Rathinam, S., Almeida, P., Kim, Z., Jackson, S., Tinka, A., & Grossman, W. (2007). *Automous searching and tracking of a river using an UAV.* Paper presented at the American control conference.

Reason, J. (1997). *Managing the risks of organizational accidents.* Aldershot, UK: Ashgate.

Ruano, S., Cuevas, C., Gallego, G., & Garcia, N. (2017). Augmented Reality Tool for the Situational Awareness Improvement of UAV Operators. *Sensors (Basel), 17*(2), 297. doi:10.339017020297 PMID:28178189

Rubio, S., Diaz, E., Martin, J., & Puente, J. (2004). Evaluation of subjective mental workload: A comparison of SWAT, NASA-TLX, and workload profile methods. *Applied Psychology, 53*(1), 61–86. doi:10.1111/j.1464-0597.2004.00161.x

Ruff, H. A., Narayanan, S., & Draper, M. H. (2002). Human interaction with levels of automation and decision-aid fidelity in the supervisory control of mulitple simulated unmanned air vehicles. *Presence (Cambridge, Mass.), 11*(4), 335–351. doi:10.1162/105474602760204264

Sarter, N. B., Woods, D. D., & Billings, C. E. (1999). *Automations Surprises*. Columbus, OH: The Ohio State University, Cognitive Systems Engineering Laboratory.

Sauer, J., Wastell, D., & Hockey, G. (1996). Skill maintenance in extended spaceflight: A human factors analysis of space and analogue work environments. *Acta Astronautica*, *39*(8), 579–587. doi:10.1016/S0094-5765(97)00006-4 PMID:11540781

Schutte, P. C., & Trujillo, A. C. (1996). *Flight crew task management in non-normal situations*. Paper presented at the 40th Annual Meeting of the Human Factors and Ergonomics Society, Santa Monica, CA. 10.1177/154193129604000422

Sorenson, L. J., Stanton, N. A., & Banks, A. P. (2011). Back to SA school: Contrasing three approaches to situation awareness in the cockpit. *Theoretical Issues in Ergonomics Science*, *12*(6), 451–471. doi:10.1080/1463922X.2010.491874

Stacy, N. J. S., Craig, D. W., Staromlynska, J., & Smith, R. B. (2002). *The Global Hawk UAV Australian deployment: imaging radar sensor modifications and employment for maritime surveillance*. Paper presented at the Geoscience and Remote Sensing Symposium 2002. 10.1109/IGARSS.2002.1025623

Staffan, M. (2002). Similarities and differences in psychophysiological reactions between simulated and real Air-to-ground missions. *The International Journal of Aviation Psychology*, *12*(1), 47–61.

Thomas, M. J. W. (2004). Predictors of Threat and Error Management: Identification of Core Nontechnical Skills and implications for Training Systems Design. *The International Journal of Aviation Psychology*, *14*(2), 207–231. doi:10.120715327108ijap1402_6

Tvaryanas, A. P., Thompson, B. T., & Constable, S. H. (2005). *U.S. Military Unmanned Aerial Vehicle Mishaps: Assessment of the Role of Human Factors using HFACS, Technical Report*. Brooks City-Base, TX: USAF Human Performance Directorate.

van de Merwe, K., van Dijk, H., & Zon, R. (2012). Eye movements as an indicator of situation awareness in a flight simulator experiment. *The International Journal of Aviation Psychology*, *22*(1), 78–95. doi:10.1080/10508414.2012.635129

Van Erp, J. B. F. (2000). *Controlling unmanned vehicles: The human factors solution*. Paper presented at the RTO Meeting Proceedings 44 (RTO-MP-44).

Van Erp, J. B. F., & Van Breda, L. (1999). Human factors issues and advanced interface design in maritime unmanned aerial vehicles: A Project Overview (Report TNO TM-99-A004). Soesterberg, The Netherlands.

Weimer, J. (1995). *Research techniques in human engineering*. Englewood Cliffs, NJ: Prentice Hall.

Wickens, C. D. (1999). Cognitive factors in aviation. In Handbook of applied cognition (pp. 247-282). John Wiley & Sons.

Wickens, C. D. (2002). Multiple resources and performance prediction. *Theoretical Issues in Ergonomics Science, 3*(2), 159–177. doi:10.1080/14639220210123806

Wickens, C. D. (2003). Pilot actions and tasks: Selection, execution, and control. In P. Tsang & M. Vidulich (Eds.), *Principles and Practice of Aviation Psychology* (pp. 239–263). Mahwah, NJ: Erlbaum.

Wickens, C. D., & Dixon, S. R. (2006). *Human machine interface analysis of unmanned vehicle systems* (Technical report HFD-06-1/MAAD-06-01). Savoy, IL: University of Illinois, Institute of Aviation, Aviation Human Factors Division.

Wickens, C. D., & Hollands, J. G. (2000). *Engineering psychology and human performance* (3rd ed.). Upper Saddle River, NJ: Prentice Hall.

Wiegmann, D. A., & Shappell, S. A. (2001). Human error perspectives in aviation. *The International Journal of Aviation Psychology, 11*(4), 341–357. doi:10.1207/S15327108IJAP1104_2

Wiener, E. L., Kanki, B. G., & Helmreich, R. L. (1993). *Cockpit resource management*. San Diego, CA: Academic.

Wikens, C. D., Dixon, S., & Chang, D. (2003). *Using Interference Models to Predict Performance in Multiple-Task UAV Environment - 2 UAVs: AHFD-03-9/MAAD-03-1*. Savoy, IL: Aviation Human Factors Division, Institute of Aviation, Universit of Illinois at Urbana-Champaign. doi:10.21236/ADA496812

Yeh, M., & Wickens, C. D. (2001). Display signalling in augmented reality: Effects of cue reliability and image realism on attention allocation and trust calibration. *Human Factors, 43*(3), 355–365. doi:10.1518/001872001775898269 PMID:11866192

ADDITIONAL READING

Bennet, W., Jr., Bridewell, J. B., Rowe, L. J., Craig, S. D., & Poole, H. M. (2016). Training Issues for Remotely Pilot Aircraft Sytems from a Human Systems Integration Persective. In N. J. Cooke, L. J. Rowe, W. Bennett, D. Q. Joralmon (Eds.), Remotely Piloted Aircraft Systems (pp. 163-176). John Wiley & Sons.

De la Torre, G., Ramallo, M., & Cervantes, E. (2016). Workload perception in drone flight training simulators. *Computers in Human Behavior*, *64*, 449–454. doi:10.1016/j.chb.2016.07.040

Drury, J. L., & Scott, S. D. (Producer). (2008). Awareness in UAV Operations. Retrieved from http://www.cs.uml.edu/~holly/91.550/papers/Drury_Scott_UAV-Awareness.pdf

Endsley, M. R. (2000). Theoretical Underpinnings of Situation Awareness: a Critical Review. In M. R. Endsley & D. J. Garland (Eds.), *Situation Awareness Analysis and Measurement*. Mahwah, NJ: Lawrence Erlbaum Associates.

Hing, J., & Oh, P. Y. (2009). Development of an Unmmaned Aerial Vehicle Pilot System with Integrated Motion Cueing for Training and Pilot Evaluation. *Journal of Intelligent & Robotic Systems*, *54*(1-3), 3–19. doi:10.100710846-008-9252-3

McCarley, J. S., & Wickens, C. D. (2005). *Human factors implications of UAVs in the national airspace*. Aviation Human Factors Division: University of Illinois at Urbana-Champaign.

Riganati, J. P., & Harrison, G. (2016). The Path Forward for Remotely Piloted Aircraft Systems. In N. J. Cooke, L. J. Rowe, W. Bennett, D. Q. Joralmon (Eds.), Remotely Piloted Aircraft Systems (pp. 223-240): John Wiley & Sons. doi:10.1002/9781118965900.ch10

Wiegmann, D. A., & Shappell, S. A. (2001). Human error perspectives in aviation. *The International Journal of Aviation Psychology*, *11*(4), 341–357. doi:10.1207/S15327108IJAP1104_2

KEY TERMS AND DEFINITIONS

Artefact: Shein's model of culture is well established in the aviation human factors area. Shein divides organizational culture into three levels. One level refers to artefacts, which are the visual organizational structures and processes that are easy to identify, yet are hard to understand.

Flight Management Systems (FMS): A fundamental component of a modern airliner's avionics and a specialized computer system that automates a wide variety of in-flight tasks, reducing the workload on the flight crew.

Fly-by-Wire: A semi-automatic and typically computer-regulated system for controlling the flight of an aircraft that replaces the conventional manual flight controls of an aircraft with an electronic interface.

Haptic Displays: Haptic technology or kinesthetic communication recreates the sense of touch by applying forces, vibrations, or motions to the user. Haptic devices may incorporate tactile sensors that measure forces exerted by the user on the interface.

"Inhabited" Aviation: An aircraft that is controlled by a pilot that inhabits the aircraft cockpit during operation.

Non-Technical Skills (NTS or NOTECHS): Encompasses attributes including the ability to recognize and manage human-performance limitations, to make sound decisions, communicate effectively, lead, and work as a team and maintain situation awareness. When coupled with strong technical skills, NTS is the difference between performance that is acceptable and performance that is outstanding.

Situational Awareness: The perception of environmental elements and events with respect to time or space, the comprehension of their meaning, and the projection of their future status.

Stick-and-Rudder: A colloquialism used in the aviation industry to refer to the manual manipulation of controls in the cockpit to manage the flight path of an aircraft.

Threat and Error Management: The process of identifying and responding to threats with appropriate countermeasures to reduce or remove their consequences and mitigate the probability of errors.

Vestibular Input: The sensation of any change in position, direction, or movement of the head. The receptors are located in the inner ear and are activated by the fluid in the ear canals moving as you move.

Chapter 2
Delivery of Special Cargoes Using the Unmanned Aerial Vehicles

Kristina Marintseva
National Aviation University, Ukraine

Gennadiy Yun
National Aviation University, Ukraine

Igor Vasilenko
National Aviation University, Ukraine

ABSTRACT

This chapter presents the role, functions, and prospects of civil unmanned aerial vehicles development, as well as technical and regulatory barriers to the introduction of unmanned aircraft into special cargo transportation technology. The authors' main idea is that the degree of UAV involvement in freight traffic will continue to grow rapidly as the range of UAV flight and carrying capacity increases, and the air law is liberalized. It is proposed to evaluate the economic efficiency of UAV application and their share in the market for the transportation of urgent and perishable goods using the methodology based on the principles of logistics and mathematical modeling. In the formulated model, the process of special cargoes delivery by unmanned aerial vehicles is integrated into the supply chain by all modes of transport along the set route network, taking into account the requirements formulated by the freight forwarder, carrier, and logistics company.

DOI: 10.4018/978-1-5225-7900-7.ch002

INTRODUCTION

The world economy globalization and the processes connected with this phenomenon largely influenced the overall structure of the world markets of air transportation and freight forwarding, including those requiring special delivery and storage conditions. On the one hand, the Freight Forwarding Companies (FFC) have significantly expanded the range of services. On the other hand, the client's demands for the goods delivery have increased, in particular in terms of comprehensive services, the full responsibility of the FFC for delivery management, shipment processing, formalities, cargo collection and consolidation etc.

Special, perishable and dangerous goods delivery is a popular and interesting market segment for FFC. To ensure effective delivery of special cargo, it is required to consider the whole chain of cargo delivery and take steps to minimize the time of goods delivery. This requirement is particularly important for regions with poorly developed transport connections, where it is unfeasible to use the charter flights or aviation. However, as it often happens in business, the relevant solution to the problem was found in military aviation, which has been using unmanned aerial vehicles (UAVs) in its operations for a long time. The civil UAV test flights performed in some countries of Europe, Africa, and Asia with the aim of medicines delivery to inaccessible communities and food (pizza, beer) supply to the vicinity of the city were done with minimal costs and minimal risks (Domino's, 2016, Bryan, Char, 2014, Rosen, 2017).

The objectives of this chapter are to educate the reader with information concerning theoretical and methodological approaches and practical recommendations for managing the economic efficiency of multimodal delivery of special cargoes with an emphasis on the capabilities and prospects of using UAVs. To achieve this goal, the following sequence of studies is determined:

1. About the present and future of UAV.
2. Role of UAV in forwarding operations.
3. Special cargo delivery modeling.

The authors of this work are not deluded by numerous iridescent publications in the media about the fast and widespread use of UAVs in many spheres of our life and, first of all, in the goods transportation. There are many barriers, which should be overcome in this unquestionably innovative direction, namely the technical, organizational and economic ones. The most important barrier is the regulatory one. Carriers and FFCs, apparently, should not waste time in anticipation of the final version of the legislation on UAV flights regulation, but to prepare for such a desired invasion of drones in advance.

BACKGROUND

21st century aviation is the unmanned aviation era. The impetus to the intensive development of unmanned aviation, as well as many other high-tech branches of engineering science, was the widespread use of unmanned aerial vehicles in the armed forces of the United States, Israel and other countries. More than 100 years ago, Nikola Tesla developed and demonstrated a miniature radio-controlled vessel (Cheney, 2001). Since then, the theory and practice of unmanned aviation has made a huge leap from the unmanned aerial vehicle to the unmanned aerial system. For clarity, some core definitions are specified below.

According to the ICAO Doc 9854 (2005), unmanned aerial vehicle is a pilotless aircraft in the sense of Article 8 of the Convention on International Civil Aviation, which is flown without a pilot-in-command on-board and is either remotely and fully controlled from another place (ground, another aircraft, space) or programmed and fully autonomous. An unmanned aerial system (UAS) includes not only the UAV, but also additional components, such as an autonomous or human-operated control system, a command and control system, technical personnel, software, means of integration with other systems, technical and regulatory documentation (Fetisov, 2014). Appropriately equipped UAVs determine the UAS industry specialization, for example, agrarian, transport or military UAS, etc.

UAVs are manufactured all over the world. USA, Israel, Germany, France, Japan, China and a few other countries are the leaders in this segment. According to PwC's study (2016), the commercial application of drone technology, the emerging global market for business services using drones is valued at over $127 bn. A considerable share should be attributed to the cargo unmanned aviation, namely $13 bn.

Today, over dozens of companies, from small start-ups to world-known manufacturers, are working at the creation of multi-rotor UAVs. However, the development of the cargo UAV market towards UAV integration into a single airspace is hampered by technical and regulatory barriers. These problems are not solved completely in any country in the world. For example, Ukraine has been taking just the first steps in this direction. In 2018, to ensure safe flight operations of general-purpose drones within Ukrainian airspace, the State Aviation Service of Ukraine published a new draft concept. The introduction of electronic registration and certification, as well as the right to control the unmanned systems depending on their size and destination is considered for the first time. These measures should greatly simplify the fly-over permission. However, this document should be supplemented by a number of supporting documents containing the detailed rules and guidelines. For now, unmanned systems are purchased by entities having special powers (border guards, army, the Ministry of Emergency Situations) without waiting for the creation of a regulatory framework. Currently, the responsibility for commercial UAVs flights rests with the operator. The issues of certification, insurance, and registration have not been addressed entirely yet.

Among the technical barriers, it should be noted that the most UAVs are not equipped with an obstacle recognition and collision avoidance system, which leads to an increased UAV accident rate (Brar et al., 2015).

Autopilots of some UAV models are less than perfect but used for the sake of sparing the cost and weight of the onboard equipment. There is one vulnerable link in the UAV control system. This is a need for continuous information exchange with ground control points. It is very difficult to ensure a high level of reliability of the large volume of transmitted data. In the simple version, they can be suppressed by noise.

In this chapter, the aviation transport term shall mean a system of means, including the traditional civil aircraft (airplanes and helicopters) and UAVs used to carry cargoes for certain purposes. In the course of its development, unmanned aviation will complement the manned one, and in the foreseeable future, it will be able to replace the same, where UAVs prove to be more efficient. Below, in particular, when the process of special cargo delivery is described by mathematical model, the aggregate fleet of aircraft is considered without its division into manned and unmanned ones.

ABOUT THE PRESENT AND FUTURE OF UAV

Role of Cargo UAV in Air Transportation

According to ICAO forecast, the global freight traffic is expected to grow 4.4 per cent annually by 2032 (ICAO, 2016). It would seem that there is no reason for air carriers' concern over the next 6-7 years. However, "not everything is so smooth in the Danish kingdom." The adverse factors include the peculiarities of the cargo and passenger flow formation, which almost never coincide; a wide range of parameters of the carried goods by weight, dimensions, values, urgency of delivery, packaging conditions, environmental hazards and so on. It is not always possible for carriers to solve the FFCs' tasks by means of traditional vehicles and logistics.

What to do in this difficult situation? It is not new in the history of technology development when the solution is offered by IT applications in the field of robotics and mechatronics. The idea of drone use for commercial cargo delivery is in the air. Despite the existing administrative barriers, the enthusiasts and corporations are investing heavily in the development of this direction, and it looks like it is already paying off. Only a few years have passed since the first publications on the delivery of goods by non-military UAVs were perceived almost as a sensation. And now serious projects with real results began to appear.

Here are a few examples of the most outstanding projects in which their developers have demonstrated the role of cargo UAVs in the future of air transportation not only by words, but also by products ready for serial production.

In December 2016, Amazon, the largest Internet retailer, announced its fastest way to deliver purchases by copters – Amazon Prime Air (Amazon Prime Air, 2016). According to Amazon, through this service, the goods purchased from Amazon online store can be delivered to the customers within 30 minutes, which is 4 times faster than the current fastest deliver service, Amazon Prime Now.

DHL, which is one of the largest logistics companies in the world, shows a steady interest in the use of UAVs for the delivery of goods. Flight tests were conducted in autumn 2014. Within 3 months, drones transported medicines and essentials from Nordeich, Germany, to the island of Joust, Germany. The flight route ran over the North Sea and was 12 km long (Bryan & Char, 2014).

In 2016, Zipline from California, in agreement with the government of the African state of Rwanda, began delivering donor blood and medications by UAVs (Rosen, 2017).

Modeling the delivery of medicines to the refugee camp in Haiti, Matternet from California debuted with the use of drones in 2012. Based on the data obtained during these tests, it was claimed that the cost of delivering 2 kg of cargo over a distance of 10 km was 24 cents (The Guardian, 2013).

In February 2017, the postal giant UPS announced the field trials of a postal van, which was at the same time an aerodrome for drones. The concept of using this machine is the same as in Matternet and Mercedes-Benz project. The van with parcels drives up to the village, and then the drones deliver parcels to the addressees (Stewart, 2017).

The Domino's Pizza projects of pizza delivery by drones are impressive in their depth and diversity (Domino's, 2016). In June 2013, the company introduced a conceptual model for the delivery of pizza by DomiCopter UAVs.

The precedent has already been established; UAVs are now used for air cargo transportation. However, before the expected future of universal UAV application comes, a number of challenges would need to be overcome. The most important of them is to find a balance the interests of all parties who use the UAVs. FFC are interested in effective delivery, citizens need a comfortable environment, and the state needs to maintain the security of the country. Only if such balance is maintained, the desired efficiency of UASs can be achieved.

Functional Capability of Cargo UAV

The main characteristics by which the cargo UAV is evaluated include the load-carrying capacity. Not every drone can be considered a cargo UAV. Not only the total load capacity of the cargo UAV (which can be quite significant), but also its ability to carry cargo of different sizes and different weights is important.

Cargo UAVs must have a load gripper, to which a transported object can be hooked, or which can capture the same. The simplest example is a hook, but it can be slings, chains, special grippers, manipulators, cargo compartments and so on. If the cargo UAV design does not provide for the possibility of installing a load-handling device, the transportation of the goods can be seriously hampered or even impossible. In professional models, capable of lifting several kilograms of cargo into the air, the fixtures should not only ensure the required reliability, but also allow fast fastening and unhooking the cargo.

The issue of the Cargo UAV reliability is equally important. Some of the models are equipped with parachutes, allowing a smooth landing even in the event of failure of all engines. Almost always, there is a video camera on board. It is usually used for the First-Person View (FPV) control. Cargo UAV with a large load capacity also have a high degree of autonomy and are able to travel long distances on one battery charge or on one fuel tank refueling. The visual contact with the cargo UAV can be lost quickly, while the camera allows the operator to maneuver the drone. Video broadcast to the console or to the screen of the mobile device helps controlling.

Another main limiting factor is the flight duration, which is greatly influenced by the energy consumption per unit of flight time of the cargo UAV, all other conditions being equal. Despite the general fascination with cargo copters and convertiplanes, the UAVs of the traditional airplane scheme are undeservedly ignored. If you compare cargo copters and Cargo Unmanned Aerial Plane (UAP), to date, the copter design is inferior to the aircraft design by a number of characteristics. Difficulties begin when trying to transport a light, but space-occupying cargo. The increase in sailings leads to the copter diversion from the course and may even cause an accident.

Powerful brushless engines installed on copters are able to lift a significant weight in the air, but they consume a lot of energy, like the unmanned helicopters powered by batteries. The larger the battery capacity, the more expensive, larger, and heavier it is, and therefore, it proportionally consumes a part of the copter payload. The flight time on one battery rarely exceeds 40 minutes. A light cargo of up to 1 kg can be raised by many amateur copters. The best professional copters raise 225 kg or more in the air. For example, a Norwegian start-up, Griff Aviation, has developed a radio-controlled copter, Griff 300 (Coxworth, 2016), which is capable of carrying a cargo of 225 kg and fly over 30 minutes. American Advanced Tactics introduced

AT Transporter, a multi-purpose ATC capable of carrying cargo up to 544 kg at a speed of 113 km h at an altitude of 3 km (Blain, 2017).

Weather conditions are one of the significant problems for copters. Unlike UAVs of the aircraft type, copters barely tolerate light rain, but they cannot fly in a downpour, snowfall, strong wind, fog, or heavy snow. Negative temperatures have a negative effect on batteries, which quickly lose capacity. The only, perhaps, advantage of copters as compared to UAVs of the aircraft type is the possibility of vertical takeoff and landing; in many cases it makes it allows delivering cargo "door to door", although in the logistics supply chain, this advantage can be utilized only in part because of the limited flight range of the modern copters.

At the same time, cargo UAPs are able to raise tens and hundreds of kilograms, and in the near future even tons of payload. Let's start with a small UAP DR-60 (Figure 1), developed in Ukraine by the head of AeroDrone, Y. Pederi. This multi-purpose UAP develops a cruising speed up to 100 km/h and can carry 60 kg of payload at the distance of 150 km; the maximum flight time is 90 minutes; the maximal altitude is 2000 m. The takeoff and landing of this UAP requires a ground only 150 × 15 m.

In 2017, the Chinese cargo drone AT200 (Figure 2), developed based on a light multipurpose New Zealand aircraft P750XL, made its first flight (Fisher, 2017). The aircraft can carry 1500 kg of payload at a distance of 2,180 km with a maximum cruising speed of 300 km/h, the duration of the flight is 8 hours, the practical maximal altitude is 6100 m, the cargo compartment volume is 10 cubic meters, and the runway length is 200 m.

The history of the cargo UAP AT200 creation can be considered as a practical implementation of a theoretical approach to aircraft design, published by Belan, Yun, and Gnashuk (2018). It is obvious that any aircraft, including UAV, initially developed to address specific tasks, will be more efficient than an upgraded, re-equipped manned aircraft similar to a New Zealand aircraft P750XL. However, it should be borne in mind that a successful selection of a manned prototype can make an unmanned conversion preferable to the originally designed, generally accepted version. This approach to the creation of a new UAV has a number of advantages, such as cost savings for the design, production and testing phases; availability and logistics of spare parts; possibility of using the existing service base; reduced UAV project development and serial production launch timelines. The authors use the "conversion" term to denote the process of transforming the manned aircraft into UAV for their subsequent economically and technically efficient use in various economy sectors. The effectiveness of such UAVs will largely be determined by the quality of conversion as mentioned above. To solve the conversion problem, a corresponding theoretical base, some combination of algorithmic and software tools, as well as technical means ensuring a complete cycle of testing the UAV control systems in all flight modes are required.

Figure 1. Unmanned aerial plane DR-60
(https://smartdrones.ua/products/dr-60, 2018)

Perspectives of Cargo UAV

In fact, one can imagine many scenarios in which the use of UAVs for cargo transportation will be justified. They are able to do a lot, from pizza delivery to the transportation of special cargoes in cities, at the regional level and even within the country. Most of the routine tasks will be performed without an operator. The international transportation is still a dream to be gradually prepared for. Artificial intelligence will take care of paving the optimal route, takeoff and landing, as well as aircraft return to the base.

To date, the most promising direction is the use of cargo UAVs in the delivery of urgent and perishable goods, which the authors call special cargo. One of the barriers (an, probably, the principal one), constraining the use of freight UAVs, as the authors see it, is the state regulation, or rather, an almost complete lack of understanding of how the industry of unmanned transport should work. It is difficult to engage in commercial cargo transportation where the rules of work are unknown and it is not clear whether the permission to use UAVs will be obtained and who should issue it.

Of course, there are certain security requirements. Accidental or deliberately directed drop of the drone with the cargo can lead to injuries and even death of people, fires, and destruction of strategically important objects. Only the correct models should rise in the air, and their routes should pass away from the crowded

Figure 2. Unmanned aerial plane AT200
(https://www.janes.com/article/75309/china-s-at200-cargo-uav-to-supply-military-installations-in-south-china-sea, 2017)

places. Otherwise, the reliability of drones should be so high that they could be safely run over the city.

The State Aviation Service of Ukraine intends to regulate the use of UAVs in the country. The office has developed a draft concept to ensure the safety of general-purpose, sports and amateur, UAV flights. UAV certification is suggested. The certificate will cover all the components required to perform the controlled flight. To determine the UAV classification, the following criteria will be evaluated: flight weight, range of action and method of control (within visibility, outside visibility and along the programmed trajectory), as well as kinetic energy upon UAV collision with an obstacle.

Also, the draft regulation suggests defining the drone operators ("external pilots") as a new category of pilot. Their duties are proposed to be equated to the duties of pilots of the manned aircraft; for this purpose, it is proposed to develop the certificate of the external pilot, in which the required qualification notes and restrictions will be specified.

The integration of drones into the air traffic environment will be implemented in two phases. Prior to the introduction of a special regulatory framework, it is proposed to allow the drone flights only in reserved airspace or at small altitudes (150 m outside the settlements and crowded places) at the weather safe for visual flights (the height of the lower border of the clouds min 450 m and the horizontal visibility min 5,000 m). Within settlements, the drones will run only based on separate permits, issued by the aviation and local authorities. It is proposed to introduce the obligatory insurance of commercial UAVs.

Today, all UAV developers, operators and amateur enthusiasts are outraged by the slow solution of problems so important for many sectors of the economy and citizens of the country by the state regulators. Still, there is hope, the ice has moved. The first documents appeared, in which the regulator tries to define the rights and obligations of those who are not indifferent to the fate of unmanned aviation.

Role of UAV in Forwarding Operations

The attraction of unmanned aircraft to the transportation of goods, in particular special ones, according to the authors, will occur in several stages. At the initial stage, they will be used to transport the goods with small weight and volume on short routes, mainly in regions barely accessible for the other types of vehicles. As the carrying capacity and the range of UAV flight increase, and the regulatory legislation, including the international one, is liberalized, the share of UAV freight traffic will only grow.

The global economy continues to develop and form the global supply chains. This process is actively promoted by the IT penetration and development of theoretical foundations of the network economy (Lu, Wang, 2008; Velthuijsen, Frank, 2014). Based on the network economy, an appropriate scientific direction is formed, studying the benefits of merging into various networks, including the transportation and supply networks. At the same time, the role of not only the FFC, but also the carrier in the global supply networks is changing, since now it must not only ensure the transportation of cargo, but also take an active part in the formation of supply chains.

The processes of general and special cargo transportation have significant differences and require special regulation. When transporting special cargo, the risks are inherent in the cargo nature. The likelihood of losses during the transportation of special cargo, in contrast to the general cargo, is also greater. However, for the customers, an important issue is not the possibility of compensation for damages by the carrier, but the delivery of goods in good quality at an acceptable time, especially those with specific properties. According to the IATA definition, special cargoes are the goods requiring the most urgent delivery and observance of special transportation conditions (IATA, 2018).

For the successful transport of special cargoes using the aviation transport, it is also required to consider the fact that the aviation component in the delivery chain requires qualitative interaction with the other modes of transport. Therefore, the actual task for the FFC in the matter of organization of cargo transportation using UAV is to ensure the safety of transportation, safety of cargo, and economic efficiency.

The steps to prevent the emergence of risks certainly require costs, so in order to ensure the economic efficiency of the delivery of special cargoes, it is required to determine the cost price of technologies for servicing all links in the cargo delivery chain. Optimization of insurance coverage for the special cargo transportation is only one of the steps to minimize risks. The core actions should be the risk prevention and control.

For example, such a measure may be procedures for licensing and certification of carriers for the delivery of special categories of goods using UAV, as well as licensing and certification of handling companies for performance of the relevant loading and unloading operations. Control by the authorized bodies should be aimed at ensuring the safety of transportation, which is especially important for the delivery of dangerous goods and goods, which are simultaneously dangerous and perishable.

The cargo safety depends on the coordination of organizational, economic, technical and technological measures implemented by the FFC, airlines, airports and other stakeholders, which, of course, significantly complicates the delivery process and requires the development of multimodal technologies, appropriate software products, considering the features of using the UAVs. Increased demand for air cargo transportation has led to an increase in the average lifting capacity of the cargo UAVs, the average length of flight, and focus on development of autonomous flight control systems.

For pharmaceutical companies, the criteria determining the delivery chain effectiveness may include minimum costs, flexible and timely supply, as well as risk minimization (Jaberidoost et al., 2013; Scott et al. 2017).

Consequently, there arises the problem of accounting for multiple criteria when delivering the products of pharmaceutical and biomedical industry, which is supposed to be one of the main customers of special cargo transportation services using UAVs.

To develop the recommendations of achieving the economic efficiency of UAS, we also define the main internal and external factors which, in the opinion of the authors, should be considered in planning and optimizing the delivery of special cargoes. The internal factors include the costs of the delivery organization, the technology of cargo consignment formation, the choice of the optimal transportation route, the mode of transport, the carrier, and the cargo handling technology. The external factors include potential and effective demand, requirements for transport safety, cargo safety and transportation quality. Internal factors depend on the goods properties and technologies used in FFC, while the external ones are determined by the requirements and constraints established by the authorized organizations (for example, aviation authorities, ICAO, IATA) and customers.

Figure 3, representing a simplified process of arranging the delivery of special cargo using the air transport provides an understanding of the need for a central warehouse (this may be the airport cargo terminal) to ensure the turnover of special

Figure 3. Scheme of organization of special cargo transportation services using air transport

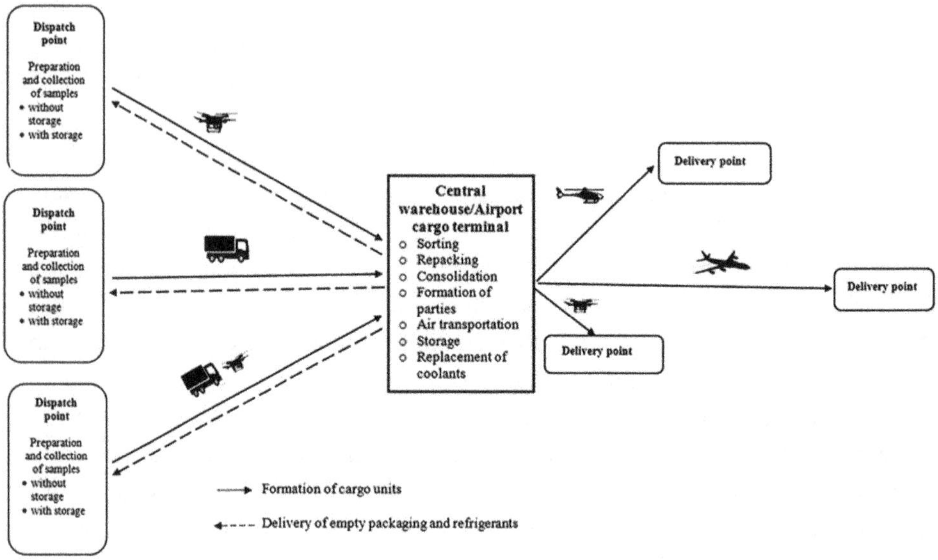

containers, storage of empty containers, sorting, repacking, consolidation of cargo for air transportation, storage (if required) and replacement of refrigerants.

On almost all continents, there are territories where the points for collection of special cargoes are located in small and medium-sized towns, where it is impossible, and sometimes unfeasible to arrange the regular and charter flights of traditional aircrafts.

UAVs may be involved in the process of transporting a special cargo from the shipment point to the delivery point both at the collection point, i.e. the central warehouse in the aviation terminal, and at the central warehouse of the delivery point. It is impossible to say a priori which parts of the routes will be assigned to certain types of UAVs, and which ones - to the ground vehicles. The answer can be obtained only by solving an optimization problem, the mathematical model of which is set out herein below. Obviously, the solution will depend on the technical and economic parameters of the UAVs (payload weight, flight range, cruise speed, tariffs and fees, requirements for take-off and landing characteristics, etc.), the number of UAVs of each type in the fleet, as well as competing modes of transport.

Considering the efficiency criteria listed above, factors and general ideas about the delivery of special cargoes using the aviation transport, we will determine the stages of FFC decision-making for process optimization.

Step 1: The sub-problems for delivery of consignments of special cargoes by road and/or cargo UAVs are solved by the criterion of the minimum total costs at the stage of assembly of individual samples, packaging, provision with refrigerants, formation of cargo units and consignments for delivery. The result of solving such a sub-problem is the optimal plan for the delivery of special cargoes, that is, the optimal number of cargo units is determined by the types of packaging and cargo consignments delivered to each collection point to the central warehouse in the aviation terminal.

Step 2: The coordinating task of delivering consignments of special cargoes in land-based aviation is being solved. Synchronization of the input and output freight flows of special cargoes to the central warehouse is carried out and the optimal plan for the delivery of special cargoes in the land-aviation communication is established. The optimization criterion is the minimum total costs for delivery in the automobile-aviation communication over the scheduled period. The problem solution will result in determination of the structure of FFC operating expenses.

Step 3: The incomes and profits of FFC are calculated. The revenues are calculated depending on the volume of cargo transportation and FFC tariffs. The indicator of the FFC performance is profit. To determine the FFC performance in the delivery of special cargoes, it is required to assess whether the FFC is satisfied by the level of profit at the existing tariffs.

Step 4: Where the level of profit is unsatisfactory for the FFC, the further optimization is possible by determining the FFC tariffs at the desired level of profitability, that is, the inverse problem is solved at this stage. The result of solving this problem is the FFC tariff system, at which the desired level of profitability is achieved. FFC efficiency is significantly affected by fluctuations in demand over the scheduled period. Ignoring the possible changes in demand in the tariff formation can lead to periodic losses of FFC during the scheduled period.

When searching for the optimal plan of the special cargo delivery, the consignments are sorted and re-formed, the number of consignments sent and those remaining for temporary storage is calculated, the urgency of dispatches, storage at departure points, selection of air carriers, the schedule of flights, ground transportation schedule and carrying capacity of the cargo UAV is considered.

SIMULATION OF THE PROCESS OF SPECIAL CARGOES DELIVERY

Building Models for Efficient Delivery of Special Cargoes

Recent publications offer various models considering the use of UAVs for the goods delivery at short distances, mainly within the city limits and its suburbs. Murray and Chu (2015) are developing two models for the delivery of parcels by unmanned aerial vehicles. These authors developed a model for the parcels delivery by unmanned aerial vehicles, in which the delivery process is similar to the delivery of fighters to the place where the combat operation is carried out by an aircraft carrier.

herein this case, the truck with the UAV on board passes most of the way, and when the distance from the truck to the cargo delivery point becomes less than the range of the UAV flight, the latter is launched from the truck. Such a tandem strategy (truck-UAV), apparently, will be quite effective, provided that the UAV can be safely returned to its launch site, that is, subject to the UAV circuit use.

An interesting model was proposed by Scott and Scott (2017), which considers the scenario of delivering the emergency medical supplies to remote areas with bad roads, but too far for the delivery by UAV. Here, too, a tandem strategy is used. However, in this case, the ground vehicles deliver the goods without the UAV to the "location", from where the cargo is further transported by the UAVs. The considered territory (region) is divided into a number of demand clusters, representing a group of villages. Each cluster will contain a demand socket. Two types of models with different optimization criteria are considered.

In this chapter, the process of special cargoes delivery is described using a model with discrete and continuous variables and non-linear cost functions, taking into account the non-delivery of intact specimens of special cargoes to the destination point at a certain time for various reasons. Transportation of the clinical trial results is associated with the cost of storage facilities for samples at low temperature, i.e. special containers, dry ice, outer packaging. The weight of the gross cargo is determined not so much by the weight of the samples, as by the weight of the additional equipment and materials. The total weight of the cargo depends to some extent on the number of samples simultaneously prepared for transportation and packaged. Then the cargo prepared for transportation is sent by land or by air to the central warehouse. In the central warehouse, samples from different departure points are sorted and repacked, depending on the destination. Transportation to destinations is performed by air transport at the tariffs meeting the special cargo transportation rules.

The process is monitored over time. The planning period can be one month or several months. The planning period is divided into individual days t.

For each day t, the need to send samples of drugs is set, where h_{ij}^{t} means the number of the samples, $i \in I$ means the index of the customer (shipment point), $j \in J$ means the index of sample destination. Shipment is performed by one batch once a day only but can be performed not on all days of the week. Accepted and not dispatched samples can be stored for 24 hours. Thus, the need for shipment of h_{ij}^{t} on day t may not be equal to the dispatch of d_{ij}^{t} no this day, which allows more efficient transportation.

However, a part of the samples $h_{ij}^{Qt} \leq h_{ij}^{t}$ must be sent urgently and cannot be kept for shipping on the next day. The sample type and the analysis to be performed does not matter, since the weight and volume of the sample are not significant. Depending on the number of samples d_{ij}^{t} sent from point i no day t, a sample packaging option is selected. The chosen packaging option is described by the weight g_{i}^{t}, volume v_{i}^{t} and cost c_{i}^{t} of packing the samples.

Where the number of samples d_{ij}^{t} exceeds the capacity of the maximum package, then the question arises as to the best distribution of the total number of samples d_{ij}^{t} between the packaging options. Namely, let the number of the packaging options be equal to P. Each option $p \in P$ is characterized by the maximum number of samples $\bar{d}_{p}$ it can accommodate, external dimensions v_{p}, the weight of the package along with the dry ice $\underline{g}_{p}$ and the cost $\underline{c}_{p}$ of packing along with the dry ice.

Here we allow a slight simplification, considering that the amount of ice does not depend on the number of samples in the package. This simplification is not significant, since such a dependence is not significant. The amount of ice can fluctuate depending on the season, the expected time of delivery to the destination etc.

The task for the entire planning period T is to determine for each day $t \in T$ and each shipping point $i \in I$ how many samples d_{ij}^{t} will be sent on that day to the central warehouse, how this quantity will be distributed by packages, what carrier to choose, based on the criterion of the minimum total packaging and dry ice, additional dry ice and storage cost and, in fact, the transportation over the scheduled period.

Formally, the problem is written as follows:

$$\min_{d_{ij}^{t}, k_{ijp}^{t}, \pi_{i}^{t}} \sum_{t \in T} \left(c_{i}^{t} + S_{i}(\Delta h_{i}^{t}) + C_{i}^{\pi_t}(v_{i}^{t}, g_{i}^{t}) \right), \tag{1}$$

$$d_{ij}^{t} = h_{ij}^{t} - \Delta h_{ij}^{t} + \Delta h_{ij}^{t-1},\ d_{ij}^{t} \geq \Delta h_{ij}^{t-1} + h_{ij}^{Qt},\ j \in J,\ i \in I,\ t \in T, \tag{2}$$

$$\sum_{p \in P} \bar{d}_p k_{ijp}^t \ge d_{ij}^t, \; j \in J, \, i \in I, \, t \in T, \tag{3}$$

$$\sum_{p \in P} \bar{d}_p k_{ijp}^{Qt} \ge h_{ij}^{Qt}, \; j \in J, \, i \in I, \, t \in T, \tag{4}$$

$$k_{ijp}^t \ge k_{ijp}^{Qt}, \; p \in P, \, j \in J, \, i \in I, \, t \in T, \tag{5}$$

$$c_i^t = \sum_{p \in P} \underline{c}_p \sum_{j \in J} k_{ijp}^t, \qquad i \in I, \, t \in T, \tag{6}$$

$$\Delta h_i^t = \sum_{j \in J} \Delta h_{ij}^t, \qquad i \in I, \, t \in T, \tag{7}$$

$$v_i^t = \sum_{p \in P} v_p \sum_{j \in J} k_{ijp}^t, \qquad i \in I, \, t \in T, \tag{8}$$

$$g_i^t = \sum_{p \in P} \underline{g}_p \sum_{j \in J} k_{ijp}^t + g_1 \sum_{j \in J} d_{ij}^t, \qquad i \in I, \, t \in T, \tag{9}$$

$$\Delta h_{ij}^t = 0, \; 1, \; 2, \ldots \text{integer}, \; j \in J, \, i \in I, \, t \in T, \tag{10}$$

$$k_{ijp}^t = 0, 1, 2, \ldots \text{integer}, \; k_{ijp}^{Qt} = 0, \; 1, \; 2, \ldots \text{integer}, \; p \in P, \, j \in J, \, i \in I, \, t \in T, \tag{11}$$

$$\Delta h_{ij}^0 = 0, \; \Delta h_{ij}^{T+1} = 0, \quad i \in I, \, j \in J, \tag{12}$$

$$\pi_i^t \in \Pi_i, \quad i \in I, \tag{13}$$

where

$i \in I$ means index of sample shipment point;
J means the set of sample destinations;
h_{ij}^t means the number of samples taken on day t to be shipped to j;
h_{ij}^{Qt} means the number of samples to be sent urgently, $h_{ij}^{Qt} \leq h_{ij}^t$;
c_i^t means the total cost of packing and dry ice at the standard transportation timelines;
Δh_i^t means the total number of samples left for storage after day t;
Δh_{ij}^t means the number of samples left for storage after the day t to be shipped to j;
v_i^t means the total volume of packages shipped on day t;
g_i^t means the total weight of the packages shipped on day t;
S_i means the function of storage costs (mainly the cost of dry ice), depending on the number of stored samples;
C_i^π means the function of transportation cost for the carrier π, depending on the total volume and weight of the packed samples;
$\underline{c}_p$ means the cost of a packaging option $p \in P$ with dry ice;
k_{ijp}^t means the total number of packaging options p to be used at point i on day t for shipping to j;
k_{ijp}^{Qt} means the number of packaging options p to be used at point i on day t for urgent shipping to j;
v_p means the volume of the packaging option $p \in P$;
$\underline{g}_p$ means the weight of the packaging option $p \in P$ with dry ice;
g_1 means the average weight of one sample;
d_{ij}^t means the number of samples sent from point i on day t to j;
$\overline{d}_p$ means the maximum number of samples placed in the package p;
Π_i means the set of possible carriers π for the shipping point i;
π_i^t means the carrier selected for day t.

The variables in the model are $\Delta h_{ij}^t, \Delta h_i^t, k_{ijp}^t, k_{ijp}^{Qt}, c_i^t, v_i^t, g_i^t, d_{ij}^t$ и π_i^t.

That is, for each day, a decision is made on how many samples and where will be sent, how they will be packed, and which carrier will be used. The variables affecting the decision are $d_{ij}^{t}, k_{ijp}^{t}, k_{ijp}^{Qt}, \pi_{i}^{t}$. The problem (1) - (13) is written for each shipping point i separately, and conditions (2) - (13) are established for each planning period t. (1) is the optimization criterion. The constraints (2) - (13) have the following meaning: (2) means the conditions for the number of samples shipped on day t, taking into account those not shipped the day before and left for the following day, as well as the conditions for the number of samples requiring urgent shipment; (3) means the condition determining the sufficiency of packages to accommodate all samples by destination; (4) means the condition determining the sufficiency of packages to accommodate all samples designated as urgent by destination ; (5) means the condition of consistency between the number of urgent packages and the total number of packages; (6) means the cost of packaging and ice; (7) means the number of samples left to be shipped on the next day; (8) means the total cargo volume; (9) means the total cargo weight; (10), (11) means the condition for integer variables (the number of samples) kept for storage and the number of packages; (12) means the threshold conditions of the dynamic process; (13) means the condition for choosing a ground carrier.

As a rule, trucking is carried out by evening and night flights. Therefore, the batches of samples sent on day t arrive at the central warehouse on day $t+1$, sorted by the shipping directions, and cargo is formed for shipment to the destination points. Thus, for each day $t+1$ and destination $j \in J$, the weight and volume of cargo received and subject to shipping are determined. The number of samples by destination is determined as follows:

$$d_{j}^{t+1} = \sum_{i \in I} d_{ij}^{t}, \quad \forall j \in J, \tag{14}$$

cargo weight by destination:

$$g_{j}^{t+1} = \sum_{p \in P} \underline{g}_{p} \sum_{i \in I} k_{ijp}^{t} + g_{1} \sum_{i \in I} d_{ij}^{t}, \tag{15}$$

volume by destination:

$$v_{j}^{t+1} = \sum_{p \in P} v_{p} \sum_{i \in I} k_{ijp}^{t}, \tag{16}$$

number of packages by type of packages and destination:

$$k_{jp}^{t+1} = \sum_{i \in I} k_{ijp}^{t}, \tag{17}$$

number of packages requiring urgent shipping by type of packages and destination:

$$k_{jp}^{Qt+1} = \sum_{i \in I} k_{ijp}^{Qt}. \tag{18}$$

The further transportation is carried out by air transport. The cost of transportation is calculated based on the tariffs for transportation of dangerous goods by regular and charter flights of airplanes, helicopters and UAVs. These tariffs are based on general tariffs with additional fees and charges. The above model (1) - (13) concerns the optimization of sample transportation from one clinic (shipping point) to the central warehouse. Where this transportation is organized by the same FFC, then it may agree on the receipt of packages with samples from different points and their shipment by air. In this case, problems (1) - (13), formulated for each separate shipping point, must be coordinated by a higher-level problem, which we formulate as follows: it is required to organize the receipt of packages with samples at the central warehouse in such a way as to minimize the total costs of FFC for delivery of samples by all shipping points, the cost of the sample air transportation and storage in clinics and in the central warehouse. Subject to the coordinated receipt of samples from different clinics at the central warehouse, the number of samples to be stored in clinics should be determined as a result of solving the coordination problem. In this case, the variables Δh_{ij}^{t} become the variables of the coordination problem, and criterion (1) for the sub-problems of individual shipping points i changes to the following one:

$$Z_i^{t+1}(\Delta h^{ti}) = S_i(\Delta h_i^t) + \min_{k_{ijp}^t, \pi_t}\left(c_i^t + C_i^{\pi_t}(v_i^t, g_i^t)\right), \tag{19}$$

provided that the conditions of (2) - (13) are met, where Δh^{ti} means vector of variables Δh_{ij}^{t}, Δh_{ij}^{t-1}, $j \in J$. That is, problem (19) is solved for each day separately. The coordination problem determines the total costs associated with the sample transportation and is recorded as follows:

$$V_T = \min_{\Delta h_{ij}^t, \Delta k_{jp}^t, a_j^t} \sum_{t \in T}\left(\sum_{j \in J} C^{a_j^t}(g_j^t, v_j^t) + \sum_{p \in P} s_p \Delta k_{jp}^t + \sum_{i \in I} Z_i^t(\Delta h^{t-1i})\right), \tag{20}$$

$$k_{jp}^{t} = \sum_{i \in I} k_{ijp}^{t-1}, \; k_{jp}^{Qt} = \sum_{i \in I} k_{ijp}^{Qt-1}, \; p \in P, \; j \in J, \; t \in T, \tag{21}$$

$$q_{jp}^{t} = k_{jp}^{t} - \Delta k_{jp}^{t} + \Delta k_{jp}^{t-1}, \; q_{jp}^{t} \geq \Delta k_{jp}^{t-1} + k_{jp}^{Qt-1}, \; p \in P, \; j \in J, \; t \in T, \tag{22}$$

$$g_{j}^{t} = \sum_{p \in P} \left(\underline{g}_{p} q_{jp}^{t} + g_{1} \bar{d}_{p} \right), \; j \in J, \; t \in T, \tag{23}$$

$$v_{j}^{t} = \sum_{p \in P} v_{p} q_{jp}^{t}, \; j \in J, \; t \in T, \tag{24}$$

$$\Delta k_{jp}^{t} = 0, 1, 2, \ldots \text{integer}, \; \Delta k_{jp}^{0} = 0, \; \Delta k_{jp}^{T+1} = 0, \; p \in P, \; j \in J, \; t \in T, \tag{25}$$

$$\Delta h_{ij}^{t} = 0, 1, 2, \ldots \text{integer}, \; \Delta h_{ij}^{0} = 0, \; \Delta h_{ij}^{T+1} = 0, \; i \in I, \; j \in J, \; t \in T, \tag{26}$$

$$a_{j}^{t} \in A_{j}^{t}, \; j \in J, \; t \in T, \tag{27}$$

where

$C^{a_j^t}$ means the cost function for air transportation by air carrier a_j^t to destination j depending on the cargo weight and volume;

a_j^t means the air carrier selected for day t to destination j;

s_p means the cost of storage of the package type p overnight in the central warehouse;

Z_i^t means function of the cost of shipping the packages from clinic i, considering the storage of samples in the clinic;

Δh^{t-1i} means the vector of variables Δh_{ij}^{t-1}, Δh_{ij}^{t-2} determining the value of function Z_i^t;

k_{jp}^t means the number of packages of type p received at the central warehouse on day t for shipping to destination j;

k_{jp}^{Qt} means the number of packages of type p received at the central warehouse on day t for urgent shipping to destination j (which cannot be retained for shipping on the next day);

k_{ijp}^{t-1} means the number of packages of type p shipped on the eve of day t from the shipping point i to destination j;

Δk_{jp}^{t} means the number of packages of type p not sent to destination j on day t (retained for storage overnight) on day t;

q_{jp}^{t} means the number of packages of type p shipped by air on day t to destination j;

g_{j}^{t} means the weight of all packages shipped by air on day t to destination j;

v_{j}^{t} means the volume of all packages shipped by air transport on day t to destination j;

A_{j}^{t} means a set of air carriers flying on day t to destination j.

The variables in this problem are Δh_{ij}^{t}, Δk_{jp}^{t}, q_{jp}^{t}, g_{j}^{t}, v_{j}^{t} and a_{j}^{t}.

In model (20) - (27), function (20) is the optimization criterion, which includes the total cost of air transportation, storage of packages in the central warehouse and transportation to the central warehouse, as well as sample storage in clinics over the whole planned period; (21) means the total number of packages received on day t at the central warehouse by shipping direction, type of packages and number of urgent packages; (22) means the number of packages to be shipped by air on day t to destination j by the type of packages; (23) means the total weight of the consignment sent to destination j on day t, taking into account the urgent packages and packages retained for storage; (24) means the total volume of cargo sent to destination j on day t; (25) means the condition of the integer number of packages stored in the central warehouse and their threshold conditions over time; (26) means the condition for the integer number of samples retained for storage in clinics and their threshold conditions over time; (27) means the condition of choosing an air carrier on day t to destination j.

Since the loss of cargo can occur at different stages of transportation and for various reasons, the FFC will incur different costs at that time. Besides, the transportation can be insured, that is, the insurance premium can be paid to compensate all or some losses of the FFC. There, each risk accident is associated with certain costs and reimbursements for FFC. Let's denote the set of risks as R, and the probability of risk accident as $r \in R$ for one shipped package χ_r.

When $r \in R$ event occurs, the FFC incurs additional costs z_{rijp}, depending on the type of package and transportation destination. For urgent packages, the costs increase, as the fines increase. Let's denote such costs as z^{Q}_{rijp} . Given that the likelihood of risk occurrence is low, the mathematical expectation of FFC costs associated with transportation risks can be written as follows:

$$Z_R = \sum_{t\in T}\sum_{i\in I}\sum_{j\in J}\sum_{p\in P}\left(((k^{t}_{ijp} - k^{Qt}_{ijp})\sum_{r\in R} z_{rijp} + k^{Qt}_{ijp}\sum_{r\in R} z^{Q}_{rijp})\chi_r \right). \tag{28}$$

Let's now consider the revenue part of the FFC arranging the sample transportation, the delivery of containers and dry ice, storage of samples in the central warehouse and coordinating these processes in time.

The revenue part of the FFC operations is the product of the volume of cargo transportation and FFC tariffs. Since the weight and volume of the package does not depend on the number of samples therein, the tariff depends mainly on the type of packaging, the destination and the delivery time. Urgent delivery is more expensive. When transporting a small number of packages by one mode of transport, it is quite simple to allocate a fixed and variable part of the costs. However, the mixed delivery of cargo, re-forming of cargo batches, and different demand for each transportation cycle make this allocation very difficult. Also, FFC expenses include the cost of package, packing and refrigerants, the quantity and types of which are also varying, which makes it difficult to determine the fixed and variable part of the costs. Therefore, it is suggested to use the averaged fixed and averaged variable tariff components of FFC per package, allowing to distribute the FFC costs for the special cargo delivery system These components were searched by mathematical modeling over an array of possible delivery scenarios. 1 package as a unit for tariff determination is chosen because in the delivery of bio-based products, 1 package is the cargo unit passing all the links of the goods delivery without re-formation. Since FFC operates in a competitive environment, tariffs should be flexible and provide for incentives. Therefore, we will additionally raise the issue of creating such a tariff system that would ensure a certain predetermined level of profitability for different options of sample transportation and would provide some flexibility. Formally, we present each tariff as two parts, fixed and variable, depending on the number of packages. Let's designate b^{f}_{ijp}, b^{v}_{ijp} as the fixed and variable part of the tariff for the ordinary delivery of a batch of samples as in package p from point i to point j for a particular day of the week, and b^{Qf}_{ijp}, b^{Qv}_{ijp} as a fixed and variable part of the tariff for urgent delivery. Then the total revenue of the FFC, associated with all shipments for the entire simulation period, is equal to:

$$D_T = \sum_{t \in T} \sum_{i \in I} \sum_{j \in J} \sum_{p \in P} \left(b_{ijp}^f + b_{ijp}^v (k_{ijp}^t - k_{ijp}^{Qt}) + b_{ijp}^{Qf} + b_{ijp}^{Qv} k_{ijp}^{Qt} \right). \tag{29}$$

The total expected income from the FFC operations for period T is equal to

$$\Pi_T = D_T - V_T - Z_R . \tag{30}$$

An additional condition for finding the values of b_{ijp}^f, b_{ijp}^v and b_{ijp}^{Qf}, b_{ijp}^{Qv} factors so as to reach the set level of profitability γ, is the following:

$$D_T \geq (1+\gamma)(V_T + Z_R) . \tag{31}$$

In this case, the FFC is interested in finding the lowest values of b_{ijp}^f, b_{ijp}^v and b_{ijp}^{Qf}, b_{ijp}^{Qv} factors to find the range of tariff values at which the minimum acceptable level of profitability is provided to it.

Consequently, in this statement, the objective function (20) can be replaced by the following one:

$$D_T \to \min_{b_{ijp}^f,\, b_{ijp}^v,\, b_{ijp}^{Qf},\, b_{ijp}^{Qv},\, \Delta h_{ij}^t,\, \Delta k_{jp}^t,\, a_j^t} \tag{32}$$

provided that conditions (2) - (14), (20) – (22), (31) are met.

The proposed model allows calculating competitive tariffs for transportation of special cargoes by minimizing costs while maintaining a set level of profitability.

Simulation of the special cargo delivery process gives to the FFC a clear and detailed plan for the formation of cargo shipments by types of packages, points of cargo collection and size of cargo batches for dispatch by air at the central warehouse daily during the scheduled period. Using the cost minimization model, one can analyze different options for the delivery of special cargoes (bio products) with consolidation and without consolidation, considering that all or a part of the shipments will require urgent delivery, and determine the delivery cost of 1 sample, 1 kg or 1 package of cargo. To ensure the profitability of special cargo delivery for the FFC, it is required to determine FFC competitive prices and tariffs, which should be flexible enough. Their level depends, on the one hand, on the market situation, structure and volumes of effective demand and, on the other hand, on the size and structure of the FFC costs.

Practical Recommendations for Forwarders

When calculating the mathematical model determining the cost part, in the practical work of the FFC, the following parameters may vary. For each point of cargo collection, the types of packages, the amount of refrigerant, and the number of cargo samples transported daily may vary. It is also possible to accumulate and temporarily store the non-urgent samples at the cargo collection points. Delivery from collection points to the central warehouse can be carried out by various motor carriers in view of the air transportation schedule. For a central warehouse, the parameters of cargo consignments and storage terms may vary, depending on the shipping urgency, timetable and carrier selection.

Based on the analysis, methods and models it was determined that when the special categories of cargo are delivered, there is an increased overall level of risk associated with the fact that, together with the ordinary risks faced by the FFC, the risks of damage and loss of cargo, transportation jeopardizing and harm to the environment and stakeholders in the cargo delivery chain, associated with the cargo properties and its interaction with the environment, occur.

The interaction of the special cargo properties with the environment and the course of various biochemical processes leads to the cargo instability, a rapid deterioration of its quality and reduced time of its intended use. Especially, the presence of such risks is manifested when delivering goods having both dangerous and perishable properties. It is found that the acceptable risk zone is the time interval during which the properties of special cargoes are preserved, and the risks associated with their properties do not occur. In this case, the ordinary risks of the delivery process take place, special cargoes are in the transportable state, and no additional costs are required. A critical risk arises during the time interval in which the properties of special cargoes begin to deteriorate, but it is still possible to return them to the transportable state due to the timely handling of bio-based products.

In the practical work of the FFC, this should be ensured by the cargo maintenance in the form of its repacking, replacement of refrigerants and its placement in the controlled temperature regime zone. As special cargos enter the critical risk zone, they start losing the transportable state and a part of the cargo may be lost; the losses depend on the specific type of special cargo. It is established that the return of special cargo to a transportable state will require additional costs for handling and extra time.

Therefore, when the FFC is planning the process of special cargo delivery, it is suggested to calculate the time, place and quantity of the cargo servicing processes in advance, and to include the corresponding costs in the cost of transportation in accordance with the proposed tools. Thus, FFC, with the help of the developed tools, can develop an optimal plan for the delivery of special cargoes with details on the shipping points, types of package, the amount of refrigerant, the number of cargo units and consignments, as well as the synchronization of the cargo delivery schedule in automobile and aviation communications.

When delivering goods, the companies engaged in express delivery of goods and mail to the customers guarantee a certain period of delivery and some set of services for the transportation registration and delivery of goods from the branches of express carriers to the airport. However, the customers deliver the cargo to the branch of the express carrier by itself. That is, express carriers, as a rule, do not provide the cargo collection services to

the customer, and pack the cargo at the collection points. Also, the delivery schedule depends on the organization of the express carrier operations and does not always coincide with the wishes of the customers. The prices of express carriers' services are quite high.

Unlike express carriers operating mainly in their network, the FFC can adapt very flexibly to the needs of the customers, organize the delivery of cargo by various modes of transport, including the cargo UAVs, perform cargo collection and formation of cargo units and cargo consignments, that is, it performs the functions of a PL-provider of logistics services. When arranging the delivery of special cargo in a road-air communication, it is required to build the organizational and economic relations of the airline and the FFC properly. The airlines to be involved in the transportation of dangerous and perishable goods must be certified for such transportation in accordance with requirements of IATA, ICAO and national authorities, and possess the required technical and technological capabilities to transport such categories of goods.

It is also advisable for FFC to choose those carriers, which have a system of servicing special cargoes and ensure regular transportation. When concluding contracts with air carriers, FFCs are suggested to prioritize the service and dispatch of special cargoes by the airline, and where there is a quite large flow, to secure the tonnage quotas for the carriage of special cargo by certain flights and prescribe penalties for untimely servicing or untimely departure of bio-products by the airline.

In the presence of agreements with airlines, FFC receives discounts from the tariff, which will reduce the total cost of special cargo (bio-product) delivery. As the volumes of special cargo transportation are increasing globally, the airlines can get a highly profitable segment of the cargo transportation market, which is explained by higher tariffs for airfreight of special cargoes with small bulk-volume characteristics of cargo. UAV can become competitors of traditional air carriers in this segment, but they are still developing small light cargoes on short routes in regions hard-to-reach for the other modes of transport.

Cooperation of air carriers with FFC allows organizing a chain of special cargo (bio-product) delivery in automobile and aviation communication at competitive prices, and gaining a foothold in a certain segment of the air transportation market. The cost of handling special cargo at the airport is also a part of the shipping cost. With appropriate infrastructure, special cargo (bio-product) handling can also be a source of additional airport revenues.

FUTURE RESEARCH DIRECTIONS

According to the authors, the time when cargo transportation by manned aircraft will become exclusive and very expensive operations is not far off. Piloted aircraft may even switch places with the cargo UAV in terms of the share in the air freight market. However, these changes will occur only after overcoming the technical and regulatory barriers significantly affecting them. The latter, by the way, has been pressured by the media recently. I would like to hope that this book will make its contribution.

Despite predictable difficulties, today there are available niches in various sectors of the economy for the use of civil UAVs, subject to certain requirements of state authorization services. As for the cargo UAVs, it is the delivery of special, light, ultra-light, low-volume cargo for short distances outside densely populated areas and closed airspace areas. Considering, for the time being, the lack of the powerful and energy–efficient copters, and the dependence of the UAVs on the airplane scheme on the runways, the much-needed "door-to-door" cargo delivery is suitable in tandem with an automobile or other mode of transport.

Agriculture is another promising area for a super-active use of UAVs for field monitoring, combat of plant diseases and pests. Since the majority of technological operations in agriculture are performed by UAVs at low altitudes, it is relatively easier to obtain a permission for flights in this case. Currently, the National Aviation University is working at the development and application of UAVs for agricultural purposes. Should the editors of the book include this subject in the next edition, we would gratefully take part in the preparation of one of the chapters.

CONCLUSION

Even from this overview chapter, it becomes clear that the cargo UAVs are not exotic. They are used to solve the professional tasks of transport and logistics. In the future, the cargo UAV standard range will be significantly expanded from micro-UAV for local movements of light packages until the appearance of a long-haul

UAV of high payload, something like the AN-225 MRIJA, but without a crew. Of course, the cost of such UAV is quite high, but such is the fee for speed, range and reliability of delivery.

Manned aircraft cost much more than UAVs in terms of maintenance and production. While a normal aircraft requires a system of protection and life support for pilots, the UAV's needs are much less. Unmanned aerial vehicles spend much less fuel due to their weight, and they need a smaller runway area. A lot less time and money are needed to train a UAV operator than a pilot.

We are witnessing the beginning of a new round of technological progress. The precedent has already been created, UAVs are already being used for commercial cargo delivery. However, before this expected future becomes a reality, it is required to solve many problems. The most important of them is to find a balance of interests of all stakeholders, which may be affected by the use of UAVs in one way or another.

Business is interested in efficient cargo delivery technologies, citizens need a comfortable environment, and the state needs to maintain the security of the country. Only provided that this balance is maintained, a fair system of legislative regulation of UAVs use can be built. Legislation is the last barrier to the use of UAVs. In terms of engineering, they are ready.

The method of managing the economic efficiency of special cargo delivery using the air transport as a set of principles, methods and mathematical models allows determining the optimal plan for the phased delivery of special cargoes in the road-air communication.

The management of the economic efficiency of special cargo delivery can be ensured by implementing the simulation of the delivery process using the UAVs. In this case, the calculation of the total costs of FFCs for the multimodal transportation of special cargoes allows taking into account the package features, batch size, carrier tariffs, cargo handling cost, and payments of reimbursement for FFC related to the permissible risk.

The principles, methods and mathematical models are an integral part of the methodology to manage the economic efficiency of special cargo delivery and provide flexibility in the decision-making by the FFCs to determine the optimal special cargo delivery option in the event of a change in the competitive situation in the special cargo transportation markets.

ACKNOWLEDGMENT

We are extremely grateful to Professor Paul Bates of the University of Southern Queensland for the kind offer to take part in the writing of this manuscript, as well as Dr. Tarryn Kille from the same university for some of the thoughts she expressed in her doctoral thesis, which served as an inspiring idea for the preparation of this chapter.

REFERENCES

Amazon Prime Air. (2016). *First Prime Air Delivery*. Retrieved from https://www.amazon.com/Amazon-Prime-Air/b?ie=UTF8&node=8037720011

Belan, V., Yun, G., & Gnashuk, A. (2018). Problem of manned aircraft conversion into unmanned aerial vehicles. In *Proceedings of the National Aviation University* (N 2, pp. 57-61). Kyiv, Ukraine: National Aviation University.

Blain, L. (2017). *AT Transporter multirotor "air car" carries cargo or people in detachable pods*. Retrieved from https://newatlas.com/advanced-tactics-transporter-at-manned-multirotor/52042/

Brar, S., Rabbat, R., Raithatha, V., Runcie, G., & Yu, A. (2015). *Drones for Deliveries. Sutardja Center for Entrepreneurship & Technology Technical Report*. Retrieved from http://scet.berkeley.edu/wp-content/uploads/ConnCarProjectReport-1.pdf

Bryan, V., & Char, P. (2014). *Drone delivery: DHL 'parcelcopter' flies to German isle*. Retrieved from https://www.reuters.com/article/us-deutsche-post-drones-idUSKCN0HJ1ED20140924?feedType=RSS

Cheney, M. (2001). *Tesla: Man Out of Time*. New York: Simon &Schuster.

Coxworth, B. (2016). *Heavy-lift drone could carry a person*. Retrieved from https://newatlas.com/griff-300-heavy-lift-drone/47049/

Domino's. (2016). *Pizza-by-drone a reality with world-first customer deliveries in New Zealand*. Retrieved from https://www.dominos.com.au/inside-dominos/media/november-2016-pizza-by-drone-a-reality-with-world-first-customer-deliveries-in-new-zealand

Fetisov, V. S. (2014). *Unmanned Aviation: Terminology, Classification, Current State: Monograph*. Ufa, Russia: PHOTON.

Fisher, R. D., Jr. (2017). *China's AT200 cargo UAV to supply military installations in South China Sea.* Retrieved from https://www.janes.com/article/75309/china-s-at200-cargo-uav-to-supply-military-installations-in-south-china-sea

IATA. (2018). *IATA Cargo Strategy.* Retrieved from https://www.iata.org/whatwedo/cargo/Documents/cargo-strategy.pdf

ICAO. (2005). *Doc 9854. Global Air Traffic Management Operational Concept* (1st ed.). Montreal, Quebec, Canada: International Civil Aviation Organization.

ICAO. (2016). *Long-Term Traffic Forecasts: Passenger and Cargo.* Retrieved from https://www.icao.int/Meetings/aviationdataseminar/Documents/ICAO-Long-Term-Traffic-Forecasts-July-2016.pdf

Jaberidoost M., Nikfar S., Abdollahiasl A., & Dinarvand R. (2013). Pharmaceutical supply chain risks: a systematic review. *DARU Journal of Pharmaceutical Sciences, 1*(21), 1-7.

Lu, L., & Wang G. (2008). A study on multi-agent supply chain framework based on network economy. *Computers & Industrial Engineering, 2*(54), 288-300.

Moiseev, V.S., Gushchina, D.S., & Moiseev, G.V. (2010). *Основы теории создания и применения информационных беспилотных авиационных комплексов: Монография* [Fundamentals of Theory of Information Unmanned Aerial Systems Design and Application: Monograph]. Kazan, Russia: Publishing house of the Ministry of Education and Science of the Republic of Tatarstan.

Murray, C. C., & Chu, A. G. (2015). The flying sidekick traveling salesman problem: Optimization of drone-assisted parcel delivery. *Transportation Research Part C, Emerging Technologies*, *54*, 86–109. doi:10.1016/j.trc.2015.03.005

PricewaterhouseCoopers (PwC). (2016). *Global Market for Commercial Applications of Drone Technology Valued at over $127 bn.* Retrieved from https://press.pwc.com/News-releases/global-market-for-commercial-applications-of-drone-technology-valued-at-over--127-bn/s/ac04349e-c40d-4767-9f92-a4d219860cd2

Rosen, J. W. (2017). *Zipline's Ambitious Medical Drone Delivery in Africa.* Retrieved from https://www.technologyreview.com/s/608034/blood-from-the-sky-ziplines-ambitious-medical-drone-delivery-in-africa/

Scott, J. E., & Scott, C. H. (2017). Drone Delivery Models for Healthcare. In *Proceedings of the 50th Hawaii International Conference on System Sciences* (pp. 3297-3304). Honolulu, HI: University of Hawai'i at Mānoa.

Stewart, J. (2017). *A drone-slinging UPS van delivers the future.* Retrieved from https://www.wired.com/2017/02/drone-slinging-ups-van-delivers-future/

The Guardian. (2013). *Flying aid drones tested in Haiti and Dominican Republic.* Retrieved from https: //www.theguardian.com/global-development/2013/jan/09/flying-aid-drones-haiti-dominican-republic

VandenBos, G., Knapp, S., & Doe, J. (2001). *Role of reference elements in the selection of resources by psychology undergraduates.* Retrieved from http://jbr.org/articles.html

Velthuijsen, H., & Frank, J. G. (2014). *Entrepreneurship in the Network Economy.* Retrieved from https://www.researchgate.net/publication/267920775_Entrepreneurship_in_the_Network_Economy

ADDITIONAL READING

Civil Aviation Authority of New Zealand. (2016). *New Zealand Airspace. Includes Drone Advice.* Retrieved from https://www.caa.govt.nz/assets/legacy/safety_info/GAPs/NZ-Airspace.pdf

IATA. (2018). *Cargo Handling Manual.* Montreal: IATA.

Jha, A. R. (2016). *Theory, Design, and Applications of Unmanned Aerial Vehicles.* Boca Raton: CRC Press. doi:10.1201/9781315371191

McKeown, P. (2009). *Information Technology and the Networked Economy.* GlobalText Project. Retrieved from https://florida.theorangegrove.org/og/file/49843a6a-9a9d-4bad-b4d4-d053f9cdf73e/1/InfoTechNetworkedEconomy.pdf

Millar, M. (2017). Global Freight Forwarding. Overview of the global forwarding sector and its impact on globalization. Retrieved from http://focusedcargonetwork.com/download/Focused-Whitepaper-Global-Freight-Forwarding.pdf

PF Collins International Trade Services. (2003). *Transportation Best Practices Manual.* Retrieved from http://www.tw.gov.nl.ca/publications/bestpracticesmanual.pdf

Phillips, N., Blauvelt, C., Ziba, M., Sherman, J., Saka, E., Bancroft, E., & Wilcox, A. (2016). *Costs Associated with the Use of Unmanned Aerial Vehicles for Transportation of Laboratory Samples in Malawi.* Seattle: VillageReach.

Thompson, T. (2016). *Drones.* Farmington Hills, Mich.: Greenhaven Press.

KEY TERMS AND DEFINITIONS

Cargo Tariff: A tariff contains rates, charges, and the conditions of service, applicable to the transportation of cargo by air transport.

Chain of Cargo Delivery: Airlines, ground handlers, freight forwarders, shippers, and industry experts.

Copter: A rotorcraft that primarily depends on engine driven rotors for motion.

Freight Forwarding Company: A company that organizes shipments for individuals or corporations to get goods from the manufacturer or producer to a market, customer or final point of distribution, and acts as an expert in the logistics network.

Multimodal Delivery: It is transportation by using two or more different means of transport through the use of transshipment (intermediate handling), organized by one company (freight forwarding company, multimodal transport operator) under one contract, with one freight document, under one liability, and one tariff (price).

Perishable Goods: Shipments that will deteriorate over a given period of time or if exposed to adverse temperature, humidity, or other environmental conditions while in carrier's possession.

Special Delivery: Transport of special cargo such as perishables and pharmaceuticals, requires compliance with regulations, standards, and training. The standards that built a global industry are applicable to criteria such as the shipping, acceptance, handling, loading, transport, and documentation.

Chapter 3
The Use of Unmanned Aerial Vehicles:
A Comparison of Turkey and Croatia

Aleksandar Erceg
J. J. Strossmayer University of Osijek, Croatia

Zafer Kilic
Istanbul Arel University, Turkey

ABSTRACT

Unmanned aerial vehicles (UAVs) are present in our lives, and although they are mostly connected to military purposes, they are becoming more present in the commercial and civilian sector. Possible applications of UAVs in the commercial and civilian sector will open new possibilities for further research and development of UAVs. This movement can bring new investment and new jobs, but at the same time, it will influence the way some activities are being done now. The use of UAVs brings savings in the production cycles and improve current operations in various industrial sectors. The chapter gives a definition and explains different types and potential applications of unmanned aerial vehicles in the word as well as the potential economic impact of their development and use. In the second part, the chapter analyzes the application of drones in Turkey and Croatia. Although different in terms of their size and the number of inhabitants, both countries are at the same level in relation to UAV application. Applications in both countries are compared, and after that, a conclusion is drawn.

DOI: 10.4018/978-1-5225-7900-7.ch003

INTRODUCTION

Globalization has brought significant changes to the world and economy. Different innovations have created improvements in services and products in different industrial sectors. Improvement in the fields of technology and information and communications has enabled the use of many new different machines which will change our future significantly. One of those improvements is unmanned aerial vehicles (UAVs), commonly known as drones. Unmanned aerial vehicle systems, whose potential benefits were not visible, have spread rapidly since the early 2000s, while both their varieties and their abilities increased at the same time. Their use was from the start connected only to military purposes but in recent days they have been used in many industrial sectors.

In this research, unmanned aerial vehicles are examined; their development is explained beginning from their invention to their present form, and their types are listed. This research is trying to find answers to how the usage of UAVs has affected different industrial sectors since the drones were introduced, especially in two countries, Turkey and Croatia, which were subjects of our research.

In the first part of the chapter, we will present UAVs and give their definition, types and main application sectors. In the second part, we will show where the drones are used in Turkey and Croatia, and then we will compare their applications. Finally, we will draw conclusions and make suggestions for further research into this emerging topic.

BACKGROUND

Stipanović et al. (2004) defined an unmanned aerial vehicle (UAV) as a powered, aerial vehicle that does not carry a human operator, uses aerodynamic forces to provide vehicle lift, can fly autonomously or be piloted remotely, can be expendable or recoverable, and can carry a lethal or non-lethal payload. In the last decades, special attention has been paid to UAVs due to the advantages of not placing human life at risk and the absence of a pilot that enables longer endurance, and consequently weight savings and costs. Unmanned aerial vehicles are referred to in many ways, from remotely piloted systems (RPAS) or aircrafts (RPA), unmanned aircraft systems (UAS), to pilotless aircraft (Erceg, Činčurak Erceg and Vasilj, 2017). UAVs are most often referred to as drones, which are defined as a flying robot or aircraft that does not carry human. These aircrafts can be controlled by humans or software by communicating onboard GPS (Eisenbeiss, 2005, Estampe, 2015). In recent years, UAVs have been most frequently associated with the military; however, drones are now also used in a wide range of civilian roles ranging from search and rescue, surveillance, traffic monitoring, weather monitoring and firefighting to personal drones and business drone-based photography, as well as videography, agriculture and even delivery services. Today, there are many more drones than ever before. Throughout the world, many competitions are held, and research and development activities are encouraged to be maintained to further develop the UAV technology (Koeniger *et al., 2005*). Based on current R&D activities, Gonzales-Aguilera and Rodriguez-Gonzalves (2017, p. 1) see the main advances of drones in: *(1) the emergence of new sensors that allow the improvement of the geometric and radiometric resolution, as well as the spectral range; (2) the evolution of new platforms that improve robustness and increase autonomy; (3) the development of software, from the navigation and communication with the platform to the processing and analysis of the images captured;* and *(4) new applications in emerging sectors: logistics, disaster assistance, security and surveillance, health and marine science.*

There are many reasons why drones are increasingly being used, mainly due to an increase in efficient technology and reductions in costs of their production. Finnegan (2015) predicts that the production of drones within the next ten years will reach 14 billion USD per year.

Drone Types

Erceg, Činčurak Erceg and Vasilj (2017) found that drones are usually divided based on specifications such as range and endurance, and based on maximum take-off weight, engines, and price. Another possible classification is according to their usage, and Sudip (2017) divided them into four main types of UAVs, i.e. multi-rotor drones, fixed-wing drones, single-rotor helicopter, and hybrid VTOL.

A multi-rotor drone is very common because both professionals and hobbyists use them. As to the price, it does not require a lot, could be considered the cheapest in the sector. This type is used by photographers for taking pictures and shoot movies (Lutz, 2014). Multi-rotor drones are easiest to manufacture and tend to be versatile enough to adapt to so many different usages. On the other hand, their flying time is very limited, and their speeds are very slow. There is a strong correlation between multi-rotor drone flying time and their battery life. Thus, people generally use this type of drones for fun.

Fixed-wing drones look more like airplanes. Unlike multi-rotor UAVs, these drones cannot stay in the air and have an average flying time. Their battery lasts up to 16 hours and is used by a human or autopilot (Sandbrook, 2015). They might be more expensive than multi-rotor ones considering the facts that the design of their bodies and wings requires craftsmanship and their materials must be enduring. They need a runway and a parachute, which represents another shortcoming. These types of drones are usually used in the defense industry because of their ability to float through the air silently. A downside to this type of drone is the requirement of a runway for takeoff and landing. However, unlike multi-rotor drones which have "a rotor hum", these drones can generally operate with almost no sound.

Single-rotor drones are the ancestor of multi-rotor drones. Despite having a fan in the wing part, they are named single-rotor drones because they have a large motor that resists the air. These UAVs can fly longer than fixed-wing drones and multi-rotor drones. This type is mostly used by hobbyists.

A hybrid VTOL shares the advantages of both fixed-wing and multi-rotor drones. A vertical take-off and landing (VTOL) drone operates and flies like a fixed-wing drone in the air; however, it has the vertical take-off and landing ability, negating the use of a runway. Heutger (2014) and Kelek (2015) noted the advantages and disadvantages of different types of drones that are used worldwide. (Table 1)

Another categorization is based on launching capability, endurance, altitude capability, and the load capacity area needed for maneuvering. According to this categorization, UAVs are divided into the following groups: lighter-than-air, fixed wings, and rotary wings. All aforementioned drone groups have their own disadvantages and advantages, but as already stated, they all have huge potential for different application fields where UAVs can be used.

DRONE APPLICATION POSSIBILITIES

Today there are many sectors of the industry that drones have been proven to be useful for, including, but not limited to, military, communications, advertising, trade, logistics, health, research and development, and discovery and surveillance. Frost and

Table 1. UAV advantages and disadvantages

Types	Advantages	Disadvantages
Multicopter	Inexpensive Low weight Easy to launch	Limited payloads Susceptible to wind due to low weight
Fixed-wing	Long range Endurance	Horizontal take-off requires substantial space or support Inferior maneuverability compared to VTOL (Vertical Take-Off and Landing)
Tilt-wing	Combination of fixed-wing and VTOL advantages	Expensive Complex technology
Unmanned (Hybrid) Helicopter	VTOL Maneuverability High payloads possible	Expensive Comparably high maintenance requirements

Source: (Heutger, 2014; Kelek, 2015)

Sullivan (2007) divided civil and market applications of drones into the following segments: government, firefighting, the energy sector, agriculture, forestry and fishery, earth observation and remote sensing and communication and broadcasting. The status of different applications of drones can be seen through the value of labor and businesses in different sectors using drones. Mazur and Wisniewski (2016) stated that current drone-powered solutions in different industrial sectors amount to more than 127 billion USD. (Figure 1)

The application of drones in infrastructure has the highest potential (45 billion USD), followed by agriculture (32.4 billion USD) and transport and logistics (13 billion USD). In what follows, we will shortly present several industrial sectors where the drones are used.

Defense Industry

With the new technology, drones have been used in the defense industry in different ways. For example, Israel has begun with domestic production of drones and it imports drone engines manufactured in the United Kingdom. Israel refused to release the full list of countries to which it sold military arms (Heutger, 2014). The United States import drones from Israel for use in combat (Lorenz, 2017). Besides these countries, research has shown that many developed countries have started using drones for national defense. From China to the USA, from Russia to South Africa,

Figure 1. Drone-powered solutions in different industrial sectors
Source: (Mazur & Wisniewski, 2016: 4)

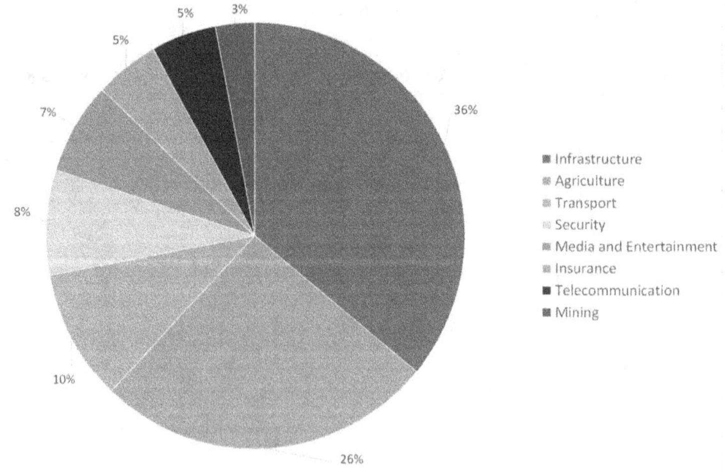

countries are using drones for, including but not limited to, border control, attack and defense, and monitoring.

The first UAVs (i.e. balloons with explosives) were used during the attack on Venice by the Austrian Habsburgs in 1849 (Consortiq, 2016). The development of UAVs started in the early 1900s and continued throughout World War I. Following the war, the development continued until World War II, when UAVs were produced and used to fly attack missions and train anti-aircraft gunners - in the 1930s the British Navy used drones for target practice (Gonzalez-Aguilera & Rodriguez-Gonzalvez, 2017). Drones are used in the military industry today for the following: air strikes, radar, rocket defense, security control, and border control systems. Most of the countries in the world developing UAV programs are developing armed UAVs, while a minority of countries are developing only unarmed UAV capabilities (Horowitz & Fuhrman, 2015). The use of drones for military purposes is expected to grow during next years (Knight, 2017).

Health Sector

Given that every second is very important for patients, drones are of great importance in the health sector. Drones are normally not late as they are not affected by roads or by traffic. In the air, there is little traffic for drones when compared to roads. Drone applications in healthcare include delivery of medicine, defibrillators, blood samples, and vaccines. Some tests have already been done with a drone carrying medicinal products. Drones can play an important role in healthcare in the future and according to Blau (2017), there are several ways how they will do that: (i) by delivering medical supplies to rural areas, (ii) by transporting blood samples to fasten different tests, (iii) by delivering relief to disaster victims, and (iv) by delivering blood units needed for surgery. The use of drones will allow more people to receive care at home for a longer period of time and potentially decrease the cost of helping people. For example, autonomous drones, such as those employed by Matternet, use GPS and other sensors to navigate between automated ground stations to deliver medications in remote locations that lack adequate roads. Matternet's drones can carry one to two kilograms and transport items about 10 km, traveling up to 40 km per hour, taking about 18 minutes including lift-off and landing (Raptopoulos, 2013).

Communication and Entertainment Sector

Drone application in the communications and entertainment sector gives users considerable advantages since many technological devices can see the pictures taken by drones. Users can see images by means of an easy application. Nowadays, many electronics companies have started to use this technology. Drones can be used for communications purposes both indoors (i.e. plants) and outdoors.

Film producers are using these UAVs for action movies, such as Marvel Studios Captain America: The Winter Soldier. Marvel Studios have brought a new dimension to recording movies by using drones to help capture footage. By using this technology both the shooting techniques have improved and new uses have been found for drones. The use of drones in communications and entertainment could contribute to steering toward a cost-effective way of maintaining critical networks. Another benefit is that drones can be used even in bad weather conditions and according to Mazur and Wisniewski (2016), a drone could even become part of communications networks as portable cell towers replacing cells on wheels at special events (i.e. sports games, concerts, etc.).

The advancing technology has changed the way of advertising a lot and an important role therein is played by drones. Advertising agencies have started to create many innovative ideas by means of drones. As a direct result of falling production costs, drones are beginning to be used for advertising.

It is rather common that even small businesses use drones to show their products to people. In 2014, drones were used for the first time as flying billboards. In Moscow, an advertising agency used drones to carry small fliers and later in Brazil several drones were seen while carrying headless mannequins. Brown (2017) stated that drones can benefit marketing in the following four ways: (1) they allow new ideas in the marketing industry; (2) they are easily obtained, so any new idea can be easily brought to the table; (3) there is no end of the content which can be created; and (4) drones can be used as the "wow" factor in adverts. Drones can be used in marketing as a tool, for quick marketing, as platforms, as actors and can have unlimited creativity potential (Agrawal, 2017).

Agriculture

Agriculture needs to accept changes to technology and groundbreaking approaches to food production. UAVs are an important part of this process together with better cooperation between industry, technology, and governments. Drones can bring important technology transformation to agriculture with plans and strategies based on real-time data collection and analysis. Mazur (2016) stated the following six fields of application in which drones can improve the agricultural sector:

- **Soil and Field Analysis**: Important for plant life cycle start since they can create precise maps needed for soil analyses which can be used for seed planting plans and for irrigation and nitrogen level management;

- **Planting**: Planting systems with drones can create a saving of up to 85% and the uptake rate of 75%. Drones will place seed pods with plants and necessary nutrients into the soil which will ensure the sustainability of the plant;
- **Crop Spraying**: Drones can increase efficiency and reduce chemicals penetrating groundwater. Drone spraying based on estimations can be completed five times quicker in relation to traditional ways;
- **Crop Monitoring**: The use of drones removes obstacles to the low efficiency of crop monitoring with time-series animations showing precise crop development and potential inefficiencies in production;
- **Irrigation**: Drones can detect which parts of the field need improvements and can calculate the vegetation index which can describe the relative density and health of the crop; and
- **Health Assessment**: Drones can detect which plant reflects a different quantity of green light and this can initiate a response to treatment the entire crop can see. Also, drones can help farmers document more precisely their potential losses for insurance.

In agriculture, UAVs could involve a fleet of drones involved in monitoring or performing different tasks. For even greater use of drones in agriculture in the future, the industry needs to develop drones which will need minimal training of users and which will be highly automated.

Transport and Logistics Sector

Logistics is generally a detailed organization and implementation of a complex operation. In a general business sense, logistics is the management of the flow of things between the point of origin and the point of consumption to meet the requirements of customers or corporations (Stroh, 2006). Logistics delivers the right product by providing high service quality and within the period of time requested by customers. What one needs when one needs it, and when one needs it can be a summary of logistics. Recently, most companies have been trying to have their orders delivered sooner to compete with other companies. This especially holds for deliveries of the so-called last mile deliveries, which Slabinac (2015) considers to be most inefficient in the whole supply chain due to smaller shipments but higher frequency. With the development of technology, UAVs are everywhere. Possible uses of drones are increasing day by day. Estampe (2015) concluded that drones will impact future transport and logistics potential in the following five ways: (i) saving money, (ii) keeping people at home, (iii) delivering parcels to places where they are not usually delivered, (iv) reducing and eliminating possible human errors, and (v) protecting and monitoring transportation paths.

Some well-known multinational companies use or test UAVs specifically because of this high level of interest. Amazon.com customers may soon be able to get their parcels in 30 minutes or less. Their Amazon Prime Air service will deliver shipments of up to 2.36 kg within 30 minutes after delivery. In the future, this option should refer to up to 80% of all Amazon deliveries (Johnson, 2017). Google is testing the possibility of using drones for deliveries. Google started the program called Project Wing, by which it plans to ship products to their customers within one to two minutes (Etherington, 2017). As one of the largest transportations and logistics companies in the world, DHL is also testing drones. They use them for shipping orders in Bavaria and for delivering medical supplies to the North Sea island of Juist. Numerous start-ups have joined the cause to gain publicity and boost sales. One of those start-ups is the Australian company Zookal, whose intention is to deliver textbooks via UAVs (Welch, 2013).

UAVs are also useful for intra-plant transportation, as well as for transportation between plants. If drones are completely operated within the plant, then there are very few regulatory issues that need to be dealt with. Another issue with drones is that currently they are limited to small form factors, as the larger ones have operating costs similar to, if not greater than, the manned options.

Advantages and Disadvantages of Drone Usage

Development predictions for the use of drones in all industrial sectors are admirable due to a significant improvement in drone technology. Currently, drones are not an integral part of the transportation and logistics sector, but they are considered because of their cost of operations and speed in relation to other potential transportation modes. Similar developments are noticed in almost all industrial sectors in which drones can be potentially used. Smith, Mazur, and Wisniewski (2017) see potential opportunities for the increased use of drones in financial and legal support, enhancing data processing and accessibility and rapid technology development, while potential challenges are aviation risk and privacy. Rosberg (2009) found the following potential setbacks for the increased use of drones in civil and commercial application: (i) non-existence of operator safety and training standards; (ii) restricted freight capacity and space limits; (iii) no protected non-military frequencies; (iv) responsibility for civil actions; (v) partial or undeveloped airspace rules that cover UAV systems; and (vi) a negative customer perception.

Although there are many advantages for drone use worldwide in different areas and industrial sectors it is needed to state that there are also challenges and disadvantages. The main challenge and disadvantage relate to security. Thus, it is necessary to create a complex air management system for the prevention of potential air collision. Besides the potential air collision, there is the issue of privacy. Since drones fly over various sites and can collect a lot of information it is important to deal with the privacy issue. Thus, another significant disadvantage is that there is no regulation regarding data collection. Additional problems can be that drones can affect navigation routine. In some area, local government restricts the use of drones (i.e. military facilities, airports, etc.) due to the safety issues for wider populations. Improdrone (2018) state four main disadvantages of drone use: (i) invasion of privacy; (ii) accidents; (iii) legal issues and (iv) potential misuse.

An important issue which will influence the future use of drones is legal regulation and this relates to some of the current disadvantages of drone use (Činčurak Erceg, Vasilj & Erceg, 2017). Not so many countries currently have legal regulation referring to drone usage for civil and commercial purposes. (Table 2)

Based on Table 2, there is a movement toward legal regulation of drone usage worldwide, which will enable the increasing use of drones and help users in different industrial sectors to use drones more frequently. Drone-related legislation is carried out on three levels, i.e. international, European and national. Work aimed at solving different legal issues which are needed is initiated through work of different aviation organizations such as the International Civil Aviation Organization, the European Aviation Safety Agency, the European Organization for the Safety of Air Navigation, Joint Authorities for Rulemaking on Unmanned Systems, etc. (Erceg, Činčurak Erceg & Vasilj, 2017)

Table 2. Legal regulation of drone usage worldwide

Country	Possibility of commercial flights	License required to fly	Possibility to perform BVLOS flights	License required for BVLOS flights	Insurance required for commercial flights	Training required for pilots to obtain a license
Poland	✓	✓	✓	✓	✓	✓
UK	✓	✓	✓	✓	✓	✓
China	✓	✓	✓	×	✓	✓
Canada	✓	✓	✓	×	✓	✓
Germany	✓	✓	×	×	✓	✓
France	✓	✓	✓	×	×	✓
South Africa	✓	✓	✓	×	×	✓
Indonesia	✓	✓	×	×	✓	✓
Australia	✓	✓	×	×	✓	✓
Brazil	✓	✓	✓	×	×	×
Mexico	✓	✓	×	×	×	✓
USA	✓	✓	×	×	×	×
Japan	✓	×	×	×	×	×
Russia	×	×	×	×	×	✓
Argentina	×	×	×	×	×	×

Source: (Maazur & Wisniewski, 2016: 21)

RESEARCH METHODOLOGY

This chapter presents the research conducted regarding use of the drones in Croatia and Turkey. For the preparation of the theoretical part of the chapter, a review of available scientific and professional papers and sources was examined. For the empirical part of the chapter i.e. review of current status of drone use in Croatia and Turkey, authors examined and collected already available data from both countries. Based on the collected data, the authors made a comparison and draw conclusions.

Authors have chosen to compare the use of drones in Croatia and Turkey since both countries are among the countries which have introduced the legal regulation of drone use. Although both countries have legal regulation of drone use, the use of drones in both countries is still insufficiently researched.

DRONES IN TURKEY AND CROATIA

Drone Usage in Turkey

Similar, to other countries in the world, Turkey is constantly developing the use of drones. Increasing numbers of unmanned aerial vehicles are used at the same time. According to the latest data released in 2016, the number of personal UAVs in Turkey amounted to 9,000, compared to 2.5 million drones in the world. It has been reported that there will be 20,000 in the coming years. Drones have become one of many different gift options for Christmas, birthday, wedding anniversary, Valentine's Day (Emir & Ere, 2016). This shows that the number of drones used for hobbies, military and industrial purposes is increasing. Teknolo (2015) stated that the most important areas of drone application in Turkey are: deliveries, first aid, wildlife conservation, search-rescue studies, preventing and detecting forest fires, crime combating, advertisement, making movies, photography, journalism, and border security.

Military capabilities that are currently available or are foreseen to be acquired in the near future are listed as follows (Türkiye İnsansız Hava Aracı Sistemleri Yol Haritası, 2011): discovery, tactical discovery and surveillance, bomb or fuzzy air attacks, advanced look-up for indirect shots, special operations and psychological operations, control and protection of borders, mine search and destruction, replenishment of health and military material, fight against smuggling, chemical, biological and radiological screening, ship identification and isolation in the maritime operations, combat search and rescue, air radio link and role assignment, point-to-point cargo delivery, and weather data collection. For military purposes, Turkey used the ANKA UAV system that was developed and produced in Turkey. In 2016,

Turkey became one of the leading military drone producing countries in the world. In Turkey, UAVs are used not only for military purposes, but also for border control and for control of many activities in the civilian sphere.

Currently, drones are mostly used in Turkey in the advertising industry and in firefighting operations. More often, drones are brought into play in areas such as woods and forests, in situations where normal interventions cannot be done effectively with firefighting equipment. Remotely controlled drones can take an active role in firefighting operations by discharging water from the special reservoir (Sahin, 2016). Necessary measurements related to forest restoration or reforestation in areas damaged in forest fires in August 2016, i.e. in Buca and Seferihisar districts of İzmir, were controlled by unmanned aerial vehicles and pictured onto maps. Debris clearing, and land preparation activities carried out by the General Directorate of Forestry Buca and Seferihisar are continuing and measurements are made in the field to carry out reforestation (Sayfa, 2016).

Another interesting field of drone application in Turkey is forensic anthropology and archeology. Due to the changes in the natural environment and vegetation, forensic archaeologists and forensic anthropologists use technologies such as magnetometer, ground penetrating radar or electrical resistivity kit (Ceker, 2014). Apart from these, UAVs have started to be used recently to detect burial places in aerodrome examinations by taking bird-eye views and to create an incident map (Ceker, 2016).

In addition to their use in different industrial sectors, in Turkey drones are also used in sports activities. Many different competitions were organized for drones and in 2017, the first official Turkish Drone Championship was held in Istanbul, in which more than 60 licensed drone pilots took part. Based on the results of this competition, the Turkish national drone competition team was established (Sayfa, 2017).

In terms of the advantages of drones, there are various legal problems. Their problematic usage is reflected in acts such as violation of the right to respect for private life, endangering people's lives, terrorism, flying in protected areas, espionage, and so on. Therefore, Turkey has UAV-related legal regulations. There are some criteria to use a drone in Turkey, such as registration, permission, insurance, and age (Cetin, 2016). For example, a person is allowed to fly a drone only in daylight and within a visual range of 2 km and he/she should not fly it higher than 120 m above the ground. The growing use of drones in Turkey in 2016 resulted in their introduction in legal regulation providing for requirements for registration and authorization of drones. According to legal regulation, those who have unmanned aerial vehicles which can be controlled by a person or institution in Turkey must primarily register with the Directorate General of Civil Aviation. When registering, criminal record and information, such as a national identification number, are required. There is also a separate rule for drones over 25 kg (Cevap, 2016). In addition, drone pilots are required to be trained according to the type of a vehicle. If UAVs are flown without a permission, a fine may be issued of up to 11,447 Turkish Lira (approximately 1,882 EUR) (Calishan, 2018). Moreover, a person who had flown an unauthorized unmanned aerial vehicle over Istanbul Atatürk Airport was arrested for jeopardizing air traffic (Directorate General of Civil Aviation). In 2018, there were 11 organizations for drone (UAV) training established before 2018. The Directorate General of Civil Aviation has recently announced that the number of drone flight training organizations has amounted to 25. There are 24,866 registered drones in Turkey, while the number of users is 31,194. When amateur devices weighing less than 500 grams are excluded, there are about 25,000 registered drones. Pursuant to Article 10, operators/owners of all UAVs and systems are liable for any damage caused to third parties.

A possible disadvantage to faster implementation of drones used in the commercial and civilian sector in Turkey is legal regulation and its changes.

Drone Usage in Croatia

In Croatia, based on the trends and development in technology, drones have already been used in different sectors. This is confirmed by more than 130 different companies using drones in their business. Drone implementation today in Croatia is in line with the current business perspective and development worldwide. This offers potential for further automation of processes in companies which will improve their efficiency in different areas (Vlahović, Knežević & Batalić, 2016). Drones are mostly used in the media and entertainment sector, surveillance and infrastructure. There are three companies producing commercial drones in Croatia (Hipersfera, Kapetaia and Tarsier drones) and this indicates an increase in drone use for civilian and commercial purposes.

Two main telecommunications companies (i.e. Hrvatski Telekom and Vipnet) evaluate the use of drones in automation of their mobile networks, which should lead to savings in operational costs. Both companies and their owners (i.e. Deutsche Telekom and A-1) also consider potential drone development for different parts of their businesses. Another field of drone application is agriculture, where the Croatian company Žito Group tests drones for collecting data from planted seeds, for farmland conditions analyses and for making digital orthophoto maps. With drones in their business, the company intends to achieve field monitoring automation and make savings in their business. Agrokor, one of the biggest Croatian companies, uses drones for their agricultural business operation (i.e. Belje and PIK Vinkovci companies), where this use resulted in savings in production that led to increased productivity. They also evaluate the possibility of using drones in the other part of their group. (Bačelić, 2016) The use of drones in agriculture was one of the main topics at the Croatian DRONEfest conference, which was held for the second time in 2017. DRONEfest is a conference for presenting the latest news about drones from different aspects (legal regulation, business possibilities, drone types, etc.) and it is a great place for further promotion of drone usage in Croatia. The conference is organized by IN2 company, which is the leading company in drone use introduction and a developer of software for drone control and use in infrastructure and agriculture. In May of the 2018 a first Drone Expo has been organized in city of Osijek. Expo had several workshops and lectures about drone use and was a host for the International Drone race cup of Croatia (Korda, 2018)

The Croatian national electricity company Hrvatska Elektroprivreda uses drones to control their power lines in remote areas. This helps them check potential malfunctions without shutting down electricity during this process.

In Croatia, drones are mostly used in the production and technology sector. Several logistics companies (e.g. Tisak and Overseas) are evaluating the use of drones for their deliveries. The Croatian Post is watching global trends, but for them, drone deliveries are not feasible due to drone reliability as delivery tools (weather conditions, short delivery range, and airspace regulation).

There is a legal framework regulating the use of unmanned aircraft vehicles in Croatia. The Air Traffic Act (Official Gazette, 2014) regulates all civil aviation activities on the territory and in the airspace of the Republic of Croatia. In accordance with the Air Traffic Act, Croatia adopted the Ordinance on unmanned aircraft systems (Official Gazette, 2015). The provisions of the Ordinance are applied to the systems with operating mass less than 150 kg. Pursuant to current legal regulation, unmanned aircraft vehicles are divided into three classes: (1) up to 5 kilograms, (2) from 5 to 25 kilograms, and (3) from 25 to (and including) 150 kilograms. Besides that, the Ordinance defines flight area classifications in relation to population and the presence of people and buildings. The category of flight operations is defined by operating mass and the flight area and there are currently four categories in accordance with the Ordinance. Regarding UAV flight safety, the aforementioned Ordinance states general conditions (i.e. flights in daylight, distance from objects, people, animals, etc.), some of which leading to the biggest restrictions - flights during daylight hours. For these reasons, the full exploitation of the technological capabilities of UAVs is disabled. One of the current problems of UAV regulation is the fact that Croatia does not have any register of UAVs. This problem becomes apparent when it is required to find an unmanned aircraft and locate a responsible person (i.e. operator, pilot, or owner). This register should be implemented based on comparative legal solutions. Mudrić and Katulić (2016) stated that similar registers of operators are currently present in the Czech Republic, Italy, the Netherlands and Sweden. Erceg, Činčurak Erceg and Vasilj (2017: 54) state that although Annex 5 of the Ordinance provides the fulfilment of certain requirements relating to flight operations, relating to the age of the operator, the psychophysical ability, knowledge of the aviation regulations and the ability to manage the system depending on the category of flight operations, they are not strict. Thus, the issue and the problems regarding safety of unmanned aircraft flights and potential issues in terms of damage liability should be emphasized and further researched. Considering the specificities of unmanned aviation, revision, improvement, and supplementing of the present legislation in Croatia is needed (Činčurak Erceg, Vasilj,& Erceg, 2017) and therefore changes to the Croatian unmanned aircraft legislation will have to consider the potential problems in current practice as well as the change of EU legislation

The main disadvantage of drones in terms of greater use in the commercial and civilian sector in Croatia is a slow introduction of legal regulation and a lack of regulation.

Comparison Between Turkey and Croatia

Based on the conducted research, it can be said that drones are becoming more and more popular in both countries, bearing in mind that Turkey is a significantly larger country than Croatia. In both countries, the number of drones is rising, especially the ones used for personal purposes (i.e. hobbies). The fields in which drones are currently applied in both countries are shown in Table 3.

The previous table shows drone application in Turkey and Croatia. Both countries have legal regulation for drone use. In Turkey, drone use is regulated by several laws. The use of drones under 4 kg of weight, with a speed less than 50 km/h and a flight altitude under 100m is not regulated and their use is allowed for sport and recreational purposes in open areas. Drones over 20 kg of weight are not allowed to be operated by civilians and are regulated by the Instructions on Procedures and Principles of UAV System Registration, Operation, Navigation, Maintenance and Flight Sufficiency Directive. Croatia regulated drone use in May 2015, when the Ordinance on unmanned aircraft systems (Official Gazette, 49/2015, 77/2015) was adopted. In accordance to the Ordinance there are several types of drones and several categories of flight operations based on the operating mass and the presence of the people and the buildings. Although both countries have legal regulation of the drone use, due to unexplored potentials and continuous development of unmanned aircraft, this area will need to be constantly advanced and legally regulation should be adjusted to the changes in use and technology development.

Both countries use drones in the military, but Turkey also produces them and is very advanced in their production with plans for further developments.

Drones are used in both countries in the commercial and civilian sector and are becoming even perfect gifts for different occasions for people using drones for hobbies. Croatia uses drones in agriculture, power grid maintenance, telecommunications

Table 3. Drone application in Turkey and Croatia

Field	Turkey	Croatia
Legal regulation	✓	✓
Military	✓	✓
Agriculture	×	✓
Logistics (deliveries)	×	×
Maintenance	×	✓
Telecommunications	✓	✓
Advertising, film making, marketing	✓	✓
Hobby	✓	✓
Drone thematic conferences	×	✓
Sports	✓	✓
Forensic anthropology and archeology	✓	×

Source: Authors' research, 2018

and the advertising, filming, and marketing sectors. In Turkey, drones are used in the telecommunications sector, advertising, filming, and marketing sectors and in forensic anthropology and archeology. In both countries, national postal companies are evaluating the use of drones and will probably soon start testing them for delivering parcels to rural and hard-to-reach areas. This is also the situation with different parcel delivery service companies (i.e. DHL, DPD, UPS, FedEx), which are still exploring possibilities in relation to legal regulation for the use of drones for last mile deliveries.

Every year, a drone thematic conference called DRONEfest is held in Croatia and it represents an excellent place for exchanging the latest news about drone development trends and information about their use in different sectors. In Turkey, Tubitak organizes the Turkish International UAV competition. The aim of the competition is to show potential applications of UAVs in the civilian sector, especially in emergencies such as fire or different accidents (Tubitak, 2017).

Finally, Turkey has started a drone racing championship that will also influence and raise the popularity of drones in the country. This is a sign that drones will have a future in our lives, especially in the logistics sector since DHL is the main sponsor for the world drone racing series (DR1 racing, 2017). In Croatia, similar sport events have been organized in last two years.

Based on the aforementioned, both countries have accepted the use of drones in the commercial and civilian sector and it is expected that drones will play an even more significant role in the future in different sectors where they will be able to contribute to optimization of business processes.

CONCLUSION

Unmanned aerial vehicles (drones) have brought significant advantages and innovations to different industrial sectors. Today they are used in many different applications, including, but not limited to, cargo, military, and advertising. Drones are increasingly becoming a staple of everyday life and a way to help productivity. On the one hand, they can help reduce shipping costs, and on the other, they can help secure national security. Usage of drones is increasing, not only due to a reduction in their prices, but also due to people's interests, and from today's perspective, drone technology will continue to be developed and changed.

Drones are used in different sectors in Turkey and Croatia. Both countries use them for military purposes since this was the intention when drones came into first use. In other sectors, drones are becoming more and more popular and their use is mainly limited to legal regulation and a slow introduction of legal regulation. This is also the main barrier to greater use of drones as legal regulation is changing and there are no standardized rules of using drones. Thus, with further development of law and drone technology, drones will be used more frequently in industrial sectors where they are not present now – e.g. in delivering parcels in last mile deliveries. Their further use will lead to further development in technology and an increase in drone production. This will result in a need for drone pilots who must be trained, and hence create drone use and maintenance job opportunities, which will at the end boost the national economy.

Since drones are currently present in some industrial sectors in both countries and their use in other sectors is being evaluated, we propose further research into the following:

- Economic benefits of drone usage in different sectors in both countries,
- Drone production and its influence on increased drone usage in both countries, and
- The potential new industrial sector where drones can be introduced.

ACKNOWLEDGMENT

This research received no specific grant from any funding agency in the public, commercial, or not-for-profit sectors.

REFERENCES

Agrawal, A. J. (2017). *5 Ways Marketers Can Take Advantage of Drone Technology*. Retrieved from: https://www.forbes.com/sites/ajagrawal/2017/06/10/5-ways-marketers-can-take-advantage-of-drone-technology/#2c3eb5e658cc

Bačelić, M. (2016, August 1). Kako se u biznisu možete koristiti bespilotnim letjelicama, Lider, 24-28.

Blau, M. (2017). *Condom drops and airborne meds: 6 ways drones could change health care*. Retrieved from: https://www.statnews.com/2017/06/13/drones-health-care/

Brown, J. (2017). *How Drones Will Revolutionize Marketing Practices*. Retrieved from: https://www.digitaldoughnut.com/articles/2017/march/how-drones-will-revolutionize-marketing-practices

Calishan, N. (2018). *İzinsiz uçuşa 11 bin 447 TL'lik ceza kesilecek*. Retrieved from: https://www.haberturk.com/izinsiz-drone-ucurana-11-bin-447-tl-lik-para-cezasi-2131702-ekonomi

Çeker, D. (2014). Adli Antropolojide Perimortem ve Postmortem Kırıkların Ayırımı ve Travma Analizlerindeki Önemi. *Ankara Üniversitesi Dil ve Tarih Coğrafya Fakültesi Antropoloji Dergisi.*, *27*, 47–64.

Ceker, D. (2016). Olay yeri inceleme ve Calismalarinda adli arkeolog ve adli antropologlarin rolu: Kuzey kibris ve Turkiyede'ki guncel durum. *AÜDTCF. Antropoloji Dergisi*, *32*, 13–21.

Cetin, O. (2016). *Türkiye'de Drone kullanmanın şartları*. Retrieved from: https://mediatrend.mediamarkt.com.tr/turkiyede-drone-kullanmanin-sartlari/

Cevap, H. (2016). *Drone kullanmak yasal mıdır?* Retrieved from: https://www.cnnturk.com/video/hazir-cevap/hukuk/drone-kullanmak-yasal-midir

Činčurak Erceg, B., Vasilj, A., & Erceg, A. (2017). Unmanned Aircrafts ('Drones') – The Need for Legal Regulation and Harmonization of Legislation in the European Union. In M. Froechlich & C. Voda (Eds.), *South Eastern Europe and the European Union - Legal Perspectives - Series of Papers* (Vol. 3, pp. 9–22). Saarbrucken: SEE EU Cluster of Excellence in European and International Law.

Consortiq. (2016). *A Short History of Unmanned Aerial Vehicles (UAVs)*. Retrieved from: https://consortiq.com/short-history-unmanned-aerial-vehicles-uavs/

DR1 Racing. (2017). *DHL Champions Series*. Retrieved from: https://www.dr1racing.tv/champions-series

Eisenbeiss, H. (2004). A mini unmanned aerial vehicle (UAV): System Overview and Image Acquisition. *Presented at International Workshop on Processing and Visualization Using High-Resolution Imagery*, 1-7.

Emir, A., & Eren, A. (2016). *Gökyüzünün Yeni Sakinleri*. Retrieved from: https//:archive.ismmmo.org.tr

Erceg, A., Činčurak Erceg, B., & Vasilj, A. (2017). Unmanned aircraft systems in logistics – legal regulation and worldwide examples toward use in Croatia. In *Proceedings of the 17th International Scientific Conference Business Logistics in Modern Management*. Osijek: Ekonomski fakultet.

Estampe, D. (2015). *Planning the Future Supply Chain Together - New Technologies: Big data, 3d Printers, Drones*. White paper. Bordeaux: Kedge Business School.

Etherington, D. (2017). *Google's Project Wing team takes a key step towards making drone delivery real*. Retrieved from: https://techcrunch.com/2017/06/07/googles-project-wing-team-takes-a-key-step-towards-making-drone-delivery-real/

Finnegan, P. (2015). *Press release: UAV Production Will Total $93 Billion*. Retrieved from: http://tealgroup.com/index.php/teal-group-news-media/item/press-release-uav-production-will-total-93-billion

Frost & Sullivan. (2007). *Study Analyzing the Current Activities in the Field of UAV*. Retrieved from: https://ec.europa.eu/home-affairs/sites/homeaffairs/files/e-library/documents/policies/security/pdf/uav_study_element_2_en.pdf

Gonzalez-Aguilera, D., & Rodriguez-Gonzalvez, P. (2017). Drones editorial. *Drones*, *1*(1), 1–5. doi:10.3390/drones1010001

Heutger, M. (2014). *Unmanned Aerial Vehicles in Logistics – A DHL Perspective on Implications and Use Cases for the Logistics Industry.* DHL Customer Solutions & Innovation. Retrieved from: http://www.dhl.com/content/dam/downloads/g0/about_us/logistics_insights/DHL_TrendReport_UAV.pdf

Horowitz, M. C., & Fuhrmann, M. (2015). *Droning On: Explaining the Proliferation of Unmanned Aerial Vehicles.* doi:10.2139srn.2514339

Improdone. (2018). *The Pros and Cons of Drones.* Retrieved from: https://improdrone.com/the-pros-and-cons-of-drones/

Johnson, T. (2017). *Amazon Prime Air: The Evolution of Amazon's Drone Delivery.* Retrieved from: http://www.cpcstrategy.com/blog/2017/07/amazon-prime-air/

Kelek, B. S. (2015). Blue Sky Birds Come to the World. *Journal of International Trade. Logistics and Law, 1*(1), 41–49.

Knight, Z. (2017). *What Germany's first armed drones could do.* Retrieved from: https://www.dw.com/en/what-germanys-first-armed-drones-could-do/a-39355009

Koeniger, N., Koeniger, G., Gries, M., & Tingek, S. (2005). Drone Competition at Drone Congregation Areas in Four Apis Species. *Apidologie, 36*(2), 211–221. doi:10.1051/apido:2005011

Korda, Z. (2018). *U Osijeku otvoren prvi sajam dronova u Hrvatskoj.* Retrieved from: https://www.tportal.hr/tehno/clanak/u-osijeku-otvoren-prvi-sajam-dronova-u-hrvatskoj-20180512

Lorenz, A., von Mittelstaedt, J., & Schmitz, G. P. (2011). *Messengers of Death - Are Drones Creating a New Global Arms Race?* Retrieved from: http://www.spiegel.de/international/world/messengers-of-death-are-drones-creating-a-new-global-arms race-a-792590-2.html

Lutz, B. (2014). *Multirotor service-drone market-leading developments.* Retrieved from: http://www.marketwired.com/press-release/multirotor-service-drone-market-leading-developments-in-may-2014-1915272.htm

Mazur, M. (2016). *Six Ways Drones Are Revolutionizing Agriculture.* Retrieved from: https://www.technologyreview.com/s/601935/six-ways-drones-are-revolutionizing-agriculture/

Mazur, M., & Wisniewski, A. (2016). *Clarity from above - PwC global report on the commercial applications of drone technology.* PricewaterhouseCoopers. Retrieved from: http://www.pwc.pl/en/publikacje/2016/clarity-from-above.htm

Mudrić, M., & Katulić, T. (2016). Regulacija sustava bespilotnih zrakoplova u hrvatskom, europskom i međunarodnom pravnom okviru. *Pravo u gospodarstvu*, *55*(2), 123-136.

Official Gazette (2014). *Air Traffic Act*. Official Gazette. 92/2014.

Official Gazette (2015). *Ordinance on unmanned aircraft systems*. Official Gazette. 77/2015.

Rosenberg, A. S. (2009). *An evaluation of a UAV guidance system with consumer grade GPS receivers* (Ph.D. dissertation). University of Arizona. Retrieved from: http://www.casa.arizona.edu/data/abigail/Abigail_Rosenberg_Final_Dissertation.pdf

Sahin, Y. E. (2016). *Drone nedir ve hangi alanlarda kullanılır?* Retrieved from: https://www.gzt.com/teknoloji/drone-nedir-ve-hangi-alanlarda-kullanilir-2576541

Sandbrook, C. (2015). The Social Implications of Using Drones for Biodiversity Conservation. *Ambio*, *44*(4), 636–647. doi:10.100713280-015-0714-0 PMID:26508350

Sayfa, A. (2016). *Orman yangını alanlarında 'drone' dönemi*. Retrieved from: http://www.trthaber.com/haber/turkiye/orman-yangini-alanlarinda-drone-donemi-286228.html

Sayfa, A. (2017). *Türkiye Drone Şampiyonası İstanbul'da nefes kesti*. Retrieved from: https://www.emlakpencerem.com/turkiye-drone-sampiyonasi-istanbul-da-nefes-kesti/97807/

Slabinac, M. (2015). Innovative Solutions for a "Last Mile" Delivery – European Experience. In *Proceedings of 15th international scientific conference Business Logistics in Modern Management*. Osijek: Ekonomski fakultet.

Smith, J., Mazur, M., & Wisniewski, A. (2017). *Clarity from above: transport infrastructure*. PricewaterhouseCoopers. Retrieved from: http://www.pwc.pl/en/publikacje/2016/clarity-from-above-transport-infrastructure.html

Stipanović, D. M., Inalhan, G., Teo, R., & Tomlin, C. J. (2004). Decentralized overlapping control of a formation of unmanned aerial vehicles. *Automatica*, *40*(8), 1285–1296. doi:10.1016/j.automatica.2004.02.017

Stroh, M. B. (2006). *A practical guide to transportation and logistics*. Dumant: Logistics Network.

Sudip, S. (2017). *For Press Release*. Retrieved from: https://forpressrelease.com/forpressrelease-195898-production-of-multirotor-drones-for-non-military-applications-has-positively-impacted-its-market.html

Teknolo. (2015). *Dronelarin Kullanim Alanlari*. Retrieved from: http://www.teknolo.com/wp-content/uploads/2015/11/DRONELARIN-KULLANIM-ALANLARI.pdf

Tubitak. (2017). *Welcome to the International UAV TURKEY Competition!* Retrieved from: https://www.tubitak.gov.tr/en/announcements/welcome-to-the-international-uav-turkey-competition

Türkiye İnsansız Hava Aracı Sistemleri Yol Haritası. (2011). *Türkiye İnsansız Hava Aracı Sistemleri Yol Haritası*. Author.

Vlahović, N., Knežević, B., & Batalić, B. (2016). Implementing Delivery Drones in Logistics Business Process: Case of Pharmaceutical Industry. *International Journal of Social, Behavioral, Educational, Economic, Business and Industrial Engineering*, *10*(12), 3948–3953.

Welch, C. (2013). *Zookal will deliver textbooks using drones in Australia next year*. Retrieved from: http://edition.cnn.com/2013/10/18/tech/innovation/zookal-will-deliver-textbooks-using-drones/

KEY TERMS AND DEFINITIONS

Automation: Creation of technology and its application in order to control and monitor the production and delivery of various goods and services.

Last Mile Delivery: Movement of goods from a transportation hub to the final delivery destination. The final delivery destination is typically a personal residence. The focus of last mile logistics is to deliver items to the end user as fast as possible.

Legal Framework: Comprise a set of documents that include the constitution, legislation, regulations, and contracts.

Logistics: Planning, execution, and control of the procurement, movement, and stationing of personnel, material, and other resources to achieve the objectives of a campaign, plan, project, or strategy.

Unmanned Aerial Vehicles: An unmanned aerial vehicle is a type of aircraft that operates without a human pilot onboard. Recent technologies have allowed for the development of many kinds of advanced unmanned aerial vehicles used for various purposes.

Chapter 4
Drones in Healthcare:
An Extended Discussion on Humanitarian Logistics

Berk Anbaroğlu
Hacettepe University, Turkey

ABSTRACT

This chapter discusses the use of drones in healthcare with a specific focus on humanitarian logistics. Drones have already been used in healthcare in different aspects, including transfer of blood products, search and rescue missions, or collecting different types of data including aerial photographs, air quality, or radiation levels. Even though the published research evidence in the area of "drones in healthcare" is almost 1% of the broader area of "drones," the progress in public acceptance, regulations, as well as technology is undeniable. This chapter summarizes the different aspects regarding the use of drones in healthcare, while specifically focusing on humanitarian logistics. The SWOT analysis indicate that the strengths and opportunities weigh more than the weaknesses and threats, suggesting that drones will revolutionize the way medical supplies are delivered within the coming years.

DOI: 10.4018/978-1-5225-7900-7.ch004

INTRODUCTION

Airspace ownership is a topic under spotlights for centuries, from the Roman times in 27BC to the present day. Each technological progress leads to a regulation structuring how that technology should be used. It is not surprising to see that after the first aircraft, which was a hot-air balloon leaving the ground in 1783, the relevant regulations followed it in 1784 (Engvers, 2001). In today's context, the debate is much more heated as the airspace could be commercialised to an extent that have never endured before, thanks to the prospects provided by the adoption of drones. Therefore, it is not surprising to hear a senior engineer from NASA stating that "we need to accommodate drones" in air traffic control systems (Schneider, 2017) or see global challenges funded by Bill & Melinda Gates Foundation to support the distribution of vaccines by low cost drones (Global Grand Challenges, 2012).

Some researchers differentiate the words "remotely piloted vehicle", "drone" or Unmanned Aerial Systems (UAS) (Rosser, Vignesh, Terwilliger, & Parker, 2018). However, this chapter does not make a distinction between them. The definition relied on throughout the chapter is: a system composed of an air vehicle and a mission planning/control station, in which the air vehicle could potentially carry a load for Beyond Visual Line Of Sight (BVLOS) distances and could be operated at different levels of automation (Fahlstrom & Gleason, 2012, p. 8).

The research and development activities surrounding this topic is wide and its implications are substantial. The renowned TIME magazine had a special issue entitled "The Drone Age", which was printed on the issue of June 11, 2018. A total of 958 drones created the logo of the magazine as well as the red border in mid-air. The cover image depicting this spectacular event was also shot with a drone (TIME, 2018).

The word "drone", which was coined due to the similarity of its sound to a male bee, has initially been associated with a military unit, thus took a negative connotation (M. Benjamin & Ehrenreich, 2013; Rosser et al., 2018). However, there is a growing number of research on small, human-friendly highly automated drones that fly in confined spaces and in close proximity to people. The emergence of drones in public use is facilitated by their miniaturisation and reduced costs. Consequently, the prototyping and commercialization of civilian drones had already happened and many different drones could be bought online.

Drones can be classified in different ways, such as their landing and take-off type (horizontal or vertical), weight (small, medium, large), propulsion (electric or internal), flight time and their specialization such as surveillance or package delivery (Fahlstrom & Gleason, 2012; Rosser et al., 2018). A thorough comparison of 30 different UAS based on their manufacturer, type (rotor or fixed wing), size, weight, flight time and cost is provided in Kim and Davidson (2015).

The integration of drones in a civilian context would add value to economy not only in terms of research and development but also in job creation. The economic impact of this endeavour is estimated to be $13.6 billion in its first three years of integration and will grow to $82 billion in the next 10 years. As a result, more than 70,000 new jobs will be created in the first three years of integration (Jenkins & Vasigh, 2013). Another article proposes a different projection, where size of the market will reach to $82.1 billion by 2025, and generating more than 103,000 jobs (Rosser et al., 2018). These values suggest an expected and natural outcome since drones can perform many different tasks ranging from logistics to crop monitoring to infrastructure surveillance, most of which can be highly automatised.

Logistics would probably be the key sector in which the utilisation of drones would revolutionise our lives (Anbaroğlu, 2017; Barmpounakis, Vlahogianni, & Golias, 2016). The advantages of using drones to transport packages include reduced congestion in urban areas, rapid transport of packages where road infrastructure is poorly developed or access to remote areas in a safe way. Researchers proposed heuristic algorithms to handle the operational challenges such as optimal routing and scheduling of drones and delivery trucks by minimizing the delivery time (Murray & Chu, 2015).

The nature of the spatial data generated by a drone is referred to as a 'trajectory', which contains the position and timestamp of the drone at specific time intervals. Therefore, a trajectory would constitute of a series of four-dimensional points (latitude, longitude, elevation and time), which necessitates the use of Geographical Information Systems (GIS) to store, query and visualise the data. Especially, legal bodies need to store such data for long times in case of an enquiry.

The growing number of resources regarding do-it-yourself drones adds further support to the vision of 'drone deliveries' (Lim, Park, Lee, & Kim, 2012). Combined with societal efforts, such as AirShare (a design concept allowing people to offer leftovers or unused food to those in need), drone delivery has much to offer (Burton, Meier, Olarte, Skeini, & Zahan, 2017).

Use of drones in healthcare has also recently attracted the interest of researchers as well as companies and non-governmental organizations. Meier (2015) provides a concise description of how drones could be adopted in this context by including information about some of the first 'use-cases'. For instance, the first aerial image of an earthquake, which happened in San Francisco, was shot in 1906. The first search-and-rescue mission with a drone took place in May 2013. Consequently, there is a growing research interest on the use of drones in healthcare (Bhatt, Pourmand, & Sikka, 2018; Sachan, 2016). This chapter contributes to the literature by providing a thorough discussion on the humanitarian logistics aspect of the use of drones in healthcare while focusing on the most recent publications. Having reviewed the literature, the findings are summarised through a SWOT analysis.

Figure 1. Number of publications on drones and their classification based on the research discipline

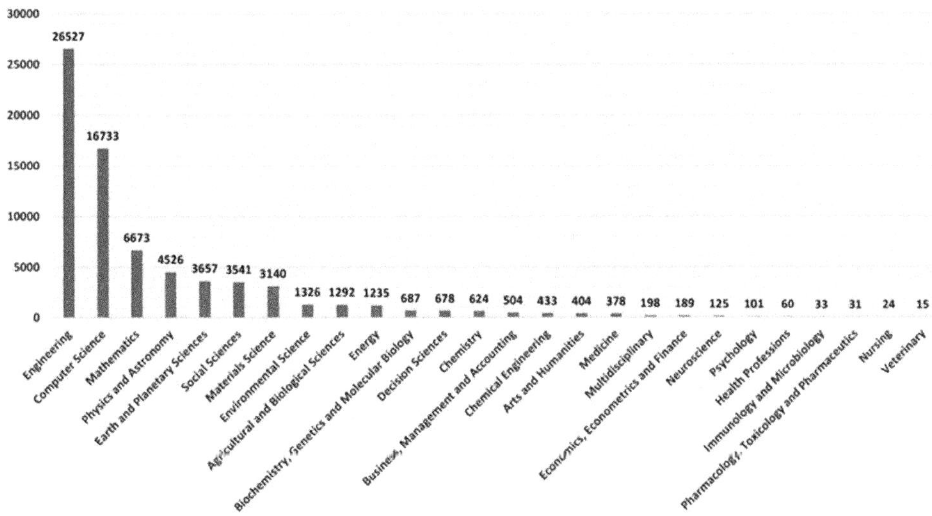

BACKGROUND

There is a growing number of research on the wide area of drones. One of the commonly used databases to search for scientific articles, SCOPUS, has been used to investigate this trend. Having noted that the word 'drone' also refers to a male bee, the papers indexed in SCOPUS having the key words {"unmanned aerial systems" or "unmanned aerial vehicles" or "drones" *and not* "bee"} are analysed on 26 December 2018. Consequently, almost 40,000 papers have been published, with almost half of it in the last four years. Since January 1st of this year [2019], over 200 papers have been indexed already. Distribution of paper's subject area is illustrated in Figure 1. This sheer scale of research interest is a clear indication of how these airborne robots are going to influence our lives in the near future.

The results indicate that "medical" sector is still in its infancy with a total of 378 publications, which corresponds to less than 1% of the total publications within the broad field of drones. A recent survey investigated the EBSCO database with a similar search pattern and a total of 202 publications are found to be containing the keywords "drone" and "medicine". The scientific publications about the use of drones in medicine are further classified into three: i) public health/disaster relief, ii) telemedicine, and iii) humanitarian logistics, where a total of 116, 8 and 78 publications have been identified corresponding to these respectively (Rosser et al., 2018).

The use of drones in healthcare would lead to a substantial impact. Humanitarian organizations or countries could immediately dispatch aid efforts by improving surveillance, search and rescue capabilities, which is also referred to as 'non-human humanitarian' effort (Meiches, 2018; Skinner et al., 2018). Emergency units such as firefighters could benefit from simulations relying on human-system interaction. In a search and rescue operation researchers have integrated inspection drones, that are capable of flying in confined spaces, and virtual reality is used to provide a better sense of the environment (Prexl, Struebig, Harder, & Hoehn, 2017). Last, as most drones rely on electric batteries, which emit less carbon than delivery trucks, drone delivery would also contribute to improving air quality. In this way, drone delivery of medical products would lead to an indirect improvement to public health (Yoo, Yu, & Jung, 2018). Empirical results also add further support by suggesting a substantial improvement regarding the "Global Warming Potential" parameter if drone delivery is utilised instead of relying on motorcycles (Park, Kim, & Suh, 2018).

Three Facades: Public Acceptance, Regulations, and Technology

Public acceptance, regulations and technology are the three key facades of seeing the effective use drones in a civilian context. As more and more research is carried out in all of these different aspects, drones carrying medical products would be a common practice in the coming years. The signs of this have already emerged in the United States with the 'UAS Integration Pilot Program' (FAA, 2018), which was launched on 7 May 2018. The pilot program matches local governments with private companies to test drone flights that are currently banned in the U.S., including flying at night, package delivery and flying over people and beyond the pilot's line of sight. Tests will occur at 10 different sites across the U.S. and the outcomes would offer valuable information to the U.S. Department of Transportation and Federal Aviation Administration (FAA), which is the aviation regulatory body to provide the regulations for drone flights.

Public Acceptance

It is not easy for the public to accept the use of civilian drones in a short span of time. Recent researches aim to understand the factors affecting public attitude to drone delivery service and prospective customers' intention to adopt the technology (Yoo et al., 2018). One of the key outcomes of Yoo et al. (2018)'s study is that customers' area of residence influence their attitude towards drone delivery. Specifically, people living in an urban environment are more interested in the safe operation of drones, whereas people living in a rural environment are interested in privacy.

Noise pollution is often an overlooked issue regarding the wide use of drones. However, due to its linkage with health problems including heart disease or lower academic performance (Passchier-Vermeer & Passchier, 2000), it is important to investigate on noise pollution to understand citizens' reactions. A recent research evidence demonstrates that the noise caused by a drone is more annoying to people than the same level of noise caused by cars or trucks (Christian & Cabell, 2017). Consequently, it is important to investigate solutions to mitigate the sound caused by a drone, especially during take-off, landing and low level flight (Miljković, 2018).

Safety is another important aspect that influence public acceptance. A drone colliding into people or properties due to malfunction, hacking or running out of battery exemplify the potential danger associated with the operation of drones in densely populated areas. Therefore, researchers aim to quantify the effect of falling drones on dummy human heads (Koh et al., 2018). The Alliance for System Safety of UAS through Research Excellence, an academic research Centre of Excellence, is also conducting research on, amongst others, quantifying the injury potential of drones resulting from collisions with public on the ground (ASSURE, 2018). Different strategies could be adopted to reduce the injury levels such as using energy-absorbing materials to cover the drone or using parachutes when loss of control occurs.

Privacy is also a topic of interest, since one of the main uses of drones is to offer enhanced vision to its customers (Zuboff, 2015). Finn and Wright (2012) argue that the "usual suspects" (i.e. the poor, people of colour and anti-government protesters) would be targeted and current legislations are inadequate as drones are complex, multimodal surveillance systems that integrate a range of technologies and capabilities. Nevertheless, regulations should be structured to protect civil liberties and freedoms and develop mechanisms to ensure that the drone is doing its intended task and not used for other purposes.

Lichtenstein, Slovic, Fischhoff, Layman, and Combs (1978), in their seminal work, highlight that the public either "overestimate small risks" or "underestimate large risks". Therefore, public awareness should be increased by elaborating on the best practices as well as the current regulations that are in practice. Therefore, understanding and constructing the public perception is a crucial step towards the acceptance of drone technologies (Clothier, Greer, Greer, & Mehta, 2015; Luppicini & So, 2016).

Regulations

In order to realise the full potential of drones, not only scientific and technological advances have to be achieved, but also a regulatory framework has to be developed bridging the gap between public and private entities. In this way, public and economic concerns could be addressed. National Aviation Authorities (NAAs), such as Federal

Aviation Administration (FAA) of U.S. or Civil Aviation Authority (CAA) of the UK are working towards such a regulatory framework. For instance, FAA requires that drones, apart from those that are under 55 pounds (~25 kg), to have on-board "detect, (sense) and avoid" capabilities (Shively, 2014). In addition, they classify drones based on their take-off weight (Oliver, 2017). On the other hand, CAA "requires drone flights to stay away from persons not actually involved in the flight operation and to be at least 150 metres from structures" which prevent the mass use of drones for logistics in an urban environment (Harrington, 2015).

Different countries have different regulations. However, for shared airspace like the one in the European Union, it would be necessary to define test conditions and performance standards in cooperation with all other regulatory bodies to evaluate the capabilities and reliability of drone technology (Boucher, 2015; Torresan et al., 2017). Consequently, European Aviation Safety Agency had already provided a technical opinion suggesting to harmonize the efforts (EASA, 2015).

The importance of following the regulations were also highlighted in the context of 2015 Nepal Earthquake. Even though there were no regulations regarding drone flights, some of the volunteers assumed that they could operate drones for humanitarian purposes without seeking permission. This act; however, backfired and some of these teams got arrested and their drones were confiscated. This suggests the importance of working together with the governmental bodies and the aviation regulators. Even if the intention might be purely humanitarian, officers could evaluate the situation differently (Meier, 2015).

Technology

Drone is a robot, meaning that it is purely a product of science and technology. Therefore, under this heading, many different topics could be discussed such as prolonging battery life through wireless charging (Choi et al., 2016), investigating different levels of autonomy (Floreano & Wood, 2015), navigating through GNSS denied environments such as urban canyons or forest canopies (Chowdhary, Johnson, Magree, Wu, & Shein, 2013) or improving the sense and avoid methods (Zhang, Cao, Ding, Zhuang, & Yao, 2016). It is not possible to cover all of these in depth under this chapter. Nevertheless, it is worth to highlight the possible security breaches. Drones relying on WiFi for communication (i.e. IEEE 802.11 standards) are vulnerable to attacks. The recently announced security protocol, WPA3, is more secure than its predecessor, since it fixes de-authentication, off-line dictionary attacks and the KRACK vulnerability (Kohlios & Hayajneh, 2018). However, developing new attacks and counter-measures are still important to assure the security of the overall system (Vanhoef & Piessens, 2018).

Path finding in an urban environment is also a recent research interest. In order to prolong battery life, hence, flight duration, solar-powered drones could be used in an urban environment. The shortest path between two points might not be optimal in terms of energy efficiency due to the differences in building heights. Therefore, researchers compared two strategies on a synthetic urban environment: time-optimal and energy-optimal paths. The energy-optimal path avoids being close to tall buildings as the total energy savings would be lower, whereas the time-optimal path finds the shortest path. The proposed methodology aims to bring a balance to both approaches (Wu et al., 2018).

Disaster response and management might necessitate multiple drones operated simultaneously, also called a swarm of drones, where each is responsible for a single task like 3D mapping of the region, locating stranded people or supplying the medical products. The communication between the drones could make the system cost-efficient rather than utilising a single drone for the same purpose (Alex & Vijaychandra, 2016). A nice overview covering the different aspects of utilising drones for disaster management, including the automated organisation between them and their safety and security, is presented by Tanzi et al. (2016).

One of the main issues to be addressed is the data scarcity. Most of the researches rely on simulated environments to generate and validate their models. Researchers investigate the effectiveness of game engines for the purpose of generating as realistic disaster data as possible (Smyth, Glavin, & Madden, 2018). As a result, drones in healthcare is a multi-stakeholder research agenda, involving governmental regulatory bodies, internationally recognised public bodies, non-governmental organisations and companies. Some of these are described in Table 1 (the stakeholder names are clickable).

DRONES IN HEALTHCARE

The growing research and development activities surrounding the topic of drones in medicine, signal their potential, and that they would play a vital role in healthcare (Balasingam, 2017; Bhatt et al., 2018). Researchers have conducted studies using drones in healthcare under three main categories: i) public health/disaster relief, ii) telemedicine, and iii) humanitarian logistics.

Public health and disaster relief related researches mainly rely on the 'enhanced data collection' aspect of using drones. In this way, high-resolution spatial data could be generated in a short span of time and at low costs compared to, for example, satellite imagery. In addition, drones could fly through hostile or inaccessible terrain or dangerous to humans due to contamination (Martin et al., 2016).

Table 1. Different stakeholders relying on drones in healthcare

Name	Explanation
Public Bodies	
European Aviation Safety Agency	Established in 2002 to ensure the protection of EU citizens as well as the environment.
International Civil Aviation Organization (ICAO)	ICAO works with the National Aviation Regulatory Authorities and industry groups from around the world to reach consensus on international civil aviation Standards and Recommended Practices (SARPs) and policies in support of a safe, efficient, secure, economically sustainable and environmentally responsible civil aviation sector.
Organizations	
UAViators	Humanitarian network consisting of 3200 members in over 120 countries. The network is dedicated to the use of drones in humanitarian efforts in a safe and effective way.
Humanitarian OpenStreetMap	A community dedicated to humanitarian action and community development through open mapping. The created maps are invaluable assets regarding disaster management.
Micromappers	A community deploying micro tasking apps specifically customized for digital humanitarian response. User-generated, multi-media content posted on social media during disasters are used to evaluate the overall situation.
Emergency Telecommunications Cluster	A global network of organizations that work together to provide shared communications services in humanitarian emergencies.
Joint Authorities for Rulemaking on Unmanned Systems	Providing support regarding technical, safety and operational requirements for the certification and safe integration of drones into airspace.
Companies	
Matternet	Swiss Post started delivering lab samples between hospitals in an urban environment by using quadrotor drones.
Zipline	Successful blood-deliveries in Rwanda that rely on fixed-wing drones since late 2016 (Ackerman & Strickland, 2018).
Flirtey	Collaboration of different partners led to the first U.S. government approved delivery of medical supplies on July 17, 2015 (Howell, 2016).
Wilstair	Within hospital delivery. Commercialization of drones delivering drugs or carrying lab samples within hospitals would be revolutionary. Costly pneumatic tubes will not be needed any more.
Vayu	Clinical lab samples are transferred from a remote village to a central lab in July 2016 in Madagascar by a VTOL capable fixed wing drone.
eHang	The designed drone (coaxial multi-rotor with VTOL) can fly with a passenger carry a passenger for approximately 10 miles through the air at speeds up to 65 mph. The aim is to automate organ transplant delivery in medical emergency situations in an urban environment.

Photogrammetric methods are used to investigate copper concentration in southern Italy (Capolupo, Pindozzi, Okello, Fiorentino, & Boccia, 2015). Photographs collected by drones are used to identify the cause of malaria in the Sabah region of Malaysia (Fornace, Drakeley, William, Espino, & Cox, 2014). Brady et. al (2016) relied on aerosol optical particle counter and a CO_2 sensor to measure aerosol and trace gas levels. In this way, they suggest, complex terrain, including urban environment, forest canopies, and even breaking surface waves could be studied.

Data collection process; however, could be hampered due to the electronic interference between the drone and sensors. Gu, Michanowicz, and Jia (2018) monitored air quality by relying on particulate matter and nitrogen dioxide observations. They emphasized the need to improve the electronic design so that the on-board sensors' readings are shielded from drone's electronic interference. Other studies relying on the 'enhanced vision' capability of a drone include the surveillance of beaches to detect drowning people (Claesson et al., 2017), detecting mosquito habitats (Hardy, Makame, Cross, Majambere, & Msellem, 2017) or locating wandering patients with dementia (Hanna, Ferworn, Lukaczyn, Abhari, & Lum, 2018). All of these studies demonstrate that relying on drones could improve the existing practice.

Telemedicine is defined as an "integrated system of healthcare delivery that employs telecommunications and computer technology as a substitute for face-to-face contact between provider and client" (Bashshur, 1995). The main use of drones in telemedicine is Out-of-Hospital Cardiac Arrests (OHCAs), which are considered to be an important health concern affecting hundreds of thousands people annually with almost a 10% survival rate (E. J. Benjamin et al., 2018).

A drone could be utilised to deliver an Automated External Defibrillator (AED) to a location close-by the patient. Instructions to use the AED are provided through the drone by either a video or an audio (Claesson et al., 2016; Zègre-Hemsey, Bogle, Cunningham, Snyder, & Rosamond, 2018). The communication could also be interactive, since a smartphone attached to the drone could be used to connect the bystander and the expert who is off-site. It is important to highlight that timely action is critical and each minute is precious to the patience having an OHCA (Larsen, Eisenberg, Cummins, & Hallstrom, 1993).

Some even referred the drones carrying an AED as the 'magic-bullet' to solve not only the long mean response times in Emergency Medical Services (EMS), but also the different issues regarding public access to defibrillation programs (Voorde et al., 2017). Delivery of an AED could be considered both as telemedicine, as an expert could be guiding a bystander from a distance by employing information technologies as well as humanitarian logistics due to the delivery of an AED.

Humanitarian Logistics

Humanitarian logistics involve the distribution of critical healthcare products to patients on a timely basis and at right conditions of the product. Vaccine transportation is the most widely studied, but the delivery of blood products, AEDs or even organs are possible. Once entered a country, the supply chain logistics of critical healthcare usually travels by road, passing through several storage locations before reaching the patient (Lee et al., 2015). Supply chain bottlenecks and inefficiencies can cause invaluable medical products to spoil, suggesting the need for innovative and lower cost methods for delivery (Haidari et al., 2016; Thiels, Aho, Zietlow, & Jenkins, 2015).

It is also equally important to demonstrate the cost-effectiveness of transporting medical products by drones. Haidari et al. (2016) demonstrated that utilising drones for vaccine delivery could be more cost-effective in a computer simulated environment compared to the delivery using road transportation. A range of conditions including the road conditions, population and vaccine schedule have been analysed to identify when and how relying on drones would be cost effective. In most of the examined scenarios, in addition to being cost-effective, vaccine availability also increases about 2%.

Delivery of medical supplies using drones would be achieved through three main steps: i) dispatch, ii) shipment and iii) delivery. During dispatch the medical supplies to be delivered are removed from the repository to be sent to the patient. Shipment is the transportation of the medical supply to the patient and delivery occurs when the package is delivered to its destination as depicted in Figure 2. Automation, in all three aspects of the process, is desired in order to provide a faster and probably a secure delivery.

During dispatch the relevant package is loaded to a drone with the destination location. There might be new generation drones that could carry several packages to be delivered to more than one destination in a single launch. The dispatch could occur through a stationary repository such as hospitals, regional medical suppliers or even pharmacies. According to Stolaroff et al. (2018), current practical range of drones is approximately 4 km, which necessitates a new network of urban warehouses. However, building such a network would be a costly process. In that case, mobile repositories could be utilised containing several drones. In this way more than one destination can be served at a time, which would still improve today's state-of-the-practice.

Shipment process is critical for several reasons. First, different drones would be sharing the same aerospace, which necessitates them to follow well-defined regulations and standards. One such regulation could be adopting an open standard that is used

Figure 2. Humanitarian logistics using drones in an urban environment (adapted from Anbaroğlu, 2017)

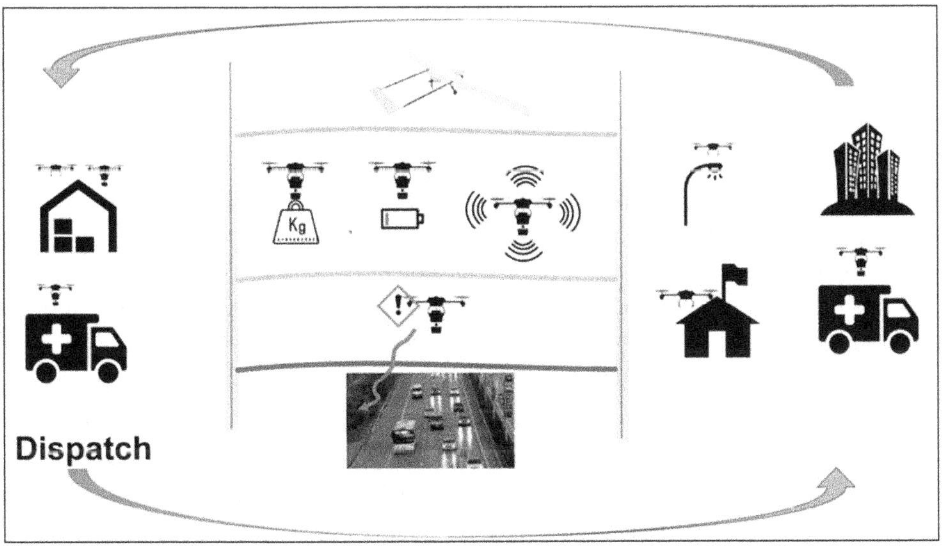

to plan non-conflicting paths for multiple drones in an obstacle rich environment. Kothari, Postlethwaite, and Gu (2009) relied on the rapidly-exploring random trees method to generate paths in near real time, by considering different obstacle types (static and dynamic) as well as the kinematic constraints of the drones such as turn radius constraints. They have assumed that drones are flying at the same height.

Different flight heights could be used to accommodate drones flying at different speeds, just like air routes of today's commercial flights as shown in different horizontal lines in Figure 2. Class G airspace, which is at 1,200 feet (~365 metres) or less to the ground, should be utilised for broad deployment of drones. Researchers investigated the use of LiDAR data for this purpose and relied on automatically generated Digital Surface Models (DSM), which incorporates ground objects that pose navigation restrictions such as airports and high-rise buildings (Feng & Yuan, 2016).

During the shipment process, all drones should employ sense-and-avoid technology so that a collision between each other, or between man-made or natural structures are prevented. In addition, drones should reach the level of 'cognitive autonomy'. A drone having cognitive autonomy could perform simultaneous localization and mapping; resolve conflicting information and plan for the future even at the risk of not accomplishing the initial goal (Floreano & Wood, 2015). For instance, due to heavy wind a drone might decide to return to a safe spot or charge its battery even though it might postpone the delivery of a medical supply. The motto here should better be "better late than never", since an uncontrolled failure of the mission would probably cause more damage.

The trajectories of the drones could also be kept for some period in case of further analysis or enquiries. Consequently, developing efficient indexing methods to retrieve the queried data in a fast manner is important. The traditional approach is to use relational database management systems, such as PostgreSQL, to store trajectory data, so that its integration with a GIS is straightforward (Mangiameli & Mussumeci, 2014). On the other hand, the emergence of NoSQL database management system, such as MongoDB, offers new horizons to store and query large trajectory data sets as well (Guan, Bo, Li, & Yu, 2017). Last, if a drone loses control and that downfall is inevitable, a parachute should open while making a loud sound to warn the people. It would also be important to aim for either natural or manufactured structures that will cause the minimum harm to people as well as the medical supply.

Finally, the medical package should be delivered. The most plausible scenario for delivery is to use "Collect and Delivery Points" (CDPs), where general-practitioners, bystanders or whoever is responsible can pick up their medical supply. There are two main types of CDPs: i) attended and ii) non-attended. In attended CDPs an officer is employed, whereas in non-attended CDPs customers use some sort of identification (e.g. PIN number) to pick up their supply. The supplies could be delivered to both types of CDPs. Indeed, Amazon has already issued a patent application to use lamp posts as a CDP, which could also be used to recharge drones (Coulton, Lindley, Sturdee, & Stead, 2017).

Apart from having a more organized way of delivery, relying on CDPs would also have the advantage of delegating the recognition of the person receiving the supply to CDPs. In this way, the privacy of the patients would be kept, which is a requirement of "Health Insurance Portability and Accountability Act (HIPAA)" of the U.S (Bhatt et al., 2018). Once the technology to deliver medical supplies to CDPs is established, the next improvement would be towards delivering supplies to residential buildings or even moving emergency vehicles in an urban road network.

Drones have already been deployed for humanitarian logistics and at different stages of disaster management including response, recovery, and mitigation (K. Kim & Davidson, 2015). It is possible to deliver essential medical products such as vaccines, blood products and diagnostic tests by drones. As Flahault et al. (2017) states "as smartphones are solving the communication needs in low and middle income countries, drone deliveries might solve transportation needs in countries where road networks are wanting".

Essential medicine delivery by drones is not limited to low and middle income countries. The Swiss Post, national postal service of Switzerland, has started using drones to carry laboratory samples between hospitals in Lugano and Bern (Swiss Post, 2018). In Australia, the prospect of 'medical maggot' delivery by long endurance drones is investigated. The authors state that "no insuperable challenges to the concept" are found, and they offer a road map to realise this economically feasible delivery method (Tatham, Stadler, Murray, & Shaban, 2017).

Safe transport of disease test samples and test kits in areas with high contagion could also be provided. Consequently, there is a growing research interest in this endeavour and the use of drones in medicine appears to be promising as it can both increase the quality and accessibility of healthcare (Bhatt et al., 2018). Companies like Google have already patented their ideas (US9307383B1) regarding the distribution of medical supplies including defibrillators, inhalers and drugs like epinephrine and insulin (Patrick, 2016).

The challenges such as the maintenance and operation of drones as well as storing the medical products at right temperatures during the delivery are eminent. For instance, Amukele et al. (2017) investigated the impact of a long (i.e. three hour flight travelling approximately 258 km) drone flight on chemistry and haematology results obtained from the flown blood samples. Apart from glucose and potassium levels being higher in the flown samples, which is attributed to temperature differences, the remaining 17 chemical and haematological tests showed similar results. This finding adds further support to transporting blood long distances using drones, but transportation environment should be carefully designed to consider temperature, vibration and acceleration. In addition, growth patterns of microbes that are transported by a drone flight are also not affected during the transportation (Amukele, Street, Carroll, Miller, & Zhang, 2016).

Search and rescue operations could also be considered under humanitarian logistics, since they might involve transportation of a product or person to improve the current situation. The duration of the process is usually critical in such missions. Drones are found to be useful to scan large areas in short time periods (Karaca et al., 2018; Pulver & Wei, 2018). Therefore, a search and rescue operation could involve the delivery of critical equipment such as a vaccine or an inflatable boat. For instance, Silvagni, Tonoli, Zenerino, and Chiaberge (2017) operated drones in mountainous areas for rescue operations. The drone is designed to meet the harsh environmental conditions including low temperatures, high altitude and strong winds. In addition, different payloads could be carried such as avalanche beacon, camera, emergency kits or even explosive material for controlled avalanche detachment.

Military could also benefit from the advancement of drones by rapidly evacuating the wounded or deceased personnel from high risk areas. The benefits include, reduced risk to human life, cost-effectiveness, ability to reach areas that are unreachable by conventional aircraft and the rapidity of transfer. However, user acceptability has to be investigated thoroughly. In addition, effective management of airspace as well as the appropriate level of care to deliver during transit are equally important (Handford, Reeves, & Parker, 2018).

Commercialization of some of these use-cases has already taken place. For instance, Rwanda launched world's first national emergency drone delivery system. The mountainous terrain of Rwanda makes it difficult to reach remote villages. The lack of proper road infrastructure further exacerbates the problem. Furthermore, some places can only be reached by boat (Gavi, 2016; Rosen, 2017). The 18-month project run by Zipline, a robotics company founded in 2011, involves two phases; the first is blood delivery to transfusion facilities to save women experiencing postpartum haemorrhaging, the second is rabies vaccine delivery. The drone is a 12 kg fixed wing that can carry almost 1.5 kg of blood or medicine and can fly 75 miles before recharging.

DISCUSSIONS

The dilemma between the advancement of technology and its implications to society is a moral and a philosophical one (Postman, 1993). Even though most of the technological advancements do occur prior to regulations, it is important to convince the public regarding the added value of the newly introduced technology. It is always necessary to learn from the previous experiences such as Los Angeles Police Department and Seattle Police Department abolishing their drone programs after public's disapproval (West, Klofstad, Uscinski, & Connolly, 2019). Therefore, decision makers need to balance the concerns of the general public, companies, drone hobbyists, and government agencies.

Three main issues have to be addressed to gain public support in this regard: privacy, safety and ethics (Luppicini & So, 2016). Considering the advancements in sensor technologies and the web technologies, it is difficult to think of a complete privacy. In addition, the number of Internet of Things (IoT) are projected to reach to 50 billion by 2020 (Isaak & Hanna, 2018). All of these developments further exacerbate the public's concern on privacy. However, researchers also identify different aspects that might change our perception on a technological product. For instance, appearance and the level of automation and as discussed, autonomy of a drone could affect the perception towards their use (Lidynia, Philipsen, & Ziefle, 2017). Different tools such as community meetings, mass media coverage of success stories and informing the public and private sector would add support to people embracing this new technology.

At the same time, regulations need to keep up with the technological progress. Indeed, a report entitled "Recurring Dilemmas: The Law's Race to Keep up with Technological Change" is dedicated for this topic and proposes to design a legal system that can cope with a rapidly changing technological environment (Moses, 2007).

Clothier, Fulton, & Walker (2008) state that "history has shown us that some of the greatest obstacles facing the realisation of a new technology are not always technical in nature but are often related to its integration into society." They further provide a thought-provoking example of the introduction of automobiles in the UK. The relevant legislation, Locomotive on Highways Act of 1865 that is also referred to as the 'Red Flag Law', might well have blocked further technological developments in the UK, leaving other countries to benefit the pioneering.

The growing research trend on the use of drones in healthcare, and specifically in humanitarian logistics signal a future where a new mode of delivery of medical supplies is introduced: drones. This new mode could potentially be applicable in both rural and urban environments. At the moment, researchers have already initiated pilot projects, and countries like Rwanda have started delivering critical medical supplies with drones. The success of such projects would not only convince the regulators but also the tax payers in terms of the societal and economic gains that the new delivery mode will bring. It is therefore important to start developing planning models in order to determine the optimum number of drone centres and the number of drones each of these centres should employ (S. J. Kim, Lim, Cho, & Côté, 2017).

The SWOT (Strengths, Weaknesses, Opportunities and Threats) analysis regarding the use of drones for humanitarian logistics is provided in Table 2. The foremost positive aspect of relying on drones in humanitarian logistics is apparent: rapid transport of medical supplies to both rural and urban areas. In rural areas the infrastructure might not be well-established, which would increase journey times by road. In urban areas traffic congestion or landscape might increase journey times. In addition, the initial test results demonstrate that the quality of the transferred medical product do not deteriorate due to its transmission with a drone. Last, relying on drones might decrease the overall costs. However, the main aspect blocking its wide use appears to be the lack of legislation. Just like in 2015 Nepal Earthquake, the use of drones could even be banned (Meier, 2015). In addition, it is important to operate each mission with care, since an incident might cause concerns amongst the public delaying its common use.

FUTURE RESEARCH DIRECTIONS

Drones would likely be used extensively in healthcare in the coming years. Countries such as Rwanda and Switzerland have already started delivering medical products through this new mode of transportation. One possible future research direction is to investigate the cost-effectiveness of this new mode compared to existing modes. Even though there is research evidence supporting the use of drones for vaccine delivery, these econometric studies have relied on computer simulations. Real-life pilot studies

Table 2. Drones in humanitarian logistics: SWOT Analysis

Strengths	Weaknesses
• Rapid transport of critical healthcare products such as AEDs . Given that the spatial distribution of drones is optimised, the majority of a 2000 km^2 landscape could be reached under a minute, which is substantially better than the traditional approach of relying on a road transport vehicle such as an ambulance or a motorcycle (Pulver & Wei, 2018). • Access to remote areas including, caves, islands, rugged terrains, mountains or rural areas. • Economically feasible (Haidari et al., 2016). • Reduced carbon emissions (Park, Kim, & Suh, 2018). • Quadrotor drones could fly and land to small designated areas, requiring low levels of capital investment. • Initial test results indicate that the quality of the transported medical supply does not deteriorate (Amukele et al., 2017, 2016).	• Lack of regulations (Tatham et al., 2017). • Lack of extensive experiments of drone flights under different climate zones, weather, landscape, urban environment or at different daylight levels. More research has to be carried to investigate the quality of the transferred medical supplies under these different conditions (Lippi & Mattiuzzi, 2016). • Lack of extensive research on the financial cost-benefit analysis. Current analyses favour delivery of medical supplies by drones (Haidari et al., 2016). However, it is important to minimize the required extra warehousing due to the limited flight range of drones (Stolaroff et al., 2018). • Necessity to understand the concerns of the public (Clothier et al., 2015; Luppicini & So, 2016).
Opportunities	**Threats**
• Existence of well-established organisations such as UAViators. Their resources are invaluable as they document the best practices as well as the lessons learned. • Research on indoor navigation and small drones could lead to intra-hospital deliveries, which might then eliminate the need of the traditional pneumatic tubes. • The benefits have already been observed. Success stories should be highlighted in media, education and governmental bodies. There are more than 25 million results for the terms: 'drone', 'saved', 'life' in Google. • Recent initiatives from regulatory bodies allowing for a variety of test scenarios (FAA, 2018).	• A legislation prohibiting drone flights. • Not being compliant with laws such as "Health Insurance Portability and Accountability Act (HIPAA)" of the U.S. that regulates the health insurance, protect the confidentiality of the patients and security of healthcare information (Bhatt et al., 2018). • Incidents cause frustration amongst the public (Solodov, Williams, Al Hanaei, & Goddard, 2018). Transferring wrong type of medical supply or drone carrying critical healthcare falling to cause another injury or even casualty would probably raise concerns.

should be carried out to draw a more realistic conclusion. In addition, each drone delivery would generate substantial amount of spatio-temporal data. Once the number of operational drones reach thousands in a single urban environment, investigating efficient ways to store and query such large trajectory datasets would be required. Societal and legal studies need to carried out to understand the implications of using drones in medical sector and develop mechanisms to solve possible controversies. Last, progress on artificial intelligence would undoubtedly advance the 'sense and avoid' capabilities of drones as well as pave the way towards autonomous delivery of medical products.

CONCLUSION

Robot automation has advanced considerably in the last decade. Drones have also advanced in many different ways including, but not limited to, their agility, speed and length of flight time. The number of publications including the keyword "drones" and its closely related terms reached 40 thousand, and intriguingly almost half of this have occurred in the last four years. There are many areas where the use of drones would bring additional benefits to current operations. Healthcare has relatively taken less attention. However, the existing literature suggests that the use of drones would bring substantial benefits to healthcare.

The first benefit is due to their advanced capabilities in terms of data collection. They could be used to quickly locate and identify people in search and rescue missions, collect test samples from contaminated areas or collect air quality data. The second benefit is telemedicine and the leading example of this is the assistance provided by the drone while delivering an AED. The drone will not only carry an AED, but also the necessary means to facilitate the communication between the bystander and the medical expert. Humanitarian logistics, which is the third benefit of drones in healthcare, would prove to be invaluable in terms of rapid transport of medical supplies to both rural and urban environments compared to the traditional land based modes. In addition, there is experimental evidence suggesting that this endeavour would also be more economically viable in the long term. The main threat to the realisation of this vision is two-fold. First, most of the developed countries' regulations do not still allow for BVLOS flights as well as flying over people, without specific approvals or waivers. In addition, the prospect of incidents might raise concerns amongst the public. Nevertheless, the initiatives such as the recently launched 'UAS Integration Pilot Program' are invaluable towards the integration of drones into civilian airspace (FAA, 2018).

ACKNOWLEDGMENT

This research is partially supported by the Scientific and Technological Research Council of Turkey (TÜBİTAK) [grant number 118Y282]. The contents of this paper reflect the views of the author, who is responsible for the facts and the accuracy of the results presented herein.

REFERENCES

Ackerman, E., & Strickland, E. (2018). Medical delivery drones take flight in east africa. *IEEE Spectrum*, *55*(1), 34–35. doi:10.1109/MSPEC.2018.8241731

Alex, C., & Vijaychandra, A. (2016). Autonomous cloud based drone system for disaster response and mitigation. In *2016 International Conference on Robotics and Automation for Humanitarian Applications (RAHA)* (pp. 1–4). Academic Press. 10.1109/RAHA.2016.7931889

Amukele, T. K., Hernandez, J., Snozek, C. L., Wyatt, R. G., Douglas, M., Amini, R., & Street, J. (2017). Drone Transport of Chemistry and Hematology Samples Over Long Distances. *American Journal of Clinical Pathology*, *148*(5), 427–435. doi:10.1093/ajcp/aqx090 PMID:29016811

Amukele, T. K., Street, J., Carroll, K., Miller, H., & Zhang, S. X. (2016). Drone Transport of Microbes in Blood and Sputum Laboratory Specimens. *Journal of Clinical Microbiology*, *54*(10), 2622–2625. doi:10.1128/JCM.01204-16 PMID:27535683

Anbaroğlu, B. (2017). Parcel Delivery in an Urban Environment Using Unmanned Aerial Systems: A Vision Paper. In ISPRS Annals of Photogrammetry, Remote Sensing and Spatial Information Sciences (Vol. IV-4-W4, pp. 73–79). Copernicus GmbH. doi:10.5194/isprs-annals-IV-4-W4-73-2017

ASSURE. (2018). *Annual Report*. Retrieved from http://www.assureuas.org/annual-reports/ASSURE_2018AR.pdf

Balasingam, M. (2017). Drones in medicine—The rise of the machines. *International Journal of Clinical Practice*, *71*(9), e12989. doi:10.1111/ijcp.12989 PMID:28851081

Barmpounakis, E. N., Vlahogianni, E. I., & Golias, J. C. (2016). Unmanned Aerial Aircraft Systems for transportation engineering: Current practice and future challenges. *International Journal of Transportation Science and Technology*, *5*(3), 111–122. doi:10.1016/j.ijtst.2017.02.001

Bashshur, R. L. (1995). On the definition and evaluation of telemedicine. *Telemedicine Journal: The Official Journal of the American Telemedicine Association*, *1*(1), 19–30. doi:10.1089/tmj.1.1995.1.19 PMID:10165319

Benjamin, E. J., Virani Salim, S., Callaway Clifton, W., Chamberlain Alanna, M., Chang Alexander, R., & Cheng, S. (2018). Heart Disease and Stroke Statistics—2018 Update: A Report From the American Heart Association. *Circulation*, *137*(12), e67–e492. doi:10.1161/CIR.0000000000000558 PMID:29386200

Benjamin, M., & Ehrenreich, B. (2013). *Drone Warfare: Killing by Remote Control (Fully revised and updated edition)*. London: Verso.

Bhatt, K., Pourmand, A., & Sikka, N. (2018). Targeted Applications of Unmanned Aerial Vehicles (Drones) in Telemedicine. *Telemedicine Journal and e-Health*, *24*(11), 833–838. doi:10.1089/tmj.2017.0289 PMID:29489441

Boucher, P. (2015). Domesticating the Drone: The Demilitarisation of Unmanned Aircraft for Civil Markets. *Science and Engineering Ethics*, *21*(6), 1393–1412. doi:10.100711948-014-9603-3 PMID:25371277

Brady, J. M., Stokes, M. D., Bonnardel, J., & Bertram, T. H. (2016). Characterization of a Quadrotor Unmanned Aircraft System for Aerosol-Particle-Concentration Measurements. *Environmental Science & Technology*, *50*(3), 1376–1383. doi:10.1021/acs.est.5b05320 PMID:26730457

Burton, E., Meier, C., Olarte, R., Skeini, H., & Zahan, F. (2017). Airshare: A Food Sharing Concept. In *Proceedings of the 29th Australian Conference on Computer-Human Interaction* (pp. 634–639). New York, NY: ACM. 10.1145/3152771.3154859

Capolupo, A., Pindozzi, S., Okello, C., Fiorentino, N., & Boccia, L. (2015). Photogrammetry for environmental monitoring: The use of drones and hydrological models for detection of soil contaminated by copper. *The Science of the Total Environment*, *514*, 298–306. doi:10.1016/j.scitotenv.2015.01.109 PMID:25668282

Choi, C. H., Jang, H. J., Lim, S. G., Lim, H. C., Cho, S. H., & Gaponov, I. (2016). Automatic wireless drone charging station creating essential environment for continuous drone operation. In *2016 International Conference on Control, Automation and Information Sciences (ICCAIS)* (pp. 132–136). Academic Press. 10.1109/ICCAIS.2016.7822448

Chowdhary, G., Johnson, E. N., Magree, D., Wu, A., & Shein, A. (2013). GPS-denied Indoor and Outdoor Monocular Vision Aided Navigation and Control of Unmanned Aircraft. *Journal of Field Robotics*, *30*(3), 415–438. doi:10.1002/rob.21454

Christian, A., & Cabell, R. (2017). *Initial investigation into the psychoacoustic properties of small unmanned aerial system noise*. Presented at the *23rd AIAA/CEAS Aeroacoustics Conference*.

Claesson, A., Fredman, D., Svensson, L., Ringh, M., Hollenberg, J., Nordberg, P., ... Ban, Y. (2016). Unmanned aerial vehicles (drones) in out-of-hospital-cardiac-arrest. *Scandinavian Journal of Trauma, Resuscitation and Emergency Medicine*, *24*. doi:10.2514/6.2017-4051

Claesson, A., Svensson, L., Nordberg, P., Ringh, M., Rosenqvist, M., Djarv, T., ... Hollenberg, J. (2017). Drones may be used to save lives in out of hospital cardiac arrest due to drowning. *Resuscitation, 114*, 152–156. doi:10.1016/j.resuscitation.2017.01.003 PMID:28110000

Clothier, R. A., Fulton, N. L., & Walker, R. A. (2008). Pilotless aircraft: The horseless carriage of the twenty-first century? *Journal of Risk Research, 11*(8), 999–1023. doi:10.1080/13669870802323353

Clothier, R. A., Greer, D. A., Greer, D. G., & Mehta, A. M. (2015). Risk Perception and the Public Acceptance of Drones. *Risk Analysis, 35*(6), 1167–1183. doi:10.1111/risa.12330 PMID:25689883

Coulton, P., Lindley, J., Sturdee, M., & Stead, M. (2017). Design Fiction as World Building. Presented at the Research Through Design.

de Voorde, P. V., Gautama, S., Momont, A., Ionescu, C. M., Paepe, P. D., & Fraeyman, N. (2017). The drone ambulance [A-UAS]: Golden bullet or just a blank? *Resuscitation, 116*, 46–48. doi:10.1016/j.resuscitation.2017.04.037 PMID:28487253

EASA. (2015). *Introduction of a regulatory framework for the operation of unmanned aircraft*. Retrieved from https://www.easa.europa.eu/sites/default/files/dfu/Introduction%20of%20a%20regulatory%20framework%20for%20the%20operation%20of%20unmanned%20aircraft.pdf

Engvers, A. (2001). *The Principle of Sovereignty in the Air - To what extent can it be upheld against aerial intruders?* Retrieved from http://lup.lub.lu.se/student-papers/record/1557232

FAA. (2018). *UAS Integration Pilot Program* [template]. Retrieved 10 December 2018, from https://www.faa.gov/uas/programs_partnerships/uas_integration_pilot_program/

Fahlstrom, P., & Gleason, T. (2012). *Introduction to UAV Systems* (4th ed.). Wiley.

Feng, D., & Yuan, X. (2016). *Automatic construction of aerial corridor for navigation of unmanned aircraft systems in class G airspace using LiDAR* (Vol. 9828, pp. 98280I-98280I – 8). doi:10.1117/12.2224359

Finn, R. L., & Wright, D. (2012). Unmanned aircraft systems: Surveillance, ethics and privacy in civil applications. *Computer Law & Security Review, 28*(2), 184–194. doi:10.1016/j.clsr.2012.01.005

Flahault, A., Geissbuhler, A., Guessous, I., Guérin, P., Bolon, I., Salathé, M., & Escher, G. (2017). Precision global health in the digital age. *Swiss Medical Weekly*, *147*, w14423. PMID:28421566

Floreano, D., & Wood, R. J. (2015). Science, technology and the future of small autonomous drones. *Nature*, *521*(7553), 460–466. doi:10.1038/nature14542 PMID:26017445

Fornace, K. M., Drakeley, C. J., William, T., Espino, F., & Cox, J. (2014). Mapping infectious disease landscapes: Unmanned aerial vehicles and epidemiology. *Trends in Parasitology*, *30*(11), 514–519. doi:10.1016/j.pt.2014.09.001 PMID:25443854

Gavi. (2016). *Rwanda launches world's first national drone delivery service powered by Zipline*. Retrieved 20 December 2018, from https://www.gavi.org/library/news/gavi-features/2016/rwanda-launches-world-s-first-national-drone-delivery-service-powered-by-zipline/

Global Grand Challenges. (2012). *On-Demand Vaccine Delivery via Low-Cost UAVs | Grand Challenges*. Retrieved 1 January 2019, from https://gcgh.grandchallenges.org/grant/demand-vaccine-delivery-low-cost-uavs

Gu, Q., Michanowicz, D. R., & Jia, C. (2018). Developing a Modular Unmanned Aerial Vehicle (UAV) Platform for Air Pollution Profiling. *Sensors (Basel)*, *18*(12), 4363. doi:10.339018124363 PMID:30544691

Guan, X., Bo, C., Li, Z., & Yu, Y. (2017). ST-hash: An efficient spatiotemporal index for massive trajectory data in a NoSQL database. In *2017 25th International Conference on Geoinformatics* (pp. 1–7). Academic Press. 10.1109/GEOINFORMATICS.2017.8090927

Haidari, L. A., Brown, S. T., Ferguson, M., Bancroft, E., Spiker, M., Wilcox, A., ... Lee, B. Y. (2016). The economic and operational value of using drones to transport vaccines. *Vaccine*, *34*(34), 4062–4067. doi:10.1016/j.vaccine.2016.06.022 PMID:27340098

Handford, C., Reeves, F., & Parker, P. (2018). Prospective use of unmanned aerial vehicles for military medical evacuation in future conflicts. *Journal of the Royal Army Medical Corps*, *164*(4), 293–296. doi:10.1136/jramc-2017-000890 PMID:29523753

Hanna, D., Ferworn, A., Lukaczyn, M., Abhari, A., & Lum, J. (2018). Using Unmanned Aerial Vehicles (UAVs) in locating wandering patients with dementia. In *2018 IEEE/ION Position, Location and Navigation Symposium (PLANS)* (pp. 809–815). IEEE. 10.1109/PLANS.2018.8373457

Hardy, A., Makame, M., Cross, D., Majambere, S., & Msellem, M. (2017). Using low-cost drones to map malaria vector habitats. *Parasites & Vectors*, *10*(1), 29. doi:10.118613071-017-1973-3 PMID:28088225

Harrington, A. (2015). Who controls the drones? *Engineering Technology*, *10*(2), 80–83. doi:10.1049/et.2015.0211

Howell, C. T. (2016). *The First Government Sanctioned Delivery of Medical Supplies by Remotely Controlled Unmanned Aerial System (UAS)*. Presented at the Xponential 2016, New Orleans, LA. Retrieved from https://ntrs.nasa.gov/search.jsp?R=20160009120

Isaak, J., & Hanna, M. J. (2018). User Data Privacy: Facebook, Cambridge Analytica, and Privacy Protection. *Computer*, *51*(8), 56–59. doi:10.1109/MC.2018.3191268

Jenkins, D., & Vasigh, B. (2013). *The Economic Impact of Unmanned Aircraft Systems Integration in the United States. Association for Unmanned Vehicle Systems International*. AUVSI.

Karaca, Y., Cicek, M., Tatli, O., Sahin, A., Pasli, S., Beser, M. F., & Turedi, S. (2018). The potential use of unmanned aircraft systems (drones) in mountain search and rescue operations. *The American Journal of Emergency Medicine*, *36*(4), 583–588. doi:10.1016/j.ajem.2017.09.025 PMID:28928001

Kim, K., & Davidson, J. (2015). Unmanned Aircraft Systems Used for Disaster Management. *Transportation Research Record: Journal of the Transportation Research Board*, *2532*(1), 83–90. doi:10.3141/2532-10

Kim, S. J., Lim, G. J., Cho, J., & Côté, M. J. (2017). Drone-Aided Healthcare Services for Patients with Chronic Diseases in Rural Areas. *Journal of Intelligent & Robotic Systems*, *88*(1), 163–180. doi:10.100710846-017-0548-z

Koh, C. H., Low, K. H., Li, L., Zhao, Y., Deng, C., Tan, S. K., ... Li, X. (2018). Weight threshold estimation of falling UAVs (Unmanned Aerial Vehicles) based on impact energy. *Transportation Research Part C, Emerging Technologies*, *93*, 228–255. doi:10.1016/j.trc.2018.04.021

Kohlios, C. P., & Hayajneh, T. (2018). A Comprehensive Attack Flow Model and Security Analysis for Wi-Fi and WPA3. *Electronics (Basel)*, *7*(11), 284. doi:10.3390/electronics7110284

Kothari, M., Postlethwaite, I., & Gu, D. (2009). Multi-UAV path planning in obstacle rich environments using Rapidly-exploring Random Trees. In *Proceedings of the 48h IEEE Conference on Decision and Control (CDC) held jointly with 2009 28th Chinese Control Conference* (pp. 3069–3074). IEEE. 10.1109/CDC.2009.5400108

Larsen, M. P., Eisenberg, M. S., Cummins, R. O., & Hallstrom, A. P. (1993). Predicting survival from out-of-hospital cardiac arrest: A graphic model. *Annals of Emergency Medicine*, *22*(11), 1652–1658. doi:10.1016/S0196-0644(05)81302-2 PMID:8214853

Lee, B. Y., Connor, D. L., Wateska, A. R., Norman, B. A., Rajgopal, J., Cakouros, B. E., ... Brown, S. T. (2015). Landscaping the structures of GAVI country vaccine supply chains and testing the effects of radical redesign. *Vaccine*, *33*(36), 4451–4458. doi:10.1016/j.vaccine.2015.07.033 PMID:26209835

Lichtenstein, S., Slovic, P., Fischhoff, B., Layman, M., & Combs, B. (1978). Judged frequency of lethal events. *Journal of Experimental Psychology. Human Learning and Memory*, *4*(6), 551–578. doi:10.1037/0278-7393.4.6.551 PMID:731196

Lidynia, C., Philipsen, R., & Ziefle, M. (2017). Droning on About Drones—Acceptance of and Perceived Barriers to Drones in Civil Usage Contexts. In *Advances in Human Factors in Robots and Unmanned Systems* (pp. 317–329). Cham: Springer; doi:10.1007/978-3-319-41959-6_26

Lim, H., Park, J., Lee, D., & Kim, H. J. (2012). Build Your Own Quadrotor: Open-Source Projects on Unmanned Aerial Vehicles. *IEEE Robotics & Automation Magazine*, *19*(3), 33–45. doi:10.1109/MRA.2012.2205629

Lippi, G., & Mattiuzzi, C. (2016). Biological samples transportation by drones: Ready for prime time? *Annals of Translational Medicine*, *4*(5), 92. doi:10.21037/atm.2016.02.03 PMID:27047951

Luppicini, R., & So, A. (2016). A technoethical review of commercial drone use in the context of governance, ethics, and privacy. *Technology in Society*, *46*, 109–119. doi:10.1016/j.techsoc.2016.03.003

Mangiameli, M., & Mussumeci, G. (2014). Real Time Integration of Field Data into a GIS Platform for the Management of Hydrological Emergencies. In ISPRS - International Archives of the Photogrammetry, Remote Sensing and Spatial Information Sciences (Vol. XL-5-W3, pp. 153–158). Copernicus GmbH. doi:10.5194/isprsarchives-XL-5-W3-153-2013

Martin, P. G., Payton, O. D., Fardoulis, J. S., Richards, D. A., Yamashiki, Y., & Scott, T. B. (2016). Low altitude unmanned aerial vehicle for characterising remediation effectiveness following the FDNPP accident. *Journal of Environmental Radioactivity, 151*, 58–63. doi:10.1016/j.jenvrad.2015.09.007 PMID:26410790

Meiches, B. (2018). Non-human humanitarians. *Review of International Studies*, 1–19. doi:10.1017/S0260210518000281

Meier, P. (2015). UAVs and Humanitarian Response. In *UAVs and Humanitarian Response. Drones and Aerial Observation: New Technologies for Property Rights, Human Rights, and Global Development. A Primer*. Academic Press. Retrieved from http://drones.newamerica.org/primer

Miljković, D. (2018). Methods for attenuation of unmanned aerial vehicle noise. In *2018 41st International Convention on Information and Communication Technology, Electronics and Microelectronics (MIPRO)* (pp. 0914–0919). Academic Press. 10.23919/MIPRO.2018.8400169

Moses, L. B. (2007). *Recurring Dilemmas: The Law's Race to Keep Up With Technological Change (SSRN Scholarly Paper No. ID 979861)*. Rochester, NY: Social Science Research Network. Retrieved from https://papers.ssrn.com/abstract=979861

Murray, C. C., & Chu, A. G. (2015). The flying sidekick traveling salesman problem: Optimization of drone-assisted parcel delivery. *Transportation Research Part C, Emerging Technologies, 54*, 86–109. doi:10.1016/j.trc.2015.03.005

Oliver, J. M. (2017). Current Status of the Use of Small Unmanned Aerial Systems for Environmental Monitoring. *Environmental Claims Journal, 29*(2), 159–170. doi:10.1080/10406026.2017.1307009

Park, J., Kim, S., & Suh, K. (2018). A Comparative Analysis of the Environmental Benefits of Drone-Based Delivery Services in Urban and Rural Areas. *Sustainability, 10*(3), 888. doi:10.3390u10030888

Passchier-Vermeer, W., & Passchier, W. F. (2000). Noise exposure and public health. *Environmental Health Perspectives, 108*(Suppl 1), 123–131. PMID:10698728

Patrick, W. G. (2016). *Request apparatus for delivery of medical support implement by UAV*. Retrieved from https://patents.google.com/patent/US9307383B1/en

Postman, N. (1993). *Technopoly: The Surrender of Culture to Technology*. New York: Vintage.

Prexl, M., Struebig, K., Harder, J., & Hoehn, A. (2017). *User studies of a head-mounted display for search and rescue teleoperation of UAVs via satellite link*. IEEE Aerospace Conference; doi:10.1109/AERO.2017.7943596

Pulver, A., & Wei, R. (2018). Optimizing the spatial location of medical drones. *Applied Geography (Sevenoaks, England)*, *90*, 9–16. doi:10.1016/j.apgeog.2017.11.009

Rosen, J. W. (2017). *Blood from the sky: an ambitious medical drone delivery system hits Rwanda*. Retrieved 20 December 2018, from https://www.technologyreview.com/s/608034/blood-from-the-sky-ziplines-ambitious-medical-drone-delivery-in-africa/

Rosser, J. C., Vignesh, V., Terwilliger, B. A., & Parker, B. C. (2018). Surgical and Medical Applications of Drones: A Comprehensive Review. *JSLS : Journal of the Society of Laparoendoscopic Surgeons*, *22*(3), e2018.00018. doi:10.4293/JSLS.2018.00018 PMID:30356360

Sachan, D. (2016). The age of drones: What might it mean for health? *Lancet*, *387*(10030), 1803–1804. doi:10.1016/S0140-6736(16)30361-0 PMID:27203491

Schneider, D. (2017). Air traffic control for delivery drones. *IEEE Spectrum*, *54*(1), 32–33. doi:10.1109/MSPEC.2017.7802742 PMID:30636790

Shively, R. J. (2014). Breaking Down Barriers to UAS Integration: What are government agencies doing? *Proceedings of the Human Factors and Ergonomics Society Annual Meeting*, *58*(1), 41–43. doi:10.1177/1541931214581009

Silvagni, M., Tonoli, A., Zenerino, E., & Chiaberge, M. (2017). Multipurpose UAV for search and rescue operations in mountain avalanche events. *Geomatics, Natural Hazards & Risk*, *8*(1), 18–33. doi:10.1080/19475705.2016.1238852

Skinner, S. W., Urdahl, S., Harrington, T., Balchanos, M. G., Garcia, E., & Mavris, D. N. (2018). UAV Swarms for Migration Flow Monitoring and Search and Rescue Mission Support. In 2018 AIAA Information Systems-AIAA Infotech @ Aerospace (Vols 1–0). American Institute of Aeronautics and Astronautics. doi:10.2514/6.2018-1489

Smyth, D. L., Glavin, F. G., & Madden, M. G. (2018). Using a Game Engine to Simulate Critical Incidents and Data Collection by Autonomous Drones. In *2018 IEEE Games, Entertainment, Media Conference (GEM)* (pp. 1–9). IEEE. 10.1109/GEM.2018.8516527

Solodov, A., Williams, A., Al Hanaei, S., & Goddard, B. (2018). Analyzing the threat of unmanned aerial vehicles (UAV) to nuclear facilities. *Security Journal, 31*(1), 305–324. doi:10.105741284-017-0102-5

Stolaroff, J. K., Samaras, C., O'Neill, E. R., Lubers, A., Mitchell, A. S., & Ceperley, D. (2018). Energy use and life cycle greenhouse gas emissions of drones for commercial package delivery. *Nature Communications, 9*(1), 409. doi:10.103841467-017-02411-5 PMID:29440638

Swiss Post. (2018). *Drone to transport laboratory samples across Lake Zurich.* Retrieved 20 December 2018, from https://www.post.ch/en/about-us/company/media/press-releases/2018/drone-to-transport-laboratory-samples-across-lake-zurich

Tanzi, T. J., Chandra, M., Isnard, J., Camara, D., Sebastien, O., & Harivelo, F. (2016). Towards 'drone-borne' disaster management: future application scenarios. In ISPRS Annals of Photogrammetry, Remote Sensing and Spatial Information Sciences (Vol. III–8, pp. 181–189). Copernicus GmbH. doi:10.5194/isprs-annals-III-8-181-2016

Tatham, P., Stadler, F., Murray, A., & Shaban, R. Z. (2017). Flying maggots: A smart logistic solution to an enduring medical challenge. *Journal of Humanitarian Logistics and Supply Chain Management, 7*(2), 172–193. doi:10.1108/JHLSCM-02-2017-0003

Thiels, C. A., Aho, J. M., Zietlow, S. P., & Jenkins, D. H. (2015). Use of Unmanned Aerial Vehicles for Medical Product Transport. *Air Medical Journal, 34*(2), 104–108. doi:10.1016/j.amj.2014.10.011 PMID:25733117

TIME. (2018). *Welcome to the Drone Age*. Retrieved 20 December 2018, from http://time.com/collection/drones/

Torresan, C., Berton, A., Carotenuto, F., Gennaro, S. F. D., Gioli, B., Matese, A., ... Wallace, L. (2017). Forestry applications of UAVs in Europe: A review. *International Journal of Remote Sensing, 38*(8–10), 2427–2447. doi:10.1080/01431161.2016.1252477

Vanhoef, M., & Piessens, F. (2018). Release the Kraken: New KRACKs in the 802.11 Standard. In *Proceedings of the 2018 ACM SIGSAC Conference on Computer and Communications Security* (pp. 299–314). New York: ACM. 10.1145/3243734.3243807

West, J. P., Klofstad, C. A., Uscinski, J. E., & Connolly, J. M. (2019). Citizen Support for Domestic Drone Use and Regulation. *American Politics Research, 47*(1), 119–151. doi:10.1177/1532673X18782208

Wu, J., Wang, H., Li, N., Yao, P., Huang, Y., & Yang, H. (2018). Path planning for solar-powered UAV in urban environment. *Neurocomputing*, *275*, 2055–2065. doi:10.1016/j.neucom.2017.10.037

Yoo, W., Yu, E., & Jung, J. (2018). Drone delivery: Factors affecting the public's attitude and intention to adopt. *Telematics and Informatics*, *35*(6), 1687–1700. doi:10.1016/j.tele.2018.04.014

Zègre-Hemsey, J. K., Bogle, B., Cunningham, C. J., Snyder, K., & Rosamond, W. (2018). Delivery of Automated External Defibrillators (AED) by Drones: Implications for Emergency Cardiac Care. *Current Cardiovascular Risk Reports*, *12*(11), 25. doi:10.100712170-018-0589-2 PMID:30443281

Zhang, Z., Cao, Y., Ding, M., Zhuang, L., & Yao, W. (2016). An intruder detection algorithm for vision based sense and avoid system. In *2016 International Conference on Unmanned Aircraft Systems (ICUAS)* (pp. 550–556). Academic Press. 10.1109/ICUAS.2016.7502521

Zuboff, S. (2015). Big other: Surveillance Capitalism and the Prospects of an Information Civilization. *Journal of Information Technology*, *30*(1), 75–89. doi:10.1057/jit.2015.5

ADDITIONAL READING

Fahlstrom, P., & Gleason, T. (2012). *Introduction to UAV Systems* (4th ed.). Wiley.

Sebbane, Y. B. (2018). *Intelligent Autonomy of UAVs: Advanced Missions and Future Use*. CRC Press.

Valavanis, K., & Vachtsevanos, G. J. (Eds.). (2015). *Handbook of Unmanned Aerial Vehicles*. Springer. doi:10.1007/978-90-481-9707-1

Završnik, A. (Ed.). (2016). *Drones and Unmanned Aerial Systems: Legal and Social Implications for Security and Surveillance*. Springer. doi:10.1007/978-3-319-23760-2

KEY TERMS AND DEFINITIONS

Autonomous Drone: A drone that is cognitively aware of its environment and itself. An autonomous drone may decide to abort a delivery mission and make a safe landing.

Collect and Delivery Point (CDP): The pickup and dropoff locations of the medical product.

Database Management System (DBMS): An information system that is used to store, query and analyze large amounts of trajectory data generated by the drones carrying medical products.

Geographic Information Systems (GIS): An information system that can be used to store, query, analyze, and visualize trajectory data and maps generated by drones.

Humanitarian Logistics: Distribution of healthcare products to patients on a timely basis and at right conditions of the product.

Operation Time: Temporal duration between the dispatch of the drone carrying a medical product and its delivery.

Trajectory Data: The location and time of each drone carrying a medical product could be sampled at regular intervals, eventually leading to a trajectory. Additional attributes such as the pitch, roll, and yaw angles as well as the speed could also be included in a trajectory.

Chapter 5
Drones to the Rescue:
A Case Study of Cyclone PAM

Peter Tatham
Griffith University, Australia

ABSTRACT

This chapter considers the potential operation of long-range drones to support the logistic response to a natural disaster using a case study of Cyclone Pam that struck Tafea Province of Vanuatu in March 2015. It provides an overview of how the core capabilities of such drones might be employed in order to overcome the key challenge facing humanitarian logisticians responding to such disasters – namely that of understanding the 6W problem of "who wants what where when and why." The chapter then discusses the people, process, and technology issues that would need to be overcome in order to operationalize the concept.

DOI: 10.4018/978-1-5225-7900-7.ch005

INTRODUCTION

Drones – technically known as Remotely Piloted Aircraft Systems (RPAS) – are frequently used in a military context for both surveillance or attack, but are increasingly being employed in a number of non-military contexts including: the provision of aerial surveillance and mapping (Nex & Remondino, 2014), in the structural evaluation of buildings (Artemenko et al., 2014), in fire detection (Huang et al., 2011), in the provision of emergency communications (Tuna et al., 2014), and in the delivery of medical supplies (Holz, 2018).

Given the broad range of potential uses of such systems, it comes as no surprise that consideration is being given to their employment as part of the response to a disaster. In this regard, the American Red Cross (2015) recently concluded that: "Aerial drones are one of the most promising and powerful new technologies to improve disaster response and relief operations. … When a disaster occurs, drones may be used to provide relief workers with better situational awareness, locate survivors … perform structural analysis of damaged infrastructure, deliver needed supplies and equipment, evacuate casualties, and help extinguish fires – among many other potential applications." (p. 4).

This Red Cross report also analyses previous cases where drones have been used in the aftermath of a disaster and suggests (p.7) that the most appropriate tasks are:

- Reconnaissance and Mapping
- Structural Assessment
- Temporary Infrastructure/Supply Delivery
- Wildfire – Detection and Extinguishing
- High-Rise Building Fire Response
- Chemical, Biological, Radiological, Nuclear, or Explosive (CBRNE) Events
- Search and Rescue Operations
- Insurance Claims Response and Risk Assessment
- Logistics Support

Furthermore, it has recently been suggested that annual sales of drones will surpass US$12 billion in 2021, which reflects a compound annual growth rate (CAGR) of 7.6% from the US$8.5 billion recorded in 2016 (Joshi, 2017). It is unsurprising, therefore, that the use of drones in support of the response to disasters is already taking place and is likely to expand. This is underlined by a recent report by the United Nations Office for the Coordination of Humanitarian Affairs (UNOCHA) in which it was observed that: "[the] move from speculation to reality raises challenging questions around … how best to integrate [drones] into humanitarian response." (UNOCHA, 2014, p. 3).

It is also important to appreciate that there are significant capability differences between what are generically referred to as drones – in particular in respect of their endurance, speed and normal operating altitude. Whilst there is, as yet, no standardised classification system, that used by the United States Department of Defense offers an overview (Table 1).

In practice, however, the use of drones in a disaster response context has mainly been limited to short range micro or mini variants (i.e. US DoD Group 1) such as those documented in a number of recently published case studies (UAViators, 2016). Whilst these case studies clearly demonstrate that such micro/mini drones have provided significant benefit to the responding agencies (and, hence, the affected populations), their endurance is typically less than 30-45 minutes. Thus such systems have a limited capability when required to transit significant distances from their base to the area of operations.

The aim of this chapter is, therefore, to consider the potential benefits and costs of the operation of *long endurance* drones in support of the logistic response to natural disasters. Such aircraft can be categorised as falling into the low end of US DoD Group 3/high end of Group 2 (see Table 1). In doing so, the events surrounding Cyclone Pam that struck Vanuatu in March 2015 will be used to demonstrate both how long endurance drones might be employed, as well as the steps that would be needed to operationalise their use in a robust way.

To achieve the chapter's aim, it will first offer a brief overview of the generic humanitarian logistic (HL) challenge. It will then provide a summary of the literature relating to drones in an HL context before discussing the capabilities of a typical long endurance example. An overview of Cyclone Pam follows, after which the chapter will outline ways in which drones could have been used to mitigate the cyclone's impact. The chapter will end with a discussion of the areas of further

Table 1. Drone Classification according to the US Department of Defense

Group	Size	Max Gross Take-Off Weight (Lbs)	Normal Operating Altitude (ft)	Airspeed (Kts)
Group 1	Small	0-20	<1,200 AGL*	<100
Group 2	Medium	21-55	<3,500	<250
Group 3	Large	<1320	<18,000 MSL**	<250
Group 4	Larger	>1320	<18,000 MSL	Any airspeed
Group 5	Largest	>1320	>18,000	Any airspeed
*AGL = Above Ground Level **MSL = Mean Sea Level				

Source: US Army (2010)

work that will be needed to underpin a broader use of drones to support the HL response to a disaster.

THE HUMANITARIAN LOGISTIC CHALLENGE

In the same way as for the commercial logistician, the challenge facing his or her humanitarian counterpart is that of matching supply with demand in an efficient and effective way. Thus, in the 'for profit' environment, the demand side of the equation becomes clear from the action of a consumer purchasing a product in a shop or via the internet. However in the aftermath of a disaster those who have survived are focussed on staying alive and minimizing the impact of the event. As a result, the process of ascertaining their requirements – usually termed 'needs assessment' – frequently has to be undertaken by a 3rd party such as staff from a government agency or from a non-government organisation (NGO).

Furthermore, this process is often challenged by the failure of communications systems (both electronic and physical) as well as a lack of knowledge relating the affected population's demographics and, hence, individuals' particular needs (Tatham & Kovács, 2010). Thus, determining the answer to the '6W question' (Who Wants What Where When and Why) can be extremely complex, particularly recognising that the price of failure is not simply a matter of reduced profits. On the other side of the equation, the physical impact of the disaster frequently disrupts re-supply routes – for example through damage to sea ports and airports, blocked roads, destroyed bridges etc., all of which reduce the speed and effectiveness of the response (Tatham et al. 2017).

The use of drones can help to mitigate both sides of the demand/supply equation. On the demand side, the use of drone-mounted cameras will help those managing the response understand the overall impact of the disaster. Furthermore, and as will be explained later in this chapter, the latest generation of drones incorporate systems that will allow communication with those on the ground in the event that the normal systems are no longer available (as was the case in the aftermath of both Cyclone Pam and also in Cyclone Winston (Fiji, 2016)). On the supply side, the ability to overfly prospective transport routes will help the logistician choose those that are least impacted by the disaster event by identifying, for example, broken bridges or blocked roads.

Whilst such capabilities are available through the use of US DoD Group 1 drones, they are typically not able to capture this information in respect of communities that are at a distance from the operating base. For example, Cyclone Pam severely impacted the 5 islands that make up the Tafea Province of Vanuatu and, in doing so, demolished the only cell phone tower covering the region (Tatham, et al., 2017).

As a result communications with the islands were extremely limited, noting that the furthest island (Anatom) is some 300km (185mi) over water from the capital (Port Vila). Thus, the use of long endurance drones would have potentially provided a significantly swifter understanding of the impact of the disaster and, hence, the logistic response requirements than was actually the case.

THE CHARACTERISTICS OF A LONG ENDURANCE DRONE TO SUPPORT HL OPERATIONS

According to OCHA (2014), drones are becoming relatively commonplace with 270 companies in 57 countries reported as manufacturing such aircraft. This figure will, unquestionably, grow as indicated by recent data from the United States Federal Aviation Administration (FAA) which predicts that annual sales will increase from an estimated 2.5 million in 2016 to some 7 million in 2020 (FAA, 2016). Such aircraft range start from very small platforms, often in a multi-rotor helicopter configuration similar to those that have recently gained significant publicity in the wake of trials by, amongst others, Amazon (Titcomb, 2016) and Domino's Pizzas (Gye, 2013). At the other end of the spectrum there are high performance fixed wing aeroplanes such as the USAF Global Hawk that is the size of a small executive jet and has a unit cost of >US$130M (GAO, 2013).

Given this potential range of drones, and putting aside cost considerations that will be discussed later, the choice of platform will reflect its desired capabilities as well as the operational environment. With this in mind it is argued that the chosen drone should meet a number of high level selection criteria:

- In light of the humanitarian context, it should not be covered by the International Traffic in Arms Regulations (ITAR) or be an otherwise restricted system.
- Ideally, it should have sufficient endurance to allow it to operate continuously throughout the hours of daylight, i.e. for some 8-12 hours.
- Whilst not essential, a night-flying capability would also be beneficial.
- It should have a relatively large payload capability which should include regular and infra-red still and video cameras. The payload should also include the ability to link into satellite networks as necessary both for command and control purposes and also to deliver the captured data to the affected country's National Disaster Management Organisation (NDMO) swiftly and efficiently.

As indicated earlier, these high level criteria would indicate the requirement for a drone that is able to operate 'beyond visual line of sight' (BVLOS). An exemplar of such a system is the Latitude HQ -160B, the details of which are summarised in Table 2:

It should be noted that the Latitude drone is launched and lands vertically (VTOL), and thus requires only a safety area of some 8m (26ft) in diameter.

LITERATURE REVIEW

A review of the literature relating to the use of drones in a disaster response context was carried out based on the methodology of Kunz and Reiner (2012) in which the following databases were searched: Proquest ABI/INFORM Complete (ABI), Business Source Complete (BSC) and Web of Science (W of S) for Academic Journals, and using the keyword and Boolean operator string:

("Unmanned Aerial Vehicle" OR "UAV" OR "Unmanned Aerial System" OR "UAS" OR "Drone" OR "Remotely Piloted Aircraft System" OR "RPAS") AND ("Disaster response" OR "Emergency Response" OR "Humanitarian Logistics")

The start point for the search timeframe (2005) was selected based on the first reported use of drones in a humanitarian context which took place in the aftermath of the Hurricane Katrina (Tatham, 2009), and the final date was the end of the calendar year 2016.

The results of this review are shown in Table 3 and, notwithstanding the search string, 41 were found to be not directly relevant. These included, for example, general discussions of the HL challenges such as those offered by Tatham and Pettit (2010) and by Tatham and Christopher (2017), both of which mention the potential use of drones without offering any specific analysis.

Table 2. Latitude HQ -160B: Key performance data

Latitude HQ -160B	
Endurance	15hrs
Cruising Speed	70kph (45 mph)
Ceiling	4,268m (14,000ft)
Wingspan	3.81m (12.6ft)
Overall Length	2.44m (8ft)
Max Gross Take-Off Weight	43kg (95lb)
Max Payload Weight	5.44kg (12lb)

(Source: Tatham et al., 2017a)

Table 3. Summary of the database search

Subject Area	ABI/ INFORM	Business Source Complete (BSC)	Web of Science (WofS)	Total
Technical operation of drone sensors and/or communications systems	5	1	12	18
Use of micro/mini drones in disaster response			10	10
Ethics/control of drone operations	7			7
Use of drones for mapping			5	5
Development of 'dextrous drones' that incorporate manipulation devices			4	4
Use of drones for post-disaster evaluation of buildings and structures	1		3	4
Use of drones for environmental analysis	2		1	3
Use of macro drones in disaster response		3		3
Safety/risk management in drones operations			1	1
Use of drones for detection of fires			1	1
Use of drones for search and rescue	1			1
Not relevant	32	2	7	41
Total	48	6	44	98

Source: Tatham et al. (2017a)

As can be seen from Table 3, a significant proportion of the literature is devoted to the actual operation of drones and their associated sensors. These papers considered ways in which the capabilities of drones could be improved through, for example, the use of particular mathematical algorithms that relate to route planning, collision avoidance techniques, and dispatching/loitering policies. In a similar way, a significant element of the literature was related to the operation of mini/micro drones, whilst a number of other themes emerged from the literature review including the use of drones for: post-disaster structural evaluation of buildings, fire detection, environmental monitoring, and search and rescue. Clearly all these are very important areas of developing research, they do not directly relate to the key focus of this paper which relates to the operation of long endurance BVLOS drones supporting HL operations.

With this latter focus in mind, a particularly important theme emerging from the literature was discussion of the ethical operation of drones as well as the implications of the restrictions imposed by various national aviation authorities. Whilst multiple agencies across the globe are actively engaged in developing regulations that provide a balance between concerns related to safety and privacy versus the potential benefits of drone operations, it is clear that the current legislative environment present in many countries severely restricts their use. As will be discussed later in this chapter, this is a key area of challenge that will need to be overcome if drones are to become part of the normal mechanism for the provision of support to the humanitarian logistician.

Within this set of papers, there was considerable discussion of the fact that drones have migrated from their original role as military weapons and the resultant ethical issues surrounding their use. In this regard Soesilo and Sandvik (2016) offer the results of a recent survey of humanitarian organisations, donors, United Nations (UN) agencies, national governments, private business and other respondents. Whilst the response rate was not provided, the survey elicited 194 inputs during the period 15 Nov 2015–15 Jan 2016, and its results of the research can be summarised in the authors' observation that:

A majority of survey respondents [66%] expressed confidence that drones have the potential to strengthen humanitarian work, and that drones can greatly enhance the speed and quality of localized needs assessments, while a significant minority [22%] viewed the use of drones in humanitarian work unfavourably. (Soesilo & Sandvik, 2016, p. 3).

Whilst the authors of this survey fully acknowledged its limitations (p. 4), they nevertheless argued that it provides a baseline against which future trends can be determined.

In particular, the survey respondents emphasised (on the positive side) the ability of drones to support various logistic processes including needs assessment and material delivery. On the negative side, however, the core concerns were related to their association with military applications and their potential to increase the 'distance' between the beneficiaries and the aid workers. Survey respondents also underlined the need for clear/improved ethical and operational guidance, whilst the development of a dedicated drone service for humanitarian operations received considerable support (61%).

A number of more relevant contributions were, however, found including the paper by Tatham (2009) in which the author discussed the suitability of drones to support the initial needs assessment in the aftermath of a rapid onset disaster and which, arguably, represents the first discussion of their use in an HL context.

Secondly, UNOCHA (2014) which surveyed the recent use of drones and suggested that: "One promising area is delivery of vaccines or other small medical supplies…" (p. 8). However, this paper went on to note that "… the range limits on small UAVs (perhaps 40km) would make them unviable where villages are too widely spaced like much of the DRC [Democratic Republic of the Congo]" (p. 8). This latter observation captures an important theme that emerged from the literature in which discussion around the use of drones in a humanitarian context generally only considers small (i.e. US DOD Group 1) aircraft.

Nevertheless, the American Red Cross (2015) also undertook a comprehensive review of the potential use of drones and their report concludes that:

Aerial drones are one of the most promising and powerful new technologies to improve disaster response and relief operations. ... When a disaster occurs, drones may be used to provide relief workers with better situational awareness, locate survivors ... perform structural analysis of damaged infrastructure, deliver needed supplies and equipment, evacuate casualties, and help extinguish fires – among many other potential applications. (p. 4).

The report went on to analyse previous cases where drones have been used and, in particular, suggested that one of appropriate tasks to be 'Supply Delivery' (p.7), with a particular emphasis on the use of large payload drones.

A further important source was the website of the UAViators organisation (www.uaviators.org) which supports a community of practitioners who are striving to develop improved ways of using drones to support those affected by disasters. Not only does this website contain a large number of case studies, but it also provides a set of Operational Guidelines (UAViators, 2018) which are currently under consideration by the UN, IFRC and NGO communities, and which are recommended for use by UNOCHA (2014).

Since the completion of the literature review summarised above, two further papers have been published. The first provides an updated analysis of the potential for the use of drones to support the humanitarian logistician (Tatham et al., 2017), whilst the second considers a specific scenario in which drones could be used to provide medical support for a remote community in Western Australia (Tatham et al., 2017a).

In summary, and drawing on the above literature and also following informal discussions with manufacturers and experts, a BVLOS drone is perceived to have particular benefit for the humanitarian logistician in the provision of the following capabilities:

- Capturing still/video photography/infra-red imagery to support the Needs Assessment process.
- Using a 'find your phone' functionality in which the drone can initiate a call to an operating cell phone from, for example, the NDMO as an extension of the Needs Assessment process.
- Acting as a temporary mobile communications system by flying in a geo-stationery orbit.

- Dropping a mobile communications device (such as a solar-powered satellite phone) into the affected area to enable direct communications with the NDMO.
- Conducting low level surveillance of prospective logistic re-supply routes to ascertain if they have been compromised by, for example, landslips or broken bridges.

The next section of this chapter is designed to demonstrate how these five core capabilities could be operationalised using the example of Cyclone Pam as an exemplar.

CYCLONE PAM: A CASE STUDY

An Overview of Cyclone PAM

The Category 5 Cyclone PAM struck Vanuatu, a nation of some 80 islands that span 1,300 km (800 mi), on the morning of 16 March 2015 with the southern group of five islands (Tafea Province) receiving a "direct hit" from one of the most intense cyclones on record. While the death toll of 15 was remarkably low, around 130,000 people (some 50 per cent of the country's population) needed emergency shelter and more than 95 per cent of the crops were destroyed (ABC, 2016).

In addition to the 6W challenge of understanding the impact of the disaster, a further complicating factor was that the cyclone damaged or destroyed much of the communications systems (including the local cell phone tower). As a result, even 48 hours after the event, it was not possible to communicate with the five islands of Tafea province which are located some 150-250 km (95-155 mi) from the capital Port Vila (United Nations Logistics Cluster, 2015).

Furthermore, the damage to the road systems (through fallen trees, landslides and general degradation of the roads themselves), together with the rough sea conditions and associated damage to boats, limited the extent to which individuals and regional authorities could self-report (United Nations Logistics Cluster, 2015a). Even ten days after the onset of the disaster, both the absence of coordinated and real-time input to the NDMO and gaps in the reporting processes severely challenged the ability of the authorities to make informed and decisive decisions during this critical initial stage of the response. This has a corresponding impact on those already affected by the disaster (United Nations Logistics Cluster, 2015b).

In summary, it is argued that the lengthy over-water transit distances from the capital to the affected islands (15-20 hours by ship each way), in combination with the impacted communications capability, underpin the potential value of using long endurance drones to support the needs assessment process. The next section will describe how this might have taken place.

An Overview of a Potential Response to Cyclone PAM Incorporating Long Endurance Drones

As discussed above, a core capability of a long endurance drone such as the Latitude HQ -160B is its capacity to undertake aerial surveillance for over 12 hours. In this regard, whilst video footage has the clear benefit of real time data capture and transmission to the NDMO, its use constrains operations to some 100km (60mi) from the operating base. Nevertheless, if this is a constraint, the capabilities of the latest generation still cameras are such that, after post-mission processing, they can provide the equivalent of a video camera output. The processing itself takes a similar time to that actual data capture – thus, for example, 3 hours of surveillance requires 3 hours of processing.

Either approach would support the Needs Assessment process by providing the NDMO with a relatively swift overview of the disaster's impact. Furthermore, the data that had been collected could be transmitted in parallel not only to the Vanuatu NDMO, but also to the equivalent organisations in those countries that assisted in the response (such as Australia and New Zealand), and to UN agencies and NGOs. This is a key benefit as it will assist all of the supporting governments and agencies in moving from a push-based response that reflects best estimates of the disaster's impact to one that is more closely driven by an improved understanding of the disaster's impact (i.e. pull-based). Such an approach clearly reflects supply chain best practice (Christopher, 2016).

The flight time for a Latitude HQ -160B drone with a high definition still camera at a cruising speed of 70kph (45mph) is some 15 hours, i.e. it has a range of around 1,000km (600mi). As shown in Figure 1 and Table 4 this would enable a comprehensive overview of the impact of the disaster on the affected areas of Tafea Province over a period of 4 days. In developing the mission profiles, the population locations on each island have been taken into account, and conservative estimates of distances and timings have been used. Notwithstanding the use of such conservative assumptions, it will be noted that in each case the transit time is 3-8.5 hours, leaving some 5-8.5 hours to conduct the oversight of the affected areas.

Figure 1. Map of Tafea Province, Vanuatu

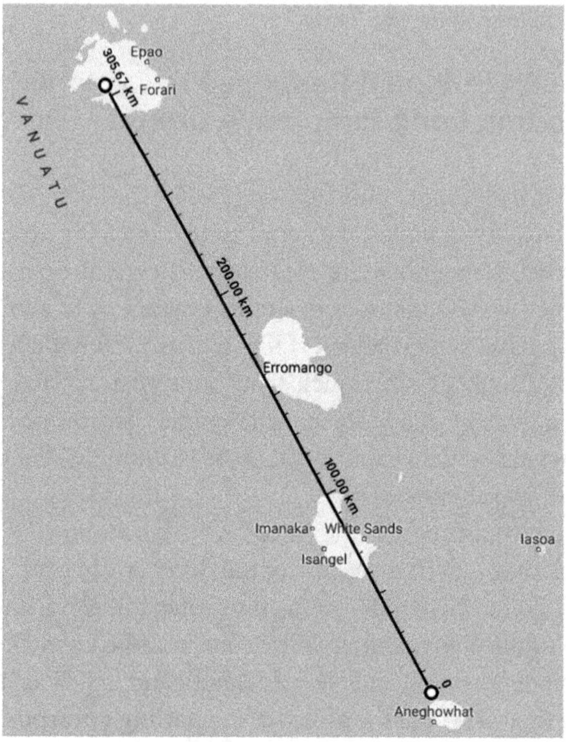

Table 4. Potential drone transit profiles

Day 1. Port Vila to Erromango to Port Vila
Total Transit Time = 3.5 Hours
Time Need on Task = 3.0 Hours
Unallocated Time = 8.5 Hours
Day 2: Port Vila to Tanna to Port Vila
Total Transit Time = 5.75 Hours
Time on Task = 3.5 Hours
Unallocated Time = 5.75 Hours
Day 3: Port Vila to Anatom to Port Vila
Total Transit Time = 8.5 Hours
Time on Task = 1.0 Hours
Unallocated Time = 5.5 Hours
Day 4: Port Vila to Aniwa to Futuna to Port Vila
Total Transit Time = 7.75 Hours
Time on Task = 1.0 Hours
Unallocated Time = 6.25 Hours

Source: The author

BROADER CONSIDERATIONS IN RELATION TO THE USE OF A LONG ENDURANCE DRONE

The use of a long endurance drone to capture the post-disaster impact is, as discussed above, potentially of significant value, however the benefits would be further improved if these results were able to be compared with a pre-disaster baseline. To an extent this could, potentially, be achieved in the typical 48 hour warning period between the clear indications that a cyclone is likely to strike a country or region and its actual onset. However, the development of a pre-planned surveillance programme that is focussed on areas of high risk is likely to provide significantly more meaningful data.

In addition to such surveillance missions, drones are also able to provide an *ad hoc* mapping service in the event that suitable information is not readily available – as was the case in the aftermath of Typhoon Haiyan that struck the Philippines in 2013 (American Red Cross, 2015, p. 16), and also as reported in a number of the recent case studies (UAViators, 2016).

It is also relevant to note that drone operations are not necessarily constrained by the high wind speeds that occur in the vicinity of a cyclone. For example in November 2007, an Aerosonde long endurance drone flew for 7.5 hours monitoring the core of the 130kph (80mph) Hurricane Noel (NASA, 2007). Whilst it is recognised that the wind speeds in the heart of Cyclone Pam were significantly higher and also that the rotor configuration of the Latitude is likely to be less robust than the regular propeller arrangement in an Aerosonde, it is probable that the drones could have safely operated in the lead up to, and aftermath of, the cyclone.

The second core capability that would have been invaluable in the response to Cyclone Pam is that of carrying and dropping a small payload (up to some 5.5kg) such as a satellite phone and associated solar powered battery systems. This would have provided a direct communications link to the NDMO thereby providing a much improved understanding of the actual post-disaster needs of the population.

Thirdly, as explained above, one of the challenges in responding to Cyclone Pam was the loss of communications to the islands that had been badly impacted. One way of overcoming this would be for the drone to fly in a geo-stationary mode and thereby act as a temporary telecommunications relay tower. In doing so, there is a clear potential that such a system would be overwhelmed by the volume of calls, however this could be overcome by limiting access to pre-determined phone numbers such as those belonging to local disaster management staff or village officials. In addition, the drone can incorporate a 'find my phone' capability which can locate cell phones on the ground. This can then be used to initiate a call to the cell phone *from* the NDMO thereby enhancing the all-important exchange of information between those affected and the responders

Finally, the ability of a drone to follow a designated route that aligns with prospective logistic routes would help the NDMO ensure the viability of a proposed re-supply route by checking that it is not blocked by fallen trees, landslips or broken bridges etc. Clearly this capability cannot provide absolute surety that a route is usable – for example a bridge or culvert may appear sound, but actually be unable to take the weight of a truck. Nevertheless it would highlight routes that are inoperable and thus avoid wasting time and effort in reaching those in need.

Importantly, however, the above mentioned benefits of the use of drones focus on the technology side of the challenge. To this must be added a 'whole-system' perspective and the other two components of the people/processes/technology triangle will be discussed in the next sections.

Overview of the Process Challenges

As discussed in UNOCHA (2014), there remains a gap in the ability of responding agencies to integrate the results of aerial observations and the associated data collection into needs and damage assessments, search and rescue, and other humanitarian functions. Overcoming this challenge is the subject of broad consideration within the sector including, for example, a recent contribution by Tatham and Spens (2016). This suggests that one resolution of the coordination challenge facing the NDMO and responding agencies might be to use a mechanism that parallels that employed by the Urban Search and Rescue (USAR) community. The International Search and Rescue Guidelines (INSARAG, 2015) provide an overarching model for the integration of the work of multiple teams from multiple countries using common processes and approaches. Importantly, this model is also being adopted by the World Health Organisation (WHO) under their Guidelines for Foreign Medical Teams (FMTs) (WHO, 2013).

In a separate, but related, contribution to assist in the resolution of the inter-agency logistic coordination challenge, Tatham et al. (2016a) recommend the adoption of a 'Common Humanitarian Logistic Picture' (CHLP) to which all agencies contribute and which can form the basis of integrated decision-making. Self-evidently, linking the drone output to the CHLP would be of considerable value in helping to ensure that response activities are appropriately prioritised and needs-based.

However, the current reality is that, although the data that are produced by drones are typically geo-tagged or referenced in some manner, there are no metadata standards. Thus the development of a robust metadata framework and its associated standards is clearly an area that requires further action in order to ensure that the technical capabilities of the drones are maximised. There is also a clear risk that the data may overwhelm existing NDMO systems, and the implications of this must be considered as part of any project to incorporate drones into the disaster response mechanisms.

The final, and key, process-related challenge is that of ensuring that the necessary permissions are in place that will allow the operation of drones in the affected country with a minimum of delay. Given that international and national air traffic management and safety authorities are struggling to achieve the appropriate balance between drone operations and the associated safety/privacy issues, this is clearly a critical area. In some cases this challenge has been mitigated by informal processes such as the decision by the Mayor of Tacloban to authorise local drone flights in the aftermath of the 2014 Typhoon Haiyan (UNOCHA, 2014, p.5). However, this is a poor substitute for the development of an agreed protocol that can be practised in advance of a disaster event and which will support, rather than impede, drone operations.

Overview of the People-Related Challenges

On the assumption that the drones are operated by a reputable commercial company, it can reasonably be assumed that the staff will have the necessary skills and expertise to conduct flying operations in a safe, effective and ethical manner, and in line with the air traffic control/safety requirements of the relevant country. There are, however, a number of people-related challenges that still need to be overcome.

The first such requirement would be for an education programme that covers a range of subjects including the risks and safety implications of drone operations, the potential benefits, and ways in which local communities can engage and support such operations. As an example, if it is planned that the drone will drop a satellite phone for use by those affected in a disaster, the appropriate protocols must be developed and practised in advance. These might include the operation of the phone as well as model through which meaningful information can be passed to the NDMO – for example by the use of a standardised question and answer system. Given that the availability of local disaster management staff may have been compromised by the cyclone, this knowledge may need to be included as part of the satellite phone delivery package – perhaps in pictorial form in the event that the operator is unable to read or uses a different language.

Secondly, given that the capability and *modus operandi* of NDMOs differ, it will be essential that the processes for capturing and integrating the drone data into the disaster management systems are prepared and exercised in advance of any actual event. In particular, the interpretation of the data provided by the drone is a sophisticated skill set and this implies that an individual versed in the operation of a particular NDMO may need to be engaged in order to ensure that the drone data is correctly interpreted and integrated with that from other sources.

COST CONSIDERATIONS

Unfortunately, as noted by Mailey (2013), comparable costs between different aircraft are not easy to calculate given there is no standardised or accepted approach to the calculation of costs/flight hour. Furthermore, to the extent that open-source literature does exist (e.g. Economist, 2009, 2011; Boyle, 2012), the focus is usually comparisons between high end military systems versus manned military aircraft. Thus, rather than attempt a potentially flawed direct comparison between alternative modes of delivering the core capabilities outlined above, the following observations in relation to the potential use of a long endurance drone are offered:

- Whilst the capital cost of the Latitude drone is not publically available, that of the Aerosonde Mk4.7 (which has a broadly similar capability (see Tatham et al., 2017a)) was reported in 2014 to be some US$100k (Corcoran, 2014), to which must be added the cost of the camera which will depend on its type (eg video v still), with the latter being around US$40k-50k. Thus, the overall capital cost of a long endurance drone with a high definition still camera is estimated to be some US$150k.
- The crew size of a Latitude consists of one pilot, one camera operator and a maintainer, and this is broadly similar to that of a small fixed or rotary wing aircraft.
- The fuel consumption of a long endurance drone is extremely low. For example, in August 1998 an Aerosonde Mk 1 flew across the Atlantic, a distance of 3,270km (2,030mi), at an altitude of 1,680m (5,500ft) in just under 27 hours consuming some 7li (1.5gal) of fuel (Barnard, 1999). In respect of the Latitude drone with a high definition still camera, a broad order consumption figure is understood to be some 0.8li/100 km at cruising speed. Thus, an indicative 15 hr mission time would use some 9li of unleaded fuel (i.e. costing around US$1-2/hour). By comparison, Robinson (2016) indicates that the two-seat Raven light helicopter uses some US$75/hour in fuel. As a result, a 15 hour flight would cost around US$1,125 compared with US$15-30 for a similar drone flight. Self-evidently, such a helicopter would not be able to conduct a 15 hour mission without re-fuelling, and so the actual time over the affected areas would be at least half, and more probably 25 per cent of that available from a drone.

CONCLUSION AND AREAS OF FURTHER RESEARCH

The above analysis of the benefits and costs of the use of an long endurance drone such as the Latitude demonstrates its potential to support logistic operations through the use of the five core capabilities (video/photography to support Needs Assessment; provision of temporary communications; 'find your phone'; the ability to drop a satellite phone; and logistic route surveillance). It is far less costly to operate than helicopters, can fly in very poor weather conditions, and can free up other assets (such as helicopters) for more appropriate activities. However, operationalising the use of such a drone will, inevitably, require considerable investment. The following represents a proposed way forward that is designed to integrate the people, process and technology requirements.

Phase 1

As summarised above, the relatively novel nature of drones in a non-military context and, in particular, the need to integrate their operations into existing disaster response systems, presents a significant challenge. To resolve this, it is believed that the most appropriate way forward is to select a country (or countries) to act as 'pathfinder(s)' in order to develop the proof of concept. The selection of the country is likely to reflect a combination of:

- Its likelihood of being impacted by a future disaster
- The potential impact of such a disaster
- The extent to which the country is likely to be supportive of the use of drones
- Its capability to make use of the resultant information in a meaningful way
- The extent to which its geography and topology are conducive to the use of drones.

Phase 2

Relevant experts from the selected country together with those skilled in the various domains (operation of the drones; data handling; air traffic control etc.) should collaborate to establish appropriate air traffic control protocols, capture base-line data to support subsequent impact analysis, and develop the systems needed to integrate the data captured by the drones into the NDMO as a basis for efficient and effective decision-making.

Once these processes and protocols have been developed, it will then be necessary to carry out appropriate training and education at all levels (community->local government->NDMO and supporting country decision-makers) in order to prepare for future drone operations. This should include the actual conduct of drone operations in order to evaluate and improve the above processes. In this was it should be possible to ensure that, in the event of a disaster, the end-to-end drone-supported operations will be conducted in a safe, ethical, effective and efficient way.

In summary, it is perceived that the use of a long endurance drone has significant potential to support the logistic response to a disaster. However, a number of important hurdles remain before the concept can be operationalised, key amongst these are the development of an air traffic control regime that supports (rather than constrains) the use of drones, and the mechanisms (both process and people-related) that translate the data captured by the drones into usable information to underpin timely and effective decision making.

REFERENCES

ABC (Australian Broadcasting Corporation). (2016). *Cyclone Pam: Vanuatu one year on*. Retrieved from: www.abc.net.au/news/2016-03-13/cyclone-pam-vanuatu-one-year-on/7242620

Ambrosia, V. G., Wegener, S., & Zajkowliski, T. (2011). The Ikhana unmanned airborne system (UAS) western states fire imaging missions: From concept to reality (2006-2010). *Geocarto International*, *26*(2), 85–101. doi:10.1080/10106049.2010.539302

American Red Cross. (2015). *Drones for Disaster Response and Relief Operations*. Retrieved from: https://www.zurichna.com/en/search?q=Drones+for+Disaster+Response+and+Relief+Operations&page=2#

Artemenko, O., Rubina, A., Golokolenko, O., Simon, T., Romisch, J., & Mitschele-Thiel, A. (2014). Validation and evaluation of the chosen path planning algorithm for localization of nodes using an unmanned aerial vehicle in disaster scenarios. *Lecture Notes of the Institute for Computer Sciences Social-Informatics and Telecommunications Engineering*, *140*, 192–203.

Barnard Microsystems. (1999). *First Atlantic crossing by an unmanned aircraft*. Retrieved from: http://www.barnardmicrosystems.com/UAV/milestones/atlantic_crossing_1.html

Boyle, A. (2012). The US and its UAVs: A Cost Benefit Analysis. *American Security Project*. Retrieved from http://www.americansecurityproject.org/the-us-and-its-uavs-a-cost-benefit-analysis/

Christopher, M. G. (2016). *Logistics and Supply Chain Management* (5th ed.). Harlow, UK: Pearson Education Ltd.

Corcoran, M. (2014). *Drone journalism: Newsgathering applications of Unmanned Aerial Vehicles (UAVs) in covering conflict, civil unrest and disaster*. Retrieved from: http://www.flinders.edu.au/ehl/fms/law_files/Drone%20Journalism%20During%20 Conflict,%20Civil%20Unrest%20and%20Disasters%20March%201%202014.pdf

Economist. (2009). *Attack of the drones*. Retrieved from: http://www.economist.com/node/14299496/print

Economist. (2011). *Flight of the drones*. Retrieved from: http://www.economist.com/node/21531433/print

FAA (Federal Aviation Administration). (2016). FAA Aerospace Forecast: Fiscal Years 2016-2036. *FAA*. Retrieved from: http://www.faa.gov/data_research/aviation/aerospace_forecasts/media/FY2016-36_FAA_Aerospace_Forecast.pdf

GAO (United States Government Accountability Office). (2013). Defense Acquisitions: Assessments of Selected Weapon Programs. *GAO*. Retrieved from: http://www.gao.gov/assets/660/653379.pdf

Gye, H. (2013). Now that's a special delivery: Domino's builds DRONE to deliver pizzas by air and beat the traffic. *Daily Mail*. Retrieved from: http://www.dailymail.co.uk/news/article-2336324/Dominos-builds-DRONE-deliver-pizzas-air-beat-traffic.html#ixzz44XJEkfMO

Holt, J., Biaz, S., Yilmaz, L., & Afi, C. A. (2014). Asymbiotic simulation architecture for evaluating UAV collision avoidance techniques. *Journal of Simulation*, *8*(1), 64–75. doi:10.1057/jos.2013.5

Holz, R. L. (2018). In Rwanda drones deliver medical supplies to remote areas. *Wall Street Journal*. Retrieved from: https://www.wsj.com/articles/in-rwanda-drones-deliver-medical-supplies-to-remote-areas-1512124200

Hong, Z., Tong, X., Cao, W., Jiang, S., Chen, P., & Liu, S. (2015). Rapid three-dimensional detection spproach for building damage due to earthquakes by the use of parallel processing of unmanned aerial vehicle imagery. *Journal of Applied Remote Sensing*, *9*(1), 097292. doi:10.1117/1.JRS.9.097292

Huang, Y., Yi, S., & Li, Z. (2011). Design of highway landslide warning and emergency response systems based on UAV. *Proceedings of the 17th Chinese conference on remote sensing*.

INSARAG. (2015). *International Search and Rescue Guidelines*. Retrieved from http://www.insarag.org/en/methodology/guidelines.html

Joshi, D. (2017). Commercial Unmanned Aerial Vehicle (UAV) Market Analysis – Industry trends, companies and what you should know. *Business Insider*. Retrieved from: http://www.businessinsider.com/commercial-uav-market-analysis-2017-8/?r=AU&IR=T

Kunz, N., & Reiner, G. (2012). A meta-analysis of humanitarian logistics research. *Journal of Humanitarian Logistics and Supply Chain Management*, *2*(2), 116–147. doi:10.1108/20426741211260723

Liu, P., Chen, A. Y., Huang, Y.-N., Han, J.-Y., Lai, J.-S., Kang, S.-C., ... Tsai, M.-H. (2014). A review of rotorcraft unmanned aerial vehicle (UAV) developments and applications in civil engineering. *Smart Structures and Systems*, *13*(6), 1065–1094. doi:10.12989ss.2014.13.6.1065

Mailey, C. (2013). Are UAS More Cost Effective that Manned Flights? *AUVSI*. Retrieved from www.auvsi.org

Milliken, J., & Linton, D. (2016). Prioritisation of citizen-centric information for disaster response. *Disasters*, *40*(3), 476–493. doi:10.1111/disa.12168 PMID:26434664

NASA. (2007). *NASA and NOAA Fly Unmanned Aircraft into Hurricane Noel*. Retrieved from: http://www.nasa.gov/centers/wallops/news/story105.html

Nedjati, A., Vizvari, B., & Izbirak, G. (2016). Post-earthquake response by small UAV helicopters. *Natural Hazards*, *80*(3), 1669–1688. doi:10.100711069-015-2046-6

Nex, F., & Remondino, F. (2014). UAV for 3D mapping applications: A review. *Applied Geomatics*, *6*(1), 1–15. doi:10.100712518-013-0120-x

Robinson. (2016). *R44 Raven II 2016 Estimated Operating Costs*. Retrieved from: https://www.helispot.be/hs/documents/heli/r44_eoc.pdf

Soeliso, D., Meier, P., Lessard-Fontaine, A., Du Plessis, J., & Stuhlberger, C. (2016). Drones in Humanitarian Action. *Swiss Foundation for Mine Action (FSD)*. Retrieved from: http://drones.fsd.ch/wp-content/uploads/2016/11/Drones-in-Humanitarian-Action.pdf

Tatham, P. H. (2009). An Initial Investigation into the Suitability of the use of Unmanned Aerial Vehicle Systems (UAVS) to Support the Emergency Assessment Process in Rapid Onset Humanitarian Disasters. *International Journal of Risk Assessment and Management*, *13*(1), 60–78. doi:10.1504/IJRAM.2009.026391

Tatham, P. H., Ball, C., Wu, Y., & Diplas, P. (2017). Remotely Piloted Aircraft Systems (RPAS) Support for Humanitarian Logistic Operations: The Current Position and A Proposed Way Ahead. *Journal of Humanitarian Logistics and Supply Chain Management*, *7*(1), 2–25. doi:10.1108/JHLSCM-05-2016-0018

Tatham, P. H., & Christopher, M. G. (2017). An Introduction to Humanitarian Logistics. In P. H. Tatham & M. G. Christopher (Eds.), *Humanitarian Logistics: Meeting the Challenge of Preparing for and Responding to Disasters* (3rd ed.). London: Kogan Page.

Tatham, P. H., & Kovács, G. (2010). The impact of gender on humanitarian logistics. *International Journal of Mass Emergencies and Disasters*, *28*(2), 148–169.

Tatham, P. H., Kovács, G., & Spens, K. M. (2016a). The humanitarian common logistic operating picture: A solution to the inter-agency coordination challenge. *Disasters: The Journal of Disaster Studies, Policy and Management*.

Tatham, P. H., & Pettit, S. J. (2010). Humanitarian Logistics: The Journey to Supply Network Management. *International Journal of Physical Distribution & Logistics Management*, *40*(8/9), 609–622. doi:10.1108/09600031011079283

Tatham, P. H., & Spens, K. M. (2016). Cracking the humanitarian logistics coordination challenge: Lessons from the urban search and rescue community. *Disasters: The Journal of Disaster Studies. Policy and Management*, *40*(2), 246–261. PMID:26283578

Tatham, P. H., Stadler, F., Murray, A., & Shaban, R. Z. (2017a). Flying maggots: A smart logistic solution to an enduring medical challenge. *Journal of Humanitarian Logistics and Supply Chain Management*, *7*(2), 172–193. doi:10.1108/JHLSCM-02-2017-0003

Titcomb, J. (2016). Amazon to step up UK tests of delivery drones. *The Telegraph*. Retrieved from: http://www.telegraph.co.uk/technology/2016/07/26/amazon-to-step-up-uk-tests-of-delivery-drones/

Tuna, G., Nezji, B., & Conte, G. (2014). Unmanned aerial vehicle-aided communications system for disaster recovery. *Journal of Network and Computer Applications*, *41*, 27–36. doi:10.1016/j.jnca.2013.10.002

UAViators. (2016). *Case Studies of drones deployed for humanitarian purposes*. Retrieved from: https://docs.google.com/document/d/1KlJyxGLJhGaCmaA8ehzO aS9Yue8zj2PcR3TlxAQME3o/edit#

UAViators. (2018). *Humanitarian UAV Code of Conduct*. Retrieved from: https:// uavcode.org/code-of-conduct/

United Nations Logistics Cluster. (2015). *Vanuatu situation updates.* Retrieved from: www.logcluster.org/document/vanuatu-situation-update-18-march-2015

United Nations Logistics Cluster. (2015a). *Vanuatu – Cyclone Pam: concept of operations.* Retrieved from: www.logcluster.org/sites/default/files/logistics_cluster_vanuatu_conops_150320.pdf

United Nations Logistics Cluster. (2015b). *Vanuatu meeting minutes.* Retrieved from: www.logcluster.org/sites/default/files/logistics_cluster_vanuatu_meeting_minutes_150331_0.pdf

UNOCHA (United Nations Office for the Coordination of Humanitarian Affairs). (2014). Unmanned aerial vehicles in humanitarian response. *UNOCHA Policy and Studies Series, Occasional Paper No 10.* Retrieved from: https://docs.unocha.org/sites/dms/Documents/Unmanned%20Aerial%20Vehicles%20in%20Humanitarian%20Response%20OCHA%20July%202014.pdf

US Army. (2010). U.S. Army Unmanned Aircraft Systems Roadmap 2010-2035. *U.S. Army Center of Excellence.* Retrieved from: http://www.rucker.army.mil/usaace/uas/US%20Army%20UAS%20RoadMap%202010%202035.pdf

WHO. (2013). *Classification and minimum standards for foreign medical teams in sudden onset disasters.* Retrieved from: http://www.who.int/hac/global_health_cluster/fmt_guidelines_september2013.pdf

Chapter 6
Using Unmanned Aerial Vehicles to Deliver Medical and Emergency Supplies to Remote Areas

Isaac Levi Henderson
Massey University, New Zealand

Savern L. Reweti
Massey University, New Zealand

Robyn Kamira
Incredible Skies Ltd, New Zealand

ABSTRACT

This chapter examines the role of unmanned aerial vehicles (UAVs) in the delivery of medical and emergency supplies to remote areas. It outlines a number of potential considerations for operators wishing to use UAVs to deliver medical and emergency supplies to remote areas. These considerations address a number of practicalities in terms of the organisation that is wishing to conduct such operations, the operations themselves, and the technology that is used for such operations. These considerations primarily stem from the nature of the international regulatory framework for unmanned aircraft operations and the peculiarities of using a UAV to deliver medical and emergency supplies. The chapter will outline some of the practicalities that have been worked through or are being worked through during a project to deliver medical and emergency supplies in Northland, New Zealand. This will provide readers with examples of some of the real-world considerations that operators face as well as outline the positive community impact that such operations can provide.

DOI: 10.4018/978-1-5225-7900-7.ch006

INTRODUCTION

Thiels, Aho, Zietlow, and Jenkins (2015) argue that the speed and ability to travel over closed roads and terrain make UAVs ideal candidates for transporting medical supplies, especially given that this can likely be done at a lower cost than conventional medical transport. According to the Swiss Foundation for Mine Action (2016), 60% of humanitarian workers favour the use of UAVs in humanitarian work, with 36% of humanitarian workers considering the use of UAVs for delivery to be of the most interest. Indeed, there have been many trials of the use of UAVs to deliver medical and emergency supplies to remote areas, such as collecting the sputum samples of patients with suspected tuberculosis from remote parts of Papua New Guinea (Médecins Sans Frontières, 2014) and reducing waiting times for human immunodeficiency virus (HIV) testing in infants in Malawi (UNICEF, 2016). Unfortunately, the academic literature on the use of UAVs to deliver medical and emergency supplies is somewhat limited, especially in terms of the practicalities that must be worked through for such an operation to take place. Accordingly, this chapter provides a foundation for practitioners and researchers by working through some of these practicalities.

Due to the lack of international standardisation and the current global regulatory environment for UAV operations, a number of practicalities result from compliance issues with the use of UAVs to deliver medical and emergency supplies (Swiss Foundation for Mine Action, 2016; Thiels et al., 2015). Article 8 of the Convention on International Civil Aviation states that any aircraft capable of being flown without a pilot shall not be flown over a State without a pilot unless special authorisation has been given by that State (CASA, 2017). The complicating factor to this is that the International Civil Aviation Organisation (ICAO) is still in the process of creating a global regulatory framework for unmanned aircraft operations. Currently, there are two major groups within ICAO working towards this end. First, there is the Unmanned Aircraft Systems Advisory Group (UAS-AG), who support the Secretariat in the development of guidance material and are currently working on a common global framework for and core boundaries of unmanned aircraft system traffic management, known as UTM (ICAO, 2018b). Second, there is the Remotely Piloted Aircraft Systems Panel (RPASP) who are developing ICAO Standards and Recommended Practices (SARPs), procedures and guidance material for unmanned aircraft so that they can be integrated within the broader aviation system (ICAO, 2018a). While this work is ongoing, ICAO signatory states are without sufficient international standards to use for creating local regulatory requirements. In terms of using UAVs to deliver emergency and medical supplies, this creates the issue of a lack of standardisation across states and the issue of individual states having regulatory authorities who may be reluctant to approve or certificate such operations because they don't fit well within the current regulatory framework (i.e., a framework oriented towards manned aircraft operations). This chapter outlines a number of common considerations that arise due to compliance issues based upon current and proposed regulations in several countries (see Appendix for a list of countries and regulatory authorities).

In addition to compliance issues, practicalities arise due to the operational/technological requirements that are intrinsic to using UAVs to deliver medical and emergency supplies. To date, UAVs have been used to deliver medicine, diagnostics/production samples, defibrillators, blood, food and water, first aid kits, organs and vaccines (Krey, 2018; Scott & Scott, 2017). Claesson et al. (2016) describe some of the difficulties that arose when testing the use of UAVs to deliver automated external defibrillators (AEDs) to patients experiencing out-of-hospital cardiac arrests in rural areas. For example, they found that parachute-release methods for delivery caused uncertainty about where the AED would land due to wind-drift, and thus they found that a latch-release method was safer and more effective (this involved the UAV flying at about 3-4m and a bystander catching the AED). Scott and Scott (2017) note that one of the issues with using battery-powered UAVs is that the range is quite limited and so they suggest a model using a land-based delivery vehicle to get nearby the drop off point, and then a UAV to do the final delivery to the area in need. They suggest that this will provide a better balance between the range and delivery time. Such operational/technological considerations, while related to compliance, are important in their own right because the UAV needs to fulfil its intended purpose.

This chapter describes a number of considerations that must be worked through by operators in terms of their organisation, their operations and the technology they use. Each consideration can be further broken down into how it relates to compliance, operations, training and/or technology. While these considerations may vary from country to country, they provide a general idea of the practicalities that must be worked through for an organisation to deliver medical and emergency supplies using a UAV. While targeted towards using UAVs for the delivery of medical and emergency supplies, this chapter provides a number of generic considerations that may be of interest to other UAV operators. In addition to providing some considerations for operators, this chapter outlines the real-world experiences of an ongoing trial to deliver medical and emergency supplies to remote areas in Northland, New Zealand. This trial is part of the Incredible Skies initiative and is called Medical Drones Aotearoa (Aotearoa is the Māori name for New Zealand). It aims to change the face of rural access to medicines, healthcare and emergency supplies using UAVs. The expected positive impacts upon the recipient community will also be discussed so that readers can understand the value that such operations may be able to provide as UAVs continue to be applied to the delivery of medical and emergency supplies.

CONSIDERATIONS FOR OPERATORS

The considerations for operators have been split into three core areas: (1) organisational, (2) operational and (3) technological. Each core area will be described in more detail in the following sections, however, the rationale for the order that they have been presented in is important. Because the use of UAVs to deliver medical and emergency supplies typically requires certification or other approvals from local regulators, individuals cannot perform such operations recreationally. This means that an organisation needs to exist. The organisation will then need to consider what it aims to achieve with the use of UAVs, and therefore what its operations need to look like. These operational requirements will then need to be related back to local regulatory requirements and may involve different solutions in terms of organisational processes, operations, and technology. Accordingly, the technological considerations should come after as these only exist in order to meet the organisation's aims and operational requirements.

The considerations in this chapter arise from a selection of sources (e.g., regulatory authorities, academic literature, UAV manufacturers) as well as the practical experience of the authors with regard to UAV operations. Therefore, it is important to caveat that this chapter is not conclusive in its considerations and recommendations, but should rather be treated as a guide of potential considerations. While every effort has been made to ensure that the considerations will be relevant to all readers, it is important that readers consider issues specific to their proposed operation, such as local regulatory requirements and recent technological developments.

Organisational Considerations

Organisational considerations are those that relate to the structure and management of the organisation that intends to carry out the delivery of medical and emergency supplies using UAVs. This section details considerations in terms what the organisation might be required to do to achieve certification and consequently which individuals will be required to make the organisation up.

The Role of UAVs in the Organisation

Organisations that want to use UAVs to deliver medical or emergency supplies need to have some rationalisation as to why UAVs should be applied to the situation. Generally, this rationale will originate from the objectives of the organisation (i.e., what the organisation is trying to achieve) and how UAVs can help an organisation achieve these objectives. For example, Zipline, a San Francisco Bay Area-based robotics start-up, had (and still has) the mission of "allowing on-demand delivery of medicines and other products at low cost without using a drop of gasoline" (Zipline, 2018, p. 3). In late 2016, Zipline (with financial backing from Google) set up distribution centres in Rwanda. Zipline's distribution centre staff prepare and pack up to 1.8 kg of medical supplies into a UAV and the mission is launched. Once the UAV is piloted to its destination, the package is dropped by parachute to within a few meters of an agreed drop-off point. The whole process from ordering to delivery is completed in 30 minutes (Dwyer, 2018). After a year and a half of successful operation its UAVs have made more than 1,400 flights carrying on-demand blood and emergency supplies over 100,000 kilometres. In 2017, Zipline expanded its medical delivery operations into a larger neighbouring country, Tanzania. (Landhuis, 2018). Here, Zipline saw that developing and operating UAVs was the best way of achieving their mission.

Similarly, in October 2017, Swiss Post launched a medical transport network in Lugano, Switzerland. The UAVs have made 350 successful deliveries, about 5 to 15 per day. Beyond, blood and medical supply deliveries, UAVs could transform another vital component of healthcare: lab tests. Timely test results assist doctors to correctly diagnose infections and eliminate guesswork in prescribing medications. Some of those decisions have life-or-death implications. For example, new-born babies produce bilirubin, a by-product formed during normal breakdown of red blood cells but high bilirubin levels can be toxic to babies so it needs to be tested for repeatedly. In most cities, millions of blood samples, urine specimens and swabs are transported to central lab facilities by motor vehicle. In remote communities, like Northland in New Zealand, these can involve a complex coordination of multiple forms of transportation such as car, plane and ferry. UAV technology offers a safer, faster, and more secure delivery mechanism. Already companies are transporting blood plasma (Palo Alto, California), defribrillators (Reno, Nevada), and vaccines (Vanuatu, South Pacific) (Landhuis, 2018). Organisations such as Médecins Sans Frontières and UNICEF (i.e., not for profits) are also involving themselves in the use of UAVs to deliver medical and emergency supplies (Médecins Sans Frontières, 2014; UNICEF, 2016).

In all successful instances, the operators had considered the role of UAVs within the supply chain and decided UAVs could contribute towards the organisation's objectives. Considering an organisation's logistics supply chain in relation to the delivery of medical and emergency supplies, UAVs are often ideal candidates due to their low costs and relatively high speeds compared to other forms of medical transport (Thiels et al., 2015). While objectives will vary markedly between organisations, a successful UAV operation will be one helps an organisation to achieve its objectives regardless of what they might be.

Certification Process

The certification process concerns how an aviation organisation becomes certificated. A certificated aviation organisation has been given an operator's certificate to conduct operations as specified in their exposition. An exposition is a document that outlines how an organisation is managed and how its operations are conducted. This document is used by regulatory authorities to determine whether an organisation should be certificated and is used following certification to audit the organisation for compliance. Accordingly, one of the primary organisational considerations is writing an exposition that complies with local regulatory requirements whilst also reflecting what is required for the day-to-day operations of the organisation. CAANZ (2015b) provide a sample exposition to help operators prepare for certification. This process is what is used in the wider aviation industry, with airlines, airports and other commercial aviation entities required to obtain certification prior to commencing operations.

Senior Persons

Certificated organisations need to identify persons who will hold several senior roles within the organisation. Depending on the size and complexity of the organisation, each of these key roles may be held by the same person, or each may be allocated to different persons. Each senior person needs to be approved into their role by the local regulator. While the terms used vary from country to country, senior persons need to show that they have the requisite knowledge and experience to be competent at their role and demonstrate a history of compliance with transport laws and other legislation to show fitness of character (e.g., CAA, 2014; CAANZ, 2015a; CASA, 2016). Table 1 shows some of the key senior roles that an organisation may need to allocate for compliance purposes. Even if the local regulator does not require the naming of senior persons for unmanned aircraft operators, each of these roles will still need to be performed by someone in the organisation to ensure a safe and efficient operation.

Training and Authorisation of Personnel

As a certificated organisation, the operator will need to have policies on what training is required for different roles and how personnel are authorised to act under the organisation's operator's certificate. With regard to UAV operations, training and authorisation processes are usually required so that personnel can act as pilot in command (PIC) and/or a visual observer (VO). Organisations are expected to keep a record of the training completed by personnel and list all personnel authorised to

Table 1. Senior persons and their responsibilities within the organisation

Senior person	Responsibilities
Chief Executive	Ensures that the operations comply with all regulatory requirements and that the organisation is suitable to carry out the operations safely. This means providing corporate and financial governance for the organisation and exercising control and/or supervision of the organisation's personnel.
Head of Flying Operations	Represents the organisation to the regulatory authority on all matters relating to flying operations.
Chief Pilot	Provides oversight of pilots and support crew during operations to ensure safe conduct of flight and safe operation of the equipment.
Training Manager	Responsible for the organisation's training standards, monitoring what training personnel have completed, when it needs to be recompleted and which authorisations they hold.
Maintenance Manager(s)	Ensures that all aircraft operated by the organisation are airworthy and maintained in accordance with the operator's maintenance manual (OMM), also known as the maintenance control manual (MCM).
Safety Manager	Manages the safety management system (SMS), facilitating risk management processes, monitoring safety performance and maintaining safety documentation.
Quality Assurance Manager	Ensures that the organisation's policies and procedures are compliant with the civil aviation rules (CARs) by conducting audits and reporting back to senior decision makers.

Sources: (CAANZ, 2015b, 2016a; CASA, 2014, 2018; ECAA, 2013; ENAC, 2016; Transport Canada, 2015)

act in the role of PIC or VO in a training and authorisations register. Table 2 shows some of the core training considerations for operators with regard to PICs and VOs.

Safety Management

As a certificated aviation organisation, the operator will also need to consider establishing a safety management system (SMS). ICAO (2013, p. xii) defines an SMS as "a systematic approach to managing safety, including the necessary organisational structures, accountability, policies and procedures". The necessary level of complexity for the SMS will depend upon the size and scope of the organisation's operations. At the very least, unmanned aircraft operators should have a hazard register, a safety reporting system, a safety policy that commits the organisation towards the core tenets of safety management and policies and procedures that mitigate risks associated with operations (CAANZ, 2015c; FAA, 2016a).

Table 2. Training considerations for personnel to become authorised as a PIC or VO

Role	Consideration type	Common examples of requirements
PIC	General	Be at least 16 years of age; and Fluency in English; and Be in safe physical and mental condition (often evidenced by a class 2 medical certificate).
	General aviation knowledge	A pilot's licence; or A remotely piloted aircraft license (or equivalent); or Passing the private pilot licence (PPL) air law exam, flight radio telephone operator (FRTO) exam, holding an FRTO rating and five hours experience focussed on airspace and flight radio use; or A certificate of achievement from a certificated aviation training organisation indicating a pass in aviation law theory, competency in operating unmanned aircraft and competency in the use of aviation radios.
	UAV-specific knowledge	A wings badge from an approved organisation relevant to the type of unmanned aircraft; or A certificate of training from the manufacturer of the unmanned aircraft to be operated; or A certificate of training from a certificated aviation training organisation authorised to conduct unmanned aircraft training.
VO	General	Be at least 16 years of age; and Fluency in English; and No visual or aural impairment that cannot be easily corrected (e.g. prescription glasses).
	Internal training	Induction into the organisation's exposition and other key documents; and Training in visual scanning techniques; and Training in standard phraseologies for communicating with the PIC and other VOs.

Sources: (CAANZ, 2015c; ENAC, 2016; FAA, 2016a; Transport Canada, 2016)

Operational Considerations

Operational considerations relate to the nature of the organisation's day-to-day operations and how these can be conducted in a safe and efficient manner. A number of operational considerations stem from regulatory restrictions, while others stem from the peculiarities of using UAVs to deliver medical and emergency supplies to remote areas.

Visual Line of Sight

One of the common requirements for unmanned aircraft operators is to maintain visual line of sight (VLOS) with their UAV at all times (e.g., CAANZ, 2015c; FAA, 2016a). This means that the PIC can see the aircraft at all times without the use of visual aids (e.g., binoculars). Even if the PIC is operating with a first-person viewer (FPV), a visual observer (VO) must have unaided VLOS. This raises some obvious issues for an operator who wants to use UAVs to deliver emergency and medical supplies to remote areas as it is unlikely that the PIC will be able to maintain VLOS. Operators can vary the VLOS requirement by carefully creating operational and training procedures and using technology to ensure that such operations can be carried out safely. In practice, there are two possible variations to the VLOS requirement. Firstly, an operator can create procedures for extended visual line of sight (EVLOS) operations. EVLOS operations involve a PIC and one or more trained VOs. The airspace is segmented and different segments are assigned to different VOs. The PIC and the VOs maintain contact via radio (civilian frequency) and at any given point in time either the PIC or one of the VOs has the aircraft within their unaided VLOS. Accordingly, EVLOS is a relatively easy variation for an operator to use as the only complications are some additional SOPs, and the creation of training and authorisation processes for visual observers. However, this still adds unnecessary operational constraints for delivering emergency and medical supplies. As the supplies are often time-sensitive, the idea of needing trained observers across the flight path creates a logistical nightmare, especially if the intended flight path is over a body of water or untraversable terrain, which are common in remote areas. This leaves such operators with the second possible variation, beyond visual line of sight (BVLOS) operations. BVLOS operations are substantially more difficult to get approval for. For example, CAANZ (2015c) say that some of the necessary features of any safety case for BVLOS operations should include identification of the airspace class to be used and how its requirements will be met, ability to provide and maintain separation from other traffic and measures to mitigate risk to persons, property and terrain. In practice this either means restricting the BVLOS operations to a restricted area (i.e., special use airspace specifically for the BVLOS operation), and/or to employ technological solutions such as ballistic parachutes, automatic recovery systems and collision avoidance systems. When the delivery of medical/ emergency supplies is to the same remote area, the special-use airspace option is not unworkable (as it is unlikely to disrupt the conventional aviation system). However, if multiple areas are being serviced, or if the area is of importance to conventional airspace users then technological solutions are necessary.

Flight Over People and Property

One of the major risks associated with unmanned aircraft operations is injury to people and/or damage to property. State regulations differ in how this risk is managed. However, most simply require that the operator has obtained permission from all affected persons and property owners unless a minimum operating distance (e.g., 100 feet) is maintained (e.g., CAANZ, 2015c; Transport Canada, 2016). Operators are also often restricted from operations over built-up areas, critical infrastructure or crowds (e.g., JCAB, 2015). While the latter point is unlikely to be of concern for delivering medical and emergency supplies to remote areas, the time-sensitive nature of the operations means that obtaining consent from all people and property owners to be overflown is impracticable. It is possible for operators to achieve variations to the requirement to obtain permission from persons and property owners by instead providing a means of notification about the operation. In terms of persons, this could involve erecting signage (especially if a road will be overflown), using employees to usher people away from the take-off/landing areas and public notification (e.g., newspaper article, Facebook posts, etc.). The idea is that people can then choose whether to enter the operating area, and by doing so they would be giving consent to being overflown. The variation for properties to be overflown is for notification to be given either pre-event (before the operation takes place) or post-event (after the operation takes place). Such notifications should include the name of the organisation, the nature of the operation (e.g., delivering medical supplies), the time and duration of the operation, and a means of contacting the PIC or organisation. ENAC (2016) suggest that the safety level to manage the risks associated with such operations is affected by the pilot (training and experience), the operational and flight management procedures, environmental conditions, the maintenance programme for the aircraft and any other elements that are significant for the safety of RPAS operations. Many of these considerations are discussed elsewhere in this chapter, so it is important for operators to have a holistic approach to the development of their exposition.

Airspace

One of the biggest safety considerations with any aircraft operation is using airspace in a safe and efficient manner. UAV operations are no different and UAV operators are expected to adhere to the same standards as manned aircraft operations. Because airspace is three-dimensional, one needs to consider not only the geographical areas of operation, but also the altitude of operations and whether they are to be conducted in uncontrolled airspace, controlled airspace, special use airspace and/or other types of airspace. Accordingly, an important part of planning any UAV operation to deliver medical or emergency supplies is for PICs to familiarise themselves with the local airspace requirements. To do this, there are several sources of information that a PIC should look at. It is also critical that these sources of information are current (as per their effective date) as airspace boundaries, classifications and requirements may be permanent or temporary. All countries use the aeronautical information regulation and control (AIRAC) effective dates, published by ICAO (n.d.). Table 3 provides a brief summary of some of the sources of airspace information that UAV operators should review and what information they will find in each one.

Because current airspace classifications and requirements are based upon manned aircraft operations, it is important that UAV operators consider how they can safely conduct operations to meet the same level of safety as a manned aircraft operating in the same airspace. Table 4 identifies some considerations in terms of areas of operation.

In additional to the considerations in Table 4, certificated unmanned aircraft operators often can receive approval from their local regulatory authorities to fly in special use airspace (SUA). SUA can be defined as airspace "in which certain activities must be confined, or where limitations may be imposed on aircraft operations that are not part of those activities" (FAA, 2016b, p. 15-3). There are several types of SUA, each with their own requirements. Essentially UAV operators who want to operate in SUA need to be able to adhere to the same requirements that would be

Table 3. Sources of airspace information and the information that they provide

Source	Information provided
Aeronautical information publication (AIP)	Aeronautical information of a lasting character essential to air navigation.
Notice to airmen (NOTAM)	Information concerning the establishment, condition or change in any aeronautical facility, service, procedure or hazard, the timely knowledge of which is essential to personnel concerned with flight operations.
AIP supplements	Temporary changes to the information published in the AIP.
Aeronautical information circulars (AIC)	Information relating to flight safety, air navigation, technical, administrative or legislative matters that do not qualify for publication in the AIP or in NOTAMs.
Designated airspace handbook (DAH)	Information about the lateral and vertical limits and other pertinent details about different airspace types. This information can be overridden by a NOTAM.
Aeronautical charts	Visual depiction of aeronautical information such as areas of designated airspace, locations of aerodromes and other information.

Sources: (Airservices Australia, 2018; CAANZ, 2018; FAA, 2018)

Table 4. Considerations based upon areas of operation

Area of operation	Considerations
Within 4km of an aerodrome	Discuss the proposed operation with the aerodrome operator to identify and manage potential hazards Monitor radio frequencies Give way to manned aircraft For controlled aerodromes, obtain ATC clearance Be aware of aerodrome information (e.g. AIPs, NOTAMs, etc.) Use of a trained VO, and only VLOS or EVLOS operations Avoid operating over active movement areas and active runway strips
Above 400ft	Operate within uncontrolled airspace (Classes F or G), otherwise see considerations for within controlled airspace Monitor radio frequencies Obtain ATC clearance Give way to manned aircraft Be aware of aerodrome information (e.g. AIPs, NOTAMs, etc.)
Within controlled airspace	Obtain ATC clearance Monitor radio frequencies Give way to manned aircraft If possible, use of a trained VO, and only VLOS or EVLOS operations, otherwise: An autonomous UAV that is fully equipped with collision avoidance and can meet the requirements to fly in that class of airspace

Sources: (CAA, 2015; CAANZ, 2015c; CASA, 2017; EASA, 2018; ENAC, 2016)

placed upon manned aircraft operating in the same airspace. Operators can usually find local SUA locations and requirements by navigating to the website of their local regulatory authority. Table 5 outlines considerations for UAV operators to comply with the requirements of SUA.

In addition to areas of operation and SUA, many countries also have other airspace areas, sometimes called non-designated airspace. Table 6 outlines some of the other types of airspace that a UAV operator may encounter.

Aircraft Maintenance

Virtually every country has requirements for the airworthiness of unmanned aircraft. An airworthy aircraft can be defined as an aircraft that conforms in all respects with its approved or properly modified type design, is maintained in accordance with the appropriate rules and is fit for flight (CAANZ, n.d.). For manned aircraft, there are clearly stipulated maintenance requirements from the manufacturer of the aircraft (for example, the aircraft maintenance manual) as well as requirements from regulatory authorities. However, many UAV manufacturers do not offer sufficient maintenance guidance for the creation of the OMM.

Table 5. Considerations for operating in special use airspace

Special use airspace	Considerations
Prohibited areas	No entry
Restricted areas	Authorisation obtained from the administering/controlling authority (usually a government agency)
Military operating areas (MOA)	Authorisation obtained from the administering/controlling authority (the corresponding military agency)
Danger/Warning areas	Consider whether the hazards pose a risk to the operation and whether the operation will pose a risk to other operators within the area
Volcanic hazard zones (VHZ)	Operations only to take place in visual meteorological conditions, so that the operators can observe any volcanic ejecta or ash plumes
Low fly zones (LFZ)	Avoid these if possible, otherwise: Obtain a briefing from the using agency (usually flight clubs or training organisations) Monitor radio frequencies Give way to manned aircraft Use of a trained VO, and only VLOS or EVLOS operations
Mandatory broadcast zones (MBZ)	Avoid operations above 400 feet if possible, otherwise: The PIC must hold an FRTO rating and therefore a pilot's licence The PIC must make regular radio calls as per the local requirements for MBZs
Alert areas	Exercise caution in alert areas due to a high volume of pilot training or an unusual operation Consider how collision avoidance can be guaranteed as part of the operation
Controlled firing areas (CFA)	There should be no need to change flight path as CFAs must be suspended when a nonparticipating aircraft might be approaching the area

Sources: (CAANZ, 2016b; FAA, 2016b)

Table 6. Considerations for operating in other airspace areas

Other airspace area	Considerations
Common frequency zones	Monitor the common frequency
Parachute landing areas	Avoid these if possible, otherwise: Consider the risks that the operation may pose to parachutists, and what risks parachutists may pose to the operation Always operate upwind of parachutists Give way to parachutists
Temporary flight restrictions (TFR)	Check NOTAMs as part of flight planning, avoid all TFRs
National security areas (NSA)	Check NOTAMs as part of flight planning, if operations are prohibited then avoid operation Consider voluntarily avoiding NSAs unless absolutely necessary

Source: (CAANZ, 2016b; FAA, 2016b)

One of the basic elements that operators may need to consider are documenting pre- and post-flight checks. Usually the user guide for a UAV contains the procedure for pre- and post-flight checks. While pilots can do simple checks like this, operators should also consider scheduled and unscheduled maintenance events that may need to be outsourced to external maintenance providers (often UAV manufacturers have recommended service centres). Scheduled maintenance events are checks, inspections, overhauls or other maintenance activities that must occur after a certain amount of flight hours, flights, calendar time or other measure (Kinnison & Siddiqui, 2013). Unscheduled maintenance events are maintenance activities that are triggered by unforeseen events such as propeller strikes, fires or hard landings (Kinnison & Siddiqui, 2013). Depending on the model of UAV, some components might also be life-limited, meaning that they must be replaced after a certain amount of flight hours, flights, calendar time, engine cycles or other measure (De Florio, 2011). Common examples of life-limited components for UAVs are propellers and batteries. A number of variations to the rules that are available to certificated organisations are contingent upon such organisations having maintenance programmes that ensure the airworthiness of their aircraft to a similar standard as manned aircraft.

Handling Cargo and Dropping Items

Handling of cargo and the dropping of items from UAVs both require the organisation to be certificated and cannot be done recreationally (SACAA, n.d.). CAANZ (2015b) suggest that the three core considerations for operators wishing to carry cargo and drop items from UAVs are how they will go about establishing a risk assessment, how they can provide assurance that there will be no premature release and how they can ensure that the actual drop will not endanger any person or property. Such procedures will vary according to the type of the UAV (e.g., fixed-wing and rotary-wing UAVs will deliver the package differently) and what is being carried (i.e., its weight, fragility, size, etc.). As medical and emergency supplies can vary significantly, it is up to the individual operators to exercise due diligence in considering the potential risk that handling cargo and dropping items may pose.

Operating Conditions

UAVs, like any aircraft, have limits on the types of conditions that they can operate in. It is critical that any operator of UAVs consults their aircraft's guidance material from the manufacturer as they will need to comply with the operating limits for the aircraft. These vary markedly between UAV types and sizes. One of the considerations for operators wishing to deliver medical and emergency supplies using UAVs would be to try and get a UAV that is versatile enough that it can operate in most conditions, unless there is a suitable back-up option using an alternate means of transport. Operators should also note that operating conditions are dynamic and subject to change. Table 7 provides some common operating conditions that UAV operators may need to consider.

Operating Hazards

In addition to operating conditions that might change during the operation, there may be hazards within the operating area that need to be considered. Table 8 outlines some common examples of hazards within the operating area and considerations for how they might be managed.

Professional Conduct

In addition to training and operational requirements for personnel, there is a certain level of professionalism that is expected of UAV operators. One of the issues taken very seriously in the aviation industry is alcohol consumption. Personnel involved in UAV operations should not be under the influence of alcohol. Alcohol impairs performance, and even if there is a blood alcohol level of zero, pilots may be

Table 7. Considerations for various operating conditions

Condition	Considerations
Wind speed	Check the wind speed operating limit of the UAV and monitor current and forecasted conditions in the area of operation to ensure that this is not exceeded.
Rain or snow	Check whether the UAV can be operated in rain and snow, and if it cannot, then avoid operating in such conditions or when such conditions are forecast to be present in the immediate future.
Fog	Consider whether VLOS can be maintained in current conditions and in the immediate future (unless the organisation has BVLOS approval).
Ice	Consider whether the conditions would allow for ice to build up on the aircraft. Do not operate if such conditions are present.
Cloud base	No operations to take place at an altitude exceeding the cloud base (unless the organisation has BVLOS approval).
Air density	Ensure that the air is dense enough for sufficient generation of lift. For rotary-wing UAVs, there is also the option of fitting high-altitude propellers.
Temperature	If the UAV is battery powered, then ensure that the batteries are not too cold or too hot.
Night time	If possible, operate indoors or within a shielded area, otherwise: Consider how VLOS will be maintained by the PIC or trained VO (unless the organisation has BVLOS approval).

Sources: (CAA, 2015; FAA, 2016a; Transport Canada, 2016)

Table 8. Possible operating hazards and considerations for their management

Hazard	Considerations
Bodies of water	Plan operations so that the UAV can always return to its landing area (i.e., allow for enough battery life after the operation is completed).
Polar areas	Check whether the UAV's control systems work in polar areas.
Roads and highways	Obtain permission from the appropriate government authority Erect signage or use a pilot vehicle to warn motorists of the operation, or (with permission from the appropriate government authority) close off the road or traffic lanes that will be operated over.
Tall structures	Avoid operating near tall structures, otherwise: Ensure that the aircraft has a control system and/or other equipment that prevent interference.
Crowds	While permission to operate above people without consent may have been obtained, usually operators cannot operate over crowds. A common definition for this is more than one person per square metre.
Obstacles	Complete a job safety assessment that takes into account any obstacles particular to an area of an operation.
High voltage power lines (>35kV)	Create and maintain a separation standard for high voltage power lines.
Trees	Create and maintain a separation standard for trees.
Electromagnetism	Avoid operating in areas of high electromagnetism, otherwise: Ensure that the aircraft has a control system and/or other equipment that prevent interference.
Birds	Consider not operating if there are large concentrations of birds in the operating area.
Other aircraft	Always give way to manned aircraft, using an air band radio and a VO where appropriate to aid situational awareness.

Sources: (CAA, 2015; FAA, 2016a; Transport Canada, 2016)

under the influence of a hangover and still not fit to fly (CAANZ, 2011). Drugs also need to be considered for their effects on performance. Illicit drug use should never be tolerated in an aviation organisation, but operators also need to be aware of the potential effects of prescription and over-the-counter drugs (Robson, 2017). Operators should ensure that their PICs consult an aviation doctor about whether any prescription or over-the-counter drugs they intend on taking are compatible with safe flight.

While not a safety issue, another area where UAV operators need to be seen to exercise professionalism is in the realm of privacy. Most UAVs, regardless of what they are being used for, are fitted with cameras and therefore collect footage that must be stored and used in a way that safeguards people's personal privacy. Most countries have a government department that manages privacy regulation. Operators should be aware of such regulations and comply with them.

Technological Considerations

Technological considerations stem from the need to meet operational and organisational requirements. Technology can either solve regulatory and practical issues or create them. Accordingly, every operator needs to carefully consider how they match technology with their operations. While the key technology to consider is the UAV to be used, there are also considerations related to the control system of the UAV and the safety redundancies that can be added to it to improve the safety of the operation.

Weight of the UAV

One consideration that seems to be universal with regulatory authorities is to have different requirements based upon the weight of the UAV to be used in the proposed operation. For example, CAANZ (2015c) uses different considerations based upon whether the aircraft weighs less than 15kg, 15 – 25kg or more than 25kg. Similarly in the United States, there are different standards based upon whether the aircraft weighs more than 25kg or not (FAA, 2016a). In Australia, they use 5 different weight classes: micro (<100g), very small (100g – 2kg), small (2kg – 25kg), medium (25kg – 150kg), and large (>150kg or >100m^3 for airships) (CASA, 2017). There are differing weight classes and corresponding regulations across the world, however, the common theme is that heavier UAVs are associated with more compliance considerations. Accordingly, while having a larger UAV may accommodate a higher payload of supplies, this will complicate obtaining regulatory approval. However, the degree of added complication will vary from state to state and may be relatively simple. Another complicating factor with the delivery of supplies is that the weight of the UAV will vary from its empty weight. Accordingly, the operator needs to set a maximum take-off weight as part of the standard operating procedures (SOPs) and carefully consider which UAV model they purchase in order to align with local regulatory requirements as well as operational requirements. Many UAV manufacturers are already aware of this issue and carefully design aircraft accordingly. For example, DJI's Agras MG-1 (used for aerial spraying) can carry a payload of up to 10kg, weighing 24.5kg when fully loaded (DJI, 2018). If it had an 11kg payload, then this would cross the 25kg for maximum take-off weight and complicate compliance for the operator in most countries.

Type of UAV

Different types of UAVs offer different advantages and disadvantages. This section outlines some of these advantages and disadvantages and applies them to what this means for the delivery of medical and emergency supplies. Example models of UAVs are provided as examples of the different types that are currently available or under development. While many of these have been developed for other purposes, there is nothing to stop them being applied to the delivery of medical and emergency supplies in the future.

There are five main types of small (2kg – 25kg) aerial UAV platforms which are similar in design to larger and heavier UAVs: (1) multi-rotor, (2) fixed-wing, (3) single rotor, (4) fixed-wing hybrid, and (5) hybrid airships (Chapman, 2016; Plimp, 2018). Multi-rotor platforms are manufactured using a number of well tested designs ranging from tricopters (3 rotors) to octocopters (8 rotors), although the most common type is the quadcopter (4 rotors).

Multi-rotor UAVs are relatively stable, easy to control and manoeuvre, have vertical take-off and landing (VTOL) capability and can hover. However, they suffer from limited endurance (approximately 25 – 30 minutes, although this is improving), relatively small payloads and are vulnerable to high wind speeds and turbulent atmospheric conditions (Olson, 2018). The rationale for constructing UAVs with 6-8 rotors is to enhance power and stability in strong winds and turbulence. In addition, the PIC can lose power in several motors on the UAV but the on-board engine management software can still maintain stability so that they can safely land the UAV. However, more rotors means higher cost and complexity as well as a reduction in range. Typical uses for multi-rotor UAVs include aerial photography and video aerial inspection (Chapman, 2016). A multi-rotor UAV (DJI Inspire 2) is popular with professional movie makers and local news stations. It weighs 3 kg and has a ceiling of 16,400 feet. It has dual-operator controls, a retractable landing gear and comes with two sophisticated 4K capable cameras. The high performance Inspire 2 model has a top speed of 111 km/hr, a range of up to 6 km, and an endurance of 20 – 25 minutes. In addition, the software system provides a range of intelligent flight modes and has a collision avoidance system that can detect objects 100 feet away (Fisher, 2017). Multi-rotor UAVs can also be fitted to deliver cargo, however, their limited payload and inability to be operated in adverse weather conditions make them less ideal for the delivery of medical and emergency supplies.

Fixed-wing UAVs work in a similar fashion to conventional aeroplanes. They have reasonable endurance (can be as high as 16 hours), can fly at high altitude, carry heavier payloads than other types of UAV and can cope better with high wind speeds and turbulence. Despite their advantages there are significant drawback's with this design, such as high cost, the need for highly trained operators (as they are more difficult to land) and the requirement to have a launcher and a capture mechanism in difficult terrain as they cannot hover (Olson, 2018). Typical uses for fixed-wing UAVs include aerial mapping, pipeline and power line inspection (Chapman, 2016). AgEagle have developed a carbon-fibre launchable fixed-wing UAV

(RX60) for agribusiness which has an autopilot and an endurance of sixty minutes. It can capture aerial maps of 300 acres of farmland per battery charge (Eckelkamp, 2018). Fixed-wing UAVs for delivering cargo also exist, however, one of the key issues is that the aircraft cannot hover and therefore the cargo needs to either be dropped from the aircraft or the aircraft needs to land for the cargo to be delivered.

Single-rotor UAVs are similar in operation to a conventional helicopter and can scale up in size and are usually powered by fossil fuels. They can have extended range, are more efficient at high speeds, have a heavy payload capability and can hover even in high winds. However, they are more difficult to fly, are noisy, are usually more expensive and are more complex in construction (Olson, 2018). Typical uses for single rotor UAVs are aerial Light Detection and Ranging (LIDAR) laser scanning and payload delivery (Chapman, 2016). NOVAerial Robotics has developed a small 12.4 kg single rotor UAV (Procyon 800E), that can hover for 40 minutes, and can carry a payload of 5 kg. It has a top speed of 90 km/hr and a range of 20 km (Blain, 2016). Single-rotor UAVs show real potential for being used to deliver medical and emergency supplies due to having less drawbacks than multi-rotor UAVs and fixed-wing UAVs. However, they are relatively uncommon and are not as cost-efficient as other types of UAVs.

Fixed-wing hybrid UAVs combine the benefits of fixed-wing UAVs with the ability to hover. Some of them are designed with tilt rotors where the wing or motors can rotate. Others are tail-sitter aircraft which rest their tails on the ground, fly vertically, and then pitch downwards to achieve level flight. The advantage of fixed-wing hybrid UAVs is the combination of VTOL capability combined with long endurance flight. However, they are aerodynamically and technically complex and many prototypes are still under development. Typical uses include payload delivery and aerial mapping (Chapman, 2016). Vertical Technologies have developed the Deltaquad which is an electric VTOL flying wing capable of fully autonomous operation. It has a flight time of 165 minutes and a range of 150 km. In addition it can achieve a maximum speed of 100 km/hr and can carry a payload of 1 kg (Vertical Technologies, 2018). As such technologies develop, they may be a good option to trial for the delivery of medical and emergency supplies.

Hybrid airships combine aerodynamic, direct and buoyant lift. The advantages of this type of UAV include quiet operation, enhanced payload capability, stability, superior loiter time and ease of control. For example, the electric powered PLIMP airship hybrid (25 kg) can fly at 48 km/hr and carry 2 kg payloads up to an altitude of 500 feet with a range of 17 nautical miles (31.5 km). Typical uses for this type of UAV include stock and crop monitoring on farms (Plimp, 2018). While not as slow as craft that only use buoyant lift, hybrid airships are still slower than other types of UAVs. This may or may not preclude them depending on the time-sensitivity of the medical and emergency supplies that are to be delivered.

Medium (25kg – 150kg), to large (>150kg or >100m^3 for airships) aerial UAV platforms share similar characteristics with the five different types of small UAVs. Capabilities such as range, endurance, ceiling, and payload are all significantly increased in the medium to large UAVs. However, larger UAVs also mean higher operating costs and greater complexities in terms of regulatory considerations.

The Russian built SKYF large multi-rotor hybrid UAV can carry up to a 400 kg payload, with a flight range of 350 km, and has a maximum endurance of 8 hours with a 50 kg payload. It has an operational ceiling of 10,000 feet and a top speed of 70 km/hr. SKYF uses a unique design which separates the function of lifting and steering. It utilises the power and endurance of gasoline engines to drive its two main lift propellers for VTOL and translational flight, and the instant torque of four sets of coaxial twin propellers located at the corners of its box frame for manoeuvring. It can be used for heavy cargo lifting, fire-fighting and automated crop dusting and costs US$150 per hour to operate which is significantly less than a small helicopter (Blain, 2017).

UAVE is a United Kingdom-based developer of the medium fixed-wing Prion Mk3 UAV. This UAV is used for scientific research, commercial applications and law enforcement. The Prion Mk3 (30kg) is an ideal platform for beyond visual line of sight (BVLOS) operations as it has a range of 1000 km and a top speed of 80 km/hr. It can carry a payload of 15 kg which can include a plethora of sensors including LIDAR, GPS, cameras and magnetometers (but could potentially be used for cargo). It is powered by a 120cc four-stroke petrol engine and is controlled via a real- time satellite link to a ground control station (GCS) which has autopilot and manual controls (UAVE, 2018).

A large single-rotor UAV has been developed by Aurora Flight Sciences. It is an autonomous LIDAR equipped autonomous aerial cargo/utility system (AACUS) which is based on the Bell UH-1H helicopter. The on-board system can plan its own flight path and even select landing sites in unmapped and hazardous regions. It has already completed cargo missions delivering gasoline, water and medical supplies to US Marines in the field (Blain, 2017).

Defiant labs has developed a new fixed-wing hybrid UAV (DX3), which features VTOL and fixed-wing flight capabilities. It has a range of 1500 km, 24 hours endurance, LIDAR for 3D mapping and can carry a 3kg payload. Its engines are powered by hydrogen fuel cell technology and is controlled via satellite link. The DX-3 UAV has been designed to monitor and inspect remote long-range infrastructure such as oil and gas pipelines, and power transmission lines. Due to its rugged construction the UAV is also suitable for disaster relief and security operations (Rees, 2016).

In the United States by 2020 there will be a US$100 billion market opportunity for UAVs, driven by increasing demand from commercial and government agencies and the retail market. Defence agencies will still be the largest market for UAVs

(US$70bn) as they are a cheaper and safer alternative to manned military aircraft. The consumer market (US$17bn) for hobby UAVs will grow exponentially and the commercial/civil market (US$13bn) is exhibiting the fastest growth (Goldman Sachs, 2018). The FAA estimates that 450,000 UAVs will be operating in United States airspace by 2022, compared to 110,000 currently. The agency also predicts the number of commercial UAV pilots will increase to 300,000 over the same period, compared to 70,000 currently. This will mean the number of commercial UAVs and UAV pilots will significantly exceed the total number of private pilots and associated manned aircraft (Pasztor, 2018). The upshot of this future growth is that there will be a greater variety of models available for a greater diversity of operations, including the delivery of medical and emergency supplies to remote areas.

Control Systems

UAVs that are controlled by a PIC should operate on a frequency acceptable to the local regulatory authority. Most UAV models utilise standard frequency ranges (2.4 GHz for remote control and 5.8 GHz for video and audio links). Some models also have the option of purchasing signal boosters and band amplifiers in order to extend the range of the aircraft. So long as these do not use alternative frequencies, then this is usually not an issue. Most UAVs use a smart phone or other system to simplify flying processes allowing for PICs to have sufficient cognitive capacity to maintain situational awareness. However, sometimes UAV operations are autonomous.

A number of commercial companies, government-sponsored research agencies and university research teams have developed autonomous systems for UAVs. Understandably these agencies are reluctant to publish technical information about their designs due to copyright and intellectual property rights which may have significant monetary value. Therefore, new entrants have to begin their project from scratch and waste a lot of time and resources duplicating the work of previous developers. One solution to this problem is an open source project called the Remotely operated Aerial Model Autopilot (RAMA), developed by the Department of Control Engineering at the Czech Technical University in Prague. A distinctive feature of the RAMA project is that it is totally open source, which means that all of the technical documentation is available at the project website. This documentation includes wiring diagrams, printed circuit boards (PCBs), software source codes, vehicle controller designs, vehicle mathematical models and real flight data (Spinka, Kroupa, & Hanzálek, 2007).

Safety Redundancies

The rapid development of more complex and expensive UAVs has enforced the need for increased reliability. At present the installation of rescue and recovery systems on many commercial and military UAVs has become standard practice to mitigate situations where the UAV suffers engine or battery power failure, fuel starvation, navigational error, or control system failure (Prisacariu, Pop, & Circiu, 2016). These systems have evolved into a range of designs to match specific UAVs and include recovery parachutes, recovery air-bags, recovery mesh nets (ground-based) and pneumatic cushion boats (water rescue). The rescue and recovery system allows for a safe landing in case of emergencies and minimises financial loss in case of extreme weather or operational events. Recovery systems have various classifications including manual and automatic command, destructive or non-destructive recovery, shock absorption, and ground mounted capture systems (e.g., nets or mesh). Recovery systems designed for fixed-wing UAVs will depend on the mission requirements of the system and must have certain characteristics to be successful. They must be safe to operate, they must protect the UAV and delicate on-board sensors from damage, they must be accurate in returning the UAV to a designated point, they must have a high degree of autonomy to reduce operator workload, and the recovery system must be reliable and its operation repeatable (Wyllie, 2001). A popular method for small to medium multi rotor UAVs is the utilisation of a ballistic recovery chute and operates in four distinctive steps: (1) deployment of drag chute, (2) deployment of main chute, (3) opening of the main chute, and (4) controlled descent of the UAV (Prisacariu et al., 2016). ParaZero is a major supplier of ballistic recovery chutes that will scale to fit any size drone. Deployment is autonomous and reacts instantly to critical failure. In addition, it can deploy successfully at low altitude and as it descends it has a broadcast system that warns people within close proximity to the landing site (ParaZero, 2018).

INCREDIBLE SKIES AND MEDICAL DRONES AOTEAROA

This section details some of the practicalities that are being worked through or have already been worked through during trials to deliver medical and emergency supplies to remote areas in Northland, New Zealand. In addition to these practicalities, this section outlines the expected community impact of the project.

Practicalities

The organisation aims to begin with 'proof of concept' operations to satisfy CAANZ and select a range of UAVs suitable for their purposes. They will then build/source the UAV and scale their operation in order to deliver medical and emergency supplies. The organisation has become a certificated aviation organisation and has been subject to the 'fit and proper' persons process to assess the competency and fitness of character of its senior persons. They have had staff sent on courses with certificated aviation training organisations as well as some who have been trained overseas and one of the staff that is a pilot (manned aircraft). They have induction and competency assessment processes that must be followed before personnel can be authorised to act in the capacity of PIC. The intended operation will test BVLOS autonomous UAV operations in a series of incremental steps in order to demonstrate the fitness of the concept. Current operations have been small tests of the aircraft's payload and connectivity that did not require approvals as a certificated organisation. The BVLOS variation, once achieved, will be subject to using restricted airspace for the purpose of research and development (with Incredible Skies as the delegated authority) as well as through the use of specialised technology (this technology is commercially-sensitive and cannot be further elaborated upon). Several aircraft types will be trialled, beginning with lightweight fixed-wing and multi-rotor aircraft and then moving towards heavier crafts and VTOL. The key considerations that the organisation is using to choose between models are payload, weight and endurance. The control systems will likely be open source. The trials will determine what safety equipment (e.g., automatic recovery systems) may be necessary to conduct operations.

Community Impact

The project recognises that people in rural areas are more likely to die from accidents and illnesses. Many choose not to pick up medicines or choose to ration them. This is because the cost of delivery, the distance, the stress and the susceptibility of remote areas to weather events causing road closures make it difficult for the local community. In the 2017 – 2018 year, access to and from the community in this trial was blocked 28 times. The project intends on operating UAVs to conduct the following operations: (1) civil defence scout operations that provide location information and deliver survival packs, (2) one-off urgent deliveries in partnership with emergency services, and (3) deliveries of scheduled repeat medicines (one flight to deliver supplies to multiple patients). Currently to deliver supplies to and from the nearest hospital or pharmacy in the trial area requires a 3-hour drive with an estimated cost of NZ$50 per trip. By using UAVs, this project intends on reducing this to 20 minutes at a cost of less than NZ$1 per trip. In addition to the reductions in delivery costs and delivery times for medical and emergency supplies, the project aims to have a positive economic impact on the local communities as well as the wider UAV sector. The project will offer training places for members of the local community, potentially leading to commercial UAV courses, further education, small business start-ups and other opportunities. Māori university students and senior high school students will also be invited to participate in lab and field activities in order to provide a practical exposure to science, technology, engineering and mathematics (STEM) disciplines. The project expects to create jobs in the local community in logistics, flying and training and the manufacture, maintenance and sale of UAV technologies. It is also hoped that these new technologies will provide opportunities for Māori land use in terms of launch sites, air strips, crew and equipment facilities, labs and telecommunications infrastructure.

CONCLUSION

This chapter has outlined a number of considerations for operators who intend on using UAVs to deliver medical and emergency supplies to remote areas. While a number of such operations have already taken place and there are organisations advocating for the use of UAVs for such purposes, the literature lacks practical guidance for potential operators about what they need to consider for such operations. This chapter separates these considerations into three areas. First, organisational considerations acknowledge that a commercial UAV operator must be a certificated aviation organisation, meaning that they need to have an exposition, senior persons, training and authorisation processes and a safety management system. Second, operational considerations outline a number of practicalities depending upon the complexity of the proposed operation, particularly with regard to regulatory requirements. Third, technological considerations acknowledge that a UAV needs to be capable of achieving the organisation's desired operations. This chapter has identified that one of the biggest issues facing potential operators is the lack of an international regulatory framework suited towards unmanned aircraft operations. This chapter summarises regulatory considerations from a number of states in order to aid potential operators in what commonalities exist internationally. In addition to this, an example of a real-world and ongoing operation to deliver medical and emergency supplies is provided along with an explanation as to the positive community impact that such operations can provide. Accordingly, this chapter provides a practical understanding to the existing literature that benefits both practitioners and researchers interested in the application of UAVs to the delivery of medical and emergency supplies.

REFERENCES

Airservices Australia. (2018). *Designated airspace handbook*. Retrieved 22 May 2018 from http://www.airservicesaustralia.com/aip/pending/dah/dah.pdf

Blain, L. (2016). *NOVAerial's helicopter drones: Better than multicopters?* Retrieved 28 May 2018 from https://newatlas.com/novaerial-procyon-800e-helicopter-drone/47013/

Blain, L. (2017). *Russian unmanned, long-range, hybrid multirotor wants to do the heavy lifting*. Retrieved 28 May 2018 from https://newatlas.com/ardn-russia-skyf-heavy-lift-hybrid-multirotor/52300/

CAA. (2014). *Criteria for an application for and grant of, or a variation to, an ATOL: Fitness, competence and accountable person*. Retrieved 17 May 2018 from https://publicapps.caa.co.uk/docs/33/Fitness%20note%2022814.pdf

CAA. (2015). *Unmanned aircraft system operations in UK airspace: Guidance*. Retrieved 18 May 2018 from http://publicapps.caa.co.uk/docs/33/CAP%20722%20Sixth%20Edition%20March%202015.pdf

CAANZ. (2011). *Alcohol consumption and medical aviation safety*. Retrieved 28 May 2018 from http://www.caa.govt.nz/assets/legacy/medical/Med_Info_Sheets/MIS014.pdf

CAANZ. (2015a). *Fit and proper person assessment handbook: Procedures and guidance*. Retrieved 17 May 2018 from https://www.caa.govt.nz/Forms/FPP_Handbook.pdf

CAANZ. (2015b). *Sample exposition: Part 102 unmanned aircraft operator's certificate*. Retrieved 15 May 2018 from https://www.caa.govt.nz/rpas/Part_102_sample_exposition.doc

CAANZ. (2015c). *Unmanned aircraft: Operator certification* [AC102-1]. Retrieved from https://www.caa.govt.nz/assets/legacy/Advisory_Circulars/AC102-1.pdf

CAANZ. (2016a). *Advisory circular: Safety management* [AC100-1]. Retrieved 17 May 2018 from https://www.caa.govt.nz/assets/legacy/Advisory_Circulars/AC100-1.pdf

CAANZ. (2016b). *Civil Aviation Rules*. Retrieved from https://www.caa.govt.nz/rules/rules.htm

CAANZ. (2018). *Civil aviation rules part 1: Definitions and abbreviations*. Retrieved 22 May 2018 from https://www.caa.govt.nz/rules/Rule_Consolidations/Part_001_Consolidation.pdf

CAANZ. (n.d.). *Part 39: Airworthiness directives*. Retrieved 29 May 2018 from https://www.caa.govt.nz/rules/part-039-brief/

CASA. (2014). *Chief pilot guide*. Retrieved 17 May 2018 from https://www.casa.gov.au/sites/g/files/net351/f/_assets/main/pilots/download/chief-pilot-guide.pdf

CASA. (2016). *Enforcement manual*. Retrieved 17 May 2018 from https://www.casa.gov.au/files/009rfullpdf

CASA. (2017). *Remotely piloted aircraft systems: Licensing and operations* [AC 101-01]. Retrieved 14 May 2018 from https://www.casa.gov.au/files/101c01pdf

CASA. (2018). *Air operator's certificate handbook: Volume 2 - flying operations*. Retrieved 17 May 2018 from https://www.casa.gov.au/file/174881/download?token=Tn6gR-ti

Chapman, A. (2016). Types of drones: Multi-rotor vs fixed-wing vs single rotor vs hybrid VTOL. *Drone Magazine*. Retrieved 28 May 2018 from https://www.auav.com.au/articles/drone-types/

Claesson, A., Fredman, D., Svensson, L., Ringh, M., Hollenberg, J., Nordberg, P., ... Ban, Y. (2016). Unmanned aerial vehicles (drones) in out-of-hospital-cardiac-arrest. *Scandinavian Journal of Trauma, Resuscitation and Emergency Medicine*, *24*(1), 124. doi:10.118613049-016-0313-5 PMID:27729058

De Florio, F. (2011). *Airworthiness: An introduction to aircraft certification* (2nd ed.). Burlington, MA: Butterworth-Heinemann. doi:10.1016/B978-0-08-096802-5.10002-9

DJI. (2018). *Agras MG-1*. Retrieved 09 May 2018 from https://www.dji.com/mg-1/info

Dwyer, F. (2018). *Medical delivery drones may soon be helping rural Americans*. Retrieved from https://www.wetalkuav.com/medical-delivery-drones-could-help-rural-americans/

EASA. (2018). *Introduction of a regulatory framework for the operation of unmanned aircraft systems in the 'open' and 'specific' categories*. Retrieved 18 May 2018 from https://www.easa.europa.eu/sites/default/files/dfu/Opinion%20No%2001-2018.pdf

ECAA. (2013). *Advisory circular: Maintenance control manual (MCM)*. Retrieved 17 May 2018 from http://www.ecaa.gov.et/upload/ac/MAINTENANCE%20CONTROL%20MANUAL(MCM).pdf

Eckelkamp. (2018). *Drone ROI realized by crop advisers*. Retrieved 29 May 2018 from https://www.agprofessional.com/article/drone-roi-realized-crop-advisers

ENAC. (2016). *Regulation: Remotely piloted aerial vehicles*. Retrieved 15 May 2018 from http://www.enac.gov.it/repository/contentmanagement/information/n1220929004/regulation_rpas_issue_2_rev%202_eng.pdf

FAA. (2016a). *Advisory circular: Small unmanned aircraft systems (sUAS)* [AC No: 107-2]. Retrieved 09 May 2018 from https://www.faa.gov/uas/media/AC_107-2_AFS-1_Signed.pdf

FAA. (2016b). *Pilot's handbook of aeronautical knowledge*. Retrieved 21 May 2018 from https://www.faa.gov/regulations_policies/handbooks_manuals/aviation/phak/

FAA. (2018). *Aeronautical information services: Aeronautical chart user's guide*. Retrieved 22 May 2018 from https://www.faa.gov/air_traffic/flight_info/aeronav/digital_products/aero_guide/media/editions/cug-complete.pdf

Fisher, J. (2017). *DJI Inspire 2*. Retrieved from http://au.pcmag.com/dji-inspire-2/44831/review/dji-inspire-2

Goldman Sachs. (2018). *Drones: Reporting for work*. Retrieved 28 May 2018 from http://www.goldmansachs.com/our-thinking/technology-driving-innovation/drones/

ICAO. (2013). *Safety Management Manual (SMM)* [Doc 9859, AN/474] (3rd ed.). Montreal, Canada: International Civil Aviation Organisation.

ICAO. (2018a). *Remotely Piloted Aircraft Systems Panel (RPASP)*. Retrieved 09 May 2018 from https://www.icao.int/safety/UA/Pages/Remotely-Piloted-Aircraft-Systems-Panel-(RPASP).aspx

ICAO. (2018b). *Unmanned Aircraft Systems Advisory Group (UAS-AG)*. Retrieved 08 May 2018 from https://www.icao.int/safety/UA/Pages/Unmanned-Aircraft-Systems-Advisory-Group-(UAS-AG).aspx

ICAO. (n.d.). *AIRAC*. Retrieved 22 May 2018 from https://www.icao.int/safety/information-management/Pages/AIRACAdherence.aspx

JCAB. (2015). *Japan's safety rules on unmanned aircraft (UA)/drone*. Retrieved 14 May 2018 from http://www.mlit.go.jp/en/koku/uas.html

Kinnison, H. A., & Siddiqui, T. (2013). *Aviation maintenance management* (2nd ed.). New York, NY: McGraw-Hill.

Krey, M. (2018). *Drones in healthcare: Application in Swiss hospitals.* Paper presented at the 51st Hawaii International Conference on System Sciences, Waikoloa, HI. Retrieved from https://scholarspace.manoa.hawaii.edu/bitstream/10125/50277/1/paper0390.pdf

Landhuis, E. (2018). *Medical cargo could be the gateway for routine drone deliveries.* Retrieved from https://www.npr.org/sections/health-shots/2018/03/10/592059175/medical-cargo-could-be-the-gateway-for-routine-drone-deliveries

Médecins Sans Frontières. (2014). *Papua New Guinea: Innovating to reach remote TB patients and improve access to treatment.* Retrieved 15 May 2018 from http://www.msf.org/article/papua-new-guinea-innovating-reach-remote-tb-patients-and-improve-access-treatment

Olson, K. (2018). *The 3 main categories of drones and their advantages and disadvantages.* Retrieved 28 May 2018 from https://www.botlink.com/blog/the-3-main-categories-of-drones-and-what-they-do

ParaZero. (2018). *ParaZero SafeAir™ Drone Safety System.* Retrieved from https://parazero.com/

Pasztor, A. (2018). FAA projects fourfold increase in commercial drones by 2022. *The Wall Street Journal.* Retrieved 28 May 2018 from https://www.wsj.com/articles/faa-projects-fourfold-increase-in-commercial-drones-by-2022-1521407110

Plimp. (2018). *Plimp airship: Revolutionizing flight.* Retrieved 28 May 2018 from http://plimp.com/

Prisacariu, V., Pop, S., & Circiu, I. (2016). Recovery system of the multi-helicopter UAV. *Review of the Air Force Academy, 1*(31), 91–98. doi:10.19062/1842-9238.2016.14.1.13

Rees, M. (2016). *New 24-hour endurance hybrid drone developed for monitoring & inspection.* Retrieved 28 May 2018 from http://www.unmannedsystemstechnology.com/2016/12/defiant-labs-announces-new-dx-3-drone-for-monitoring-inspection/

Robson, D. (2017). *Human being pilot: Human factors for aviation professionals* (2nd ed.). Hamilton, Australia: Aviation Theory Centre.

SACAA. (n.d.). *RPAS regulations (Part 101).* Retrieved 29 May 2018 from http://www.caa.co.za/Documents/RPAS/Part%20101%20-%20RPAS%20Workshops.pdf

Scott, J. E., & Scott, C. H. (2017). *Drone delivery models for healthcare*. Paper presented at the 50th Hawaii International Conference on System Sciences, Waikoloa, HI. Retrieved from http://aisel.aisnet.org/cgi/viewcontent.cgi?article=1442&context=hicss-50

Spinka, O., Kroupa, S., & Hanzálek, Z. (2007). *Control system for unmanned aerial vehicles*. Paper presented at the 5th IEEE International Conference on Industrial Informatics, Vienna, Austria. Retrieved from https://ieeexplore.ieee.org/abstract/document/4384800/

Swiss Foundation for Mine Action. (2016). *Drones in humanitarian action: A guide to the use of airborne systems in humanitarian crises*. Geneva, Switzerland: Swiss Foundation for Mine Action.

Thiels, C. A., Aho, J. M., Zietlow, S. P., & Jenkins, D. H. (2015). Use of unmanned aerial vehicles for medical product transport. *Air Medical Journal*, *34*(2), 104–108. doi:10.1016/j.amj.2014.10.011 PMID:25733117

Transport Canada. (2015). *Advisory circular: Quality assurance programs*. Retrieved 17 May 2018 from http://www.tc.gc.ca/media/documents/ca-opssvs/AC_QUA-001-ISSUE_01_-_QUALITY_ASSURANCE_PROGRAMS.pdf

Transport Canada. (2016). *Guidance material for operating unmanned air vehicle systems under an exemption*. Retrieved 14 May 2018 from http://www.tc.gc.ca/media/documents/ca-opssvs/AC-600-004.pdf

UAVE. (2018). *Large fixed wing UAVs with long flight endurance*. Retrieved 28 May 2018 from http://www.unmannedsystemstechnology.com/company/uave-ltd/

UNICEF. (2016). *Malawi tests first unmanned aerial vehicle flights for HIV early infant diagnosis*. Retrieved 15 May 2018 from https://www.unicef.org/media/media_90462.html

Vertical Technologies. (2018). *DeltaQuad rising: The next generation autonomous VTOL UAV*. Retrieved 28 May 2018 from https://www.verticaltechnologies.com/

Wyllie, T. (2001). Parachute recovery for UAV systems. *Aircraft Engineering and Aerospace Technology*, *73*(6), 542–551. doi:10.1108/00022660110696696

Zipline. (2018). *Zipline launches fastest delivery drone in the world*. Retrieved 21 September 2018 from http://www.flyzipline.com/uploads/Zipline%20Fastest%20Drone%20Press%20Release.pdf

ADDITIONAL READING

CAANZ. (2015b). Sample exposition: Part 102 unmanned aircraft operator's certificate. Retrieved 15 May 2018 from https://www.caa.govt.nz/rpas/Part_102_sample_exposition.doc

Claesson, A., Fredman, D., Svensson, L., Ringh, M., Hollenberg, J., Nordberg, P., ... Ban, Y. (2016). Unmanned aerial vehicles (drones) in out-of-hospital-cardiac-arrest. *Scandinavian Journal of Trauma, Resuscitation and Emergency Medicine, 24*(1), 124. doi:10.118613049-016-0313-5 PMID:27729058

FAA. (2016b). Pilot's handbook of aeronautical knowledge. Retrieved 21 May 2018 from https://www.faa.gov/regulations_policies/handbooks_manuals/aviation/phak/

ICAO. (n.d.). *Remotely piloted aircraft system (RPAS) concept of operations (CONOPS) for international IFR operations.* Retrieved from https://www.icao.int/safety/UA/Documents/RPAS%20CONOPS.pdf

Krey, M. (2018). Drones in healthcare: Application in Swiss hospitals. Paper presented at the 51st Hawaii International Conference on System Sciences, Waikoloa, HI. https://scholarspace.manoa.hawaii.edu/bitstream/10125/50277/1/paper0390.pdf

Medical Drones Aotearoa. (2018). Revolutionising access to crisis support + medicines for rural communities. Retrieved from https://www.medicaldrones.nz/

Scott, J. E., & Scott, C. H. (2017). *Drone delivery models for healthcare.* Paper presented at the 50th Hawaii International Conference on System Sciences, Waikoloa, HI. http://aisel.aisnet.org/cgi/viewcontent.cgi?article=1442&context=hicss-50

Swiss Foundation for Mine Action. (2016). *Drones in humanitarian action: A guide to the use of airborne systems in humanitarian crises.* Geneva, Switzerland: Swiss Foundation for Mine Action.

Thiels, C. A., Aho, J. M., Zietlow, S. P., & Jenkins, D. H. (2015). Use of unmanned aerial vehicles for medical product transport. *Air Medical Journal, 34*(2), 104–108. doi:10.1016/j.amj.2014.10.011 PMID:25733117

KEY TERMS AND DEFINITIONS

Exposition: A document that outlines how an organisation is managed and how its operations are conducted. This document is used by regulatory authorities to determine whether an organization should be certificated and is used following certification to audit the organization for compliance.

Life-Limited Parts: Parts that must be replaced at regular intervals (flight hours, flights, calendar time, or other measure) in order to ensure the airworthiness of the UAV.

Operator's Maintenance Manual: Part of the exposition that outlines how the organization will maintain the airworthiness of their aircraft through regular checks, scheduled maintenance, and procedures for managing unscheduled maintenance events.

Regulatory Authority: A government agency that is charged regulating the aviation industry by producing rules, issuing licenses and certificates, and auditing certificated aviation organizations for compliance.

Scheduled Maintenance: Regular maintenance activities to ensure that the UAV remains airworthy.

Senior Persons: People who hold managerial portfolios that involve oversight for the organization and its operations.

Unscheduled Maintenance: Maintenance activities triggered by an event that could not have been foreseen, such as propeller strikes, fires, and hard landings.

APPENDIX

Table 9. List of regulatory authorities used to form compliance considerations

Country/Entity	Regulator	Abbreviation (if any)
Australia	Civil Aviation Safety Authority	CASA
Canada	Transport Canada	-
Ethiopia	Ethiopian Civil Aviation Authority	ECAA
European Union	European Aviation Safety Agency	EASA
Italy	National Agency for Civil Aviation (Ente Nazionale per l'Aviazione Civile)	ENAC
Japan	Japan Civil Aviation Bureau (航空局)	JCAB
New Zealand	Civil Aviation Authority of New Zealand	CAANZ
South Africa	South African Civil Aviation Authority	SACAA
United Kingdom	Civil Aviation Authority	CAA
United Nations	International Civil Aviation Organisation	ICAO
United States	Federal Aviation Administration	FAA

Chapter 7
Regulations and Laws Pertaining to the use of Unmanned Aircraft Systems (UAS) by ICAO, USA, China, Japan, Australia, India, and Korea

Hwan Doo Kim
Korea Society of Air and Space Law and Policy, South Korea

ABSTRACT

The drone industry is rapidly developing around the world, and the numbers of drones are increasing. In order to maintain safety and secure stability of drone flights, regulations and laws related to drone operations are established in each country. This chapter reviews the rules and laws of drones established by the International Civil Aviation Organization, the United States, China, Japan, Australia, India, and Korea. In order to protect victims and develop the drone industry, the author proposes that it is necessary and desirable for the legislation of a unified and global "Draft Convention for the Unification of Certain Rules Relating to Drone Operations and Transport."

DOI: 10.4018/978-1-5225-7900-7.ch007

1. INTRODUCTION

Drone technologies continue to improve at a rapid pace and are slowly pushing unmanned aircraft (UAS/Drones/UAV) toward the mainstream. Companies in a variety of industries are now looking to use drones to cut costs, boost efficiencies, and create new revenue streams and business values, such as last-mile retail deliveries. An unmanned aerial vehicle (UAV), commonly known as a drone, is an aircraft without a human pilot aboard. UAVs are a component of an unmanned aircraft system (UAS); which include UAV's, and a system of communications between a ground-based controller and the Drone. The flight of UAVs may operate with various degrees of autonomy: either under remote control by a human operator or autonomously by onboard computers (ICAO, 2011).

Compared to manned aircraft, UAVs were originally used for missions too "dull, dirty or dangerous" (Tice, 1991) [1] for humans. While they originated mostly in military applications, their use is rapidly expanding to commercial, scientific, recreational, agricultural, and other applications, such as policing, peacekeeping (Franke, 2015) and surveillance, product delivery, aerial photography, agriculture, smuggling,[2] and drone racing. Civilian UAVs now vastly outnumber military UAVs, with estimates of over a million sold by 2015, so they can be seen as an early commercial application of autonomous things, to be followed by the autonomous car and home robots.

Figure 1. Many kinds of use (fire etc.) by drone

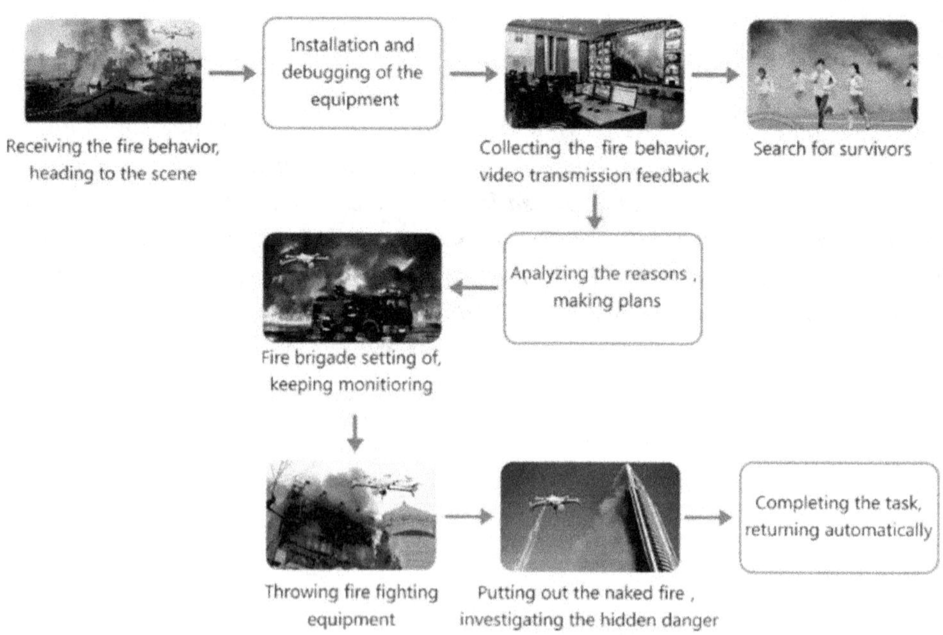

There's no arguing the drone industry growth that has occurred in the last few years, but discrepancies arise in trying to quantify that growth depending on how you define a drone, according to Recode.

Total drone unit sales climbed to 2.2 million worldwide in 2016, and revenue surged 36% to $4.5 billion, according to research firm Gartner. But the Consumer Technology Association points out that 2.4 million personal drones were sold in the U.S. alone in 2016, more than double the 1.1 million sold in 2015.

BI Intelligence, Business Insider's premium research service, defines drones as aerial vehicles that can fly autonomously or be piloted by a remote individual. Under that criterion, BI Intelligence expects sales of drones to surpass $12 billion in 2021. That's up by a compound annual growth rate (CAGR) of 7.6% from $8.5 billion in 2016. This growth will occur across the three main segments of the drone industry: Consumer Drones, Enterprise Drones (also known as Commercial Drones), and Government Drones. Consumer drones are drones purchased by individuals for noncommercial and nonprofessional purposes. BI Intelligence expects consumer drone shipments to hit 29 million in 2021, which would indicate a CAGR of 31.3%. For enterprise drones, BI Intelligence expects shipments to reach 805,000 in 2021 with a five-year CAGR of 51% from 102,600 in 2016 (Meola, 2017).

A recent Teal Group study estimates that worldwide spending in the Unmanned Aerial Vehicle (UAV) industry will increase drastically in the next decade—with expenditures of up to $11.5 billion annually. The drone industry is growing at a significant pace and is expected to contribute $82 billion to the U.S. economy within the next ten years.

Sales of drones for commercial operations are predicted to increase from 600,000 in 2016 to approximately 2.7 million by 2020. Drones have broad applicability for a wide range of industries, including energy and public utilities, agriculture, real estate, insurance, movie-making, photography, and videography. The slow progress of creating a comprehensive regulatory scheme for commercial uses of drones is delaying more extensive deployment of drones in the U.S. economy (Heffernan & Urban 2017). As the Drone industry has a bright prospect not only in the United States and the Republic of Korea but also in the whole world, it is essential to discuss the legislative examples of Drones in the International Civil Aviation Organization (ICAO), USA, China, Japan, Australia, India and South Korea.

2. REGULATION OF DRONE BY ICAO

2.1. ICAO and Unmanned Aircraft Systems (UAS)

The Unmanned Aircraft Systems Advisory Group (UAS-AG) of the ICAO (International Civil Aviation Organzation), established in 2015 to support the Secretariat in developing guidance material and expedite the development of provisions to be used by States to regulate unmanned aircraft systems (UAS), with its industry and international partners, as well as the Member States, has been instrumental in providing support to the global aviation safety collaboration.

The 39th Assembly, held from 27 September to 7 October 2016 requested that ICAO develop a global baseline of provisions and guidance material for the proper harmonization of regulations on UAS that remain outside of the international instrument flight rules (IFR) framework (ICAO, n.d.).

This request was based on three factors: the need to maintain safety for manned aircraft; the desire for harmonization of domestic UAS regulations; and the need for assurance from the leading global aviation standards-making body that the best options for UAS operations were being considered and recommended. In order to accomplish this new tasking, the UAS-AG began a second, larger phase of work (Phase II).

The UAS-AG continues to serve as a technical body, working under the management of the ICAO Secretariat. It reviewed and assessed submissions from States, industry and academia to ICAO's Requests for Information (RFIs). Recognizing that an agreed global approach will greatly assist businesses and others in launching their UAS services with suitable levels of investment confidence and operational safety, ICAO is now convening its "second DRONE ENABLE event" for 13-14 September 2018 in Chengdu, China (中国, 成都) (ICAO, 2018). Its focus will be on exploring new solutions with experts and innovators from industry, academia and other areas to help globally coordinate the development of UAS activities, and safely integrate UAS traffic management systems and existing conventional air traffic management systems. In preparation for the Chengdu event, ICAO has also issued a second request for information (RFI) to expand on the guidance material which was initiated after its "first DRONE ENABLE in 2017".

Based upon the assessments, ICAO invited various submitters to present their information to a global audience at DRONE ENABLE, "ICAO's Unmanned Aircraft Systems (UAS) Industry Symposium", was held in Montreal, Canada 22 to 23 September 2017 (ICAO, 2017). Interested parties can contribute to the 2018 ICAO RFI via the UN aviation agency's Unmanned Aviation website at www.icao.int/safety/ua.

The UAS-AG Phase II will support the Secretariat in guiding ICAO Member States with establishing a common global framework for, and core boundaries of, unmanned aircraft system traffic management (UTM), in order to allow further UTM developments to focus on better defined issues, whether technical, operational or legal.

Aimed at assisting consumers and remote pilots regardless of their skills and experience, the UN agency for civil aviation (ICAO) launched its new Unmanned Aircraft Systems (UAS) Toolkit on the occasion of International Civil Aviation Day. "The resources this new toolkit makes available are designed to help UAS operators of all ages operate their aircraft safely and responsibly," commented ICAO Council President Dr. Olumuyiwa Benard Aliu. "The importance of recognizing that these devices are aircraft, and of integrating their use safely with existing manned operations, should not be under-estimated."

Given that UAS, informally referred to as "drones," can be mistakenly and often illegally operated by less-informed pilots around airports and other areas of controlled or sensitive airspace, ICAO has been taking steps to help minimize their risks. Its new UAS Toolkit is much more than a starting point for learning the basics of their safe operation. UAS today can be flown in a variety of configurations, and outfitted with a wide range of payloads and sensors to support their increasing humanitarian and emergency operations roles.

They are also being used to monitor infrastructure and crops and for wide-ranging geological, geographical and climate-related research and development. "Given the immense economic and social potential of UAS technology, and the variety of factors at play, regulations for these systems and their use must be carefully considered," highlighted ICAO Secretary General Dr. Fang Liu. "Our new toolkit offers not only helpful information and resources, but can also serve as a platform for the exchange of global best practices, lessons learned, and effective governance approaches." Developed through ICAO's UAS Advisory Group, and in cooperation with industry and international expert partners, the ICAO UAS Toolkit can be accessed at icao.int/rpas.

Drone deliveries, drone inspections and even autonomous flying taxis are near term realities, and to make these services safe and efficient ICAO has begun the consultative work needed to establish low-altitude traffic management guidance for domestic unmanned aircraft systems (UAS). "ICAO is the natural agency to bring together the best and brightest from government and industry to define how these aircraft can be safely integrated into modern airspace, and in a way that optimizes their benefits globally for the wide range of public and private sector operators" (ICAO, n.d.).

"Multiple States and regional groups have activities underway to establish a UAS airspace management tool for lower altitudes, and ICAO's work through this RFI process will help to facilitate harmonized solutions which are safe, secure, sustainable, and most importantly globally aligned," noted ICAO Secretary General Dr. Fang Liu. "Our over-riding goal at ICAO is to better define the issues involved, whether technical, operational or legal, and also to ensure safety continues to remain our highest priority."

2.2. ICAO Regulatory Framework: Pilotless Aircraft

Article 8 of the *Convention on International Civil Aviation*, signed at Chicago on 7 December 1944 and amended by the ICAO Assembly (Doc 7300) (hereinafter referred to as "the Chicago Convention")

stipulates that: No aircraft capable of being flown without a pilot shall be flown without a pilot over the territory of a contracting State without special authorization by that State and in accordance with the terms of such authorization.... 2.1. The *Global Air Traffic Management Operational Concept* (Doc 9854) states "An unmanned aerial vehicle is a pilotless aircraft, in the sense of Article 8 of the Convention on International Civil Aviation, which is flown without a pilot-in-command on-board and is either remotely and fully controlled from another place (ground, another aircraft, space) or programmed and fully autonomous."

This understanding of UAVs was endorsed by the 35th Session of the ICAO Assembly. All UA, whether remotely-piloted, fully autonomous or a combination thereof, are subject to the provisions of Article 8. Only the remotely-piloted aircraft (RPA), however, will be able to integrate into the international civil aviation system in the foreseeable future. The functions and responsibilities of the remote pilot are essential to the safe and predictable operation of the aircraft as it interacts with other civil aircraft and the air traffic management (ATM) system. Fully autonomous aircraft operations are not being considered in this effort, nor are unmanned free balloons nor other types of aircraft which cannot be managed on a real-time basis during flight. Unmanned Aircraft Systems (UAS) will operate in accordance with ICAO Standards that exist for manned aircraft as well as any special and specific standards that address the operational, legal and safety differences between manned and unmanned aircraft operations. In order for UAS to integrate into non-segregated airspace and at non-segregated aerodromes, there shall be a pilot responsible for the UAS operation.

Pilots may utilize equipment such as an autopilot to assist in the performance of their duties; however, under no circumstances will the pilot responsibility be replaced by technologies in the foreseeable future.

UAS will operate in accordance with ICAO Standards that exist for manned aircraft as well as any special and specific standards that address the operational, legal and safety differences between manned and unmanned aircraft operations. In order for UAS to integrate into non-segregated airspace and at non-segregated

Figure 2.

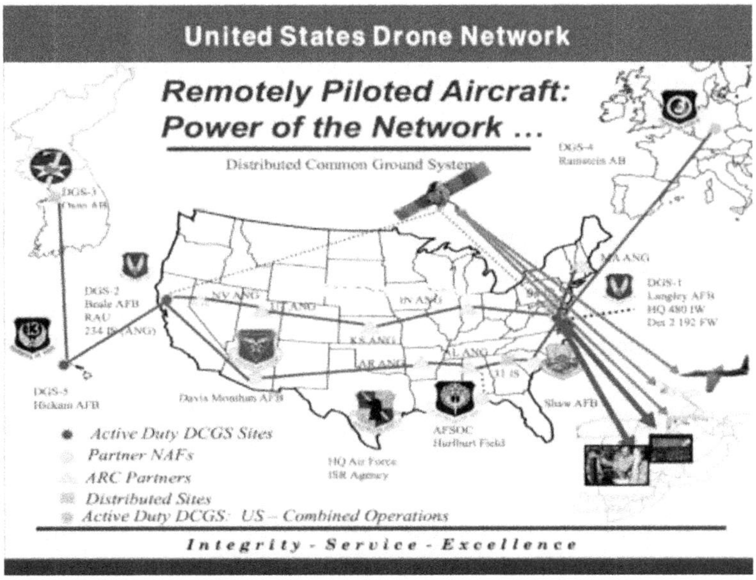

aerodromes, there shall be a pilot responsible for the UAS operation. Pilots may utilize equipment such as an autopilot to assist in the performance of their duties; however, under no circumstances will the pilot responsibility be replaced by technologies in the foreseeable future.

3. DRONE REGULATIONS AND LAWS IN THE UNITED STATES

3.1. Drone Legislation and Litigation

From legislation to litigation, 2017 was a major year for drones at the federal level, across all branches. Starting in the judiciary, a couple of major cases impacted drones in 2017. The first, Taylor v. FAA, challenged the Federal Aviation Administration's (FAA) registration requirement for all drones, including model aircraft, which Taylor believed violated language in the 2012 FAA reauthorization which prohibited FAA from regulating model aircraft. Subsequently, Congress included a provision in the 2017 National Defense Authorization Act, undoing the prohibition specifically as it relates to a registration requirement and FAA has subsequently reinstituted the requirement. Department of Transportation Secretary Elaine Chao announced in early January that FAA had registered its 1 millionth drone (NCSL, n.d.).

3.2. Definitions of UA, sUA and sUAS in the United States

- Unmanned Aircraft (UA) means an aircraft operated without the possibility of direct human intervention from within or on the aircraft.
- Small Unmanned Aircraft (sUA) means an unmanned aircraft weighing less than 55 pound (25kg) on takeoff, including everything that is onboard or attached to the aircraft.
- Small Unmanned Aircraft System (sUAS) means a small unmanned aircraft and its' associated elements (including communication links etc.) that are required for the safe and efficient operation of the small unmanned aircraft in the national airspace system.

3.3. FAA Regulation of Commercial Uses of Drones

There are many legal restrictions on commercial uses of drones, also known as unmanned aircraft systems ("UAS") or unmanned aerial vehicles ("UAV"), in the United States. Until August 2016, entities that wanted to participate in commercial drone operations had to petition the FAA for a so-called Section 333 exemption or an individual approval for specific operations. The FAA reviewed each Section 333 exemption petition on a case-by-case basis and could take months to issue an approval. On August 29, 2016, the FAA's Small UAS Rule, known as Part 107, went into effect and changed the regulatory scheme for commercial drone operations. Part 107 generally authorizes the commercial use of drones that weigh less than 55 pounds and do not fly faster than 100 miles per hour, but subject to significant conditions and limitations. Drones must:

- Remain within 400 feet of the ground (or higher, if the drone remains within 400 feet of a structure);
- Operate during daylight hours;
- Only operate when there is at least 3 miles of visibility from the control station;
- Not be flown over people;
- Operate in class g airspace absent special authorization;
- Remain within the visual line of sight of the operator at all times; and
- Not carry hazardous materials.

In order to pilot the drone, the operator must hold a Remote Pilot Airman Certificate and pass a Transportation Security Administration background check. In order to receive a Remote Pilot Airman Certificate, the applicant must pay a $150 fee, pass the initial aeronautical knowledge test (a two-hour, 60-question test) with a minimum score of 70 percent, and complete the FAA Airman Certificate and/or Rating Application. Foreign-registered drones may operate in compliance with Part 107, but the remote pilot must hold an FAA-issued Remote Pilot Airman Certificate (Herffernan & Urban, 2017). Under the Obama administration, the FAA was drafting new regulations to permit the operation of drones over people without the need for a waiver.

Federal Aviation Administration (FAA) released requirements for the registration of "hobby" drones:

- FAA requires registration for drones between 55-55 pounds
- Registration numbers linked to owners
- Registration requires name, home address, and email address
- Penalties for failure to register include civil fines of $27,500 and criminal fines of $250,000 and up to three years imprisonment
- Recent Chicago ordinance regulates where individuals can fly and use drones
- No drones within 5 miles of airports
- No drones over schools, churches, hospitals, police station
- No drones above 400 feet
- California drone bill addressed privacy concerns (vetoed by Governor)
- Recent Paradise Valley, AZ ordinance limits drone use for privacy reasons.

It remains uncertain whether such regulations will be issued under the Trump administration. The FAA convened a Drone Advisory Committee, including members from across the aviation and drone sectors, to advice on the best way to handle issues surrounding this new technology. While there have not been specific regulations implemented regarding cybersecurity or privacy issues, entities engaged in drone operations should exercise caution not to violate others' privacy rights and ensure that the drone has proper security mechanisms, such as a password-protected Wi-Fi signal.

3.4. State and Local Drone Laws and Federal Pre-Emption

Depending on the entity's location, legal advice may also be needed regarding compliance with state and local laws affecting drone operations. Whether the FAA has exclusive legal authority to regulate drones that would pre-empt State and local governments from adopting their own laws and regulations is an unresolved legal issue. The FAA, in Part 107, not only avoided asserting exclusive jurisdiction over drones, but actually encouraged state and local regulation of drones in specific respects, such as to address privacy issues arising from drone operations.

The pre-emption issue will likely be addressed through legislation and the courts. On May 25, 2017, a bipartisan group of senators introduced the "Drone Federalism Act of 2017" to "preserve State, local and tribal authorities and private property rights with respect to unmanned aircraft systems." This bill would require the FAA to define the scope of federal pre-emption, limiting it to "the extent necessary to ensure the safety and efficiency of the national airspace system for interstate commerce."[4] Drone industry stakeholders are concerned about the emergence of a patchwork quilt of potentially inconsistent state and local drone laws across the country.

3.5. FAA Regulation of Recreational Uses of Drones

The rapid expansion in the number of recreational drone users has raised significant safety concerns about intrusions into controlled airspace and interference with other aircraft as well as persons and structures on the ground. In December 2015, the FAA adopted a rule requiring small drone owners who conduct operations for recreational purposes to register their drones with the FAA.[3] On May 19, 2017, the United States Court of Appeals for the District of Columbia held that this rule "directly violates" Section 336(a)[4] of the FAA Modernization and Reform Act, which prohibited the FAA from "promulgat[ing] any rule or regulation regarding a model aircraft."[5]

The court found that drones constitute "model aircraft" and thus must not be made subject to FAA regulation.8 FAA argued that notwithstanding Section 336(a), it could impose the registration requirements pursuant to its general congressional mandate to "improve aviation safety," but the court found that Section 33(a) controlled, thereby barring the recreational drone registration rule.

The American Drone laws allow drones to ONLY be used for recreational use. They recently have begun requiring users to register their drones. "Do I Need To Register My Drone?"

When using your drone please adhere to the guidelines below.

- Maintain sufficient distance from populated areas.
- Do NOT fly your drone higher than 400 feet (120 m).
- Illegal to fly your drone in national parks, or within Washington D.C. regardless if you are flying for recreational or commercial purposes.
- There are different FAA drone regulations for commercial use and for recreational use.
- Recreational drone laws are in some ways more lax than commercial ones, but the line of sight remains pivotal (more on these laws later).

3.6. Issues for Companies to Consider Before Commencing Drone Use

Companies looking to use drones should consider whether to operate the drones themselves or hire an outside contractor do so on their behalf. If a company wants to conduct drone operations itself, it will need an employee to obtain a Remote Pilot Airman Certificate. The company also will need to purchase insurance coverage and ensure that it can conduct drone operations in a safe and compliant manner. By conducting drone operations itself, the company may reduce costs by eliminating the need to hire a contractor and avoid the delay involved in getting contractor operations established, but may increase its potential liability for its drone operations.

Alternatively, companies can contract in for drone services, which can be preferable or even necessary, particularly for more complex operations. The benefits of using an outside contractor include the ability to access a range of drone devices and the most up-to-date technology operated by a licensed, insured, experienced pilot, and to limit the company's liability. The U.S. Geological Survey National Unmanned Aircraft Systems (UAS) Project Office is leading the research and integration activities needed to make UAS data collection an efficient, safe, and cost-effective remote sensing tool for DOI and USGS scientists.

3.7. Aircraft Registration: Unmanned Aircraft (UA)

Registration is required for small Unmanned Aircraft weighing more than 0.55 pounds that do not operate exclusively under the Special Rule for Model Aircraft. Review the latest information regarding model aircraft registration in light of the May 19, 2017 decision of the U.S. Court of Appeals for the District of Columbia Circuit in *Taylor v. Huerta*. Use the streamlined online system to register small unmanned aircraft (sUAS) that weigh between 0.55 and 55 pounds.

Drone Pilot License.

- "Do I need a license to fly a drone?" It's one of the most common questions prospective drone owners ask.
- To act as a remote pilot for drones in accordance with FAA regulations, a person must obtain a remote pilot certificate.

3.8. Cause of Action for Drone in the Anglo-American Legal System

1. Basic Tort Theory Involving Duty of Care, Breach, Causation and Damages Caused by Drone

Potential negligent acts could include the following:

- Failing to maintain Drone (UAS) which may lead to mechanical failures
- Failing to account for weather conditions
- Losing control of or signal to UAS, leading to injury or damage
- Permitting someone without proper training to operate UAS
- Violation of FAA regulations (*per se*)
- Operating UASs under diminished capacity, *i.e.*,"drunk droning"
- Target Defendant: Commercial Operators/Owners, Airport

2. Causes of Action: Product Liability

The manufacturer of the Drones shall be liable for damages caused by defects in the manufacture of it;

- Design Defect:
- Conditions with the products design that makes the product inherently dangerous
    - Consumer expectations test
    - Risk utility test
- Manufacturing Defects:
- Mistakes made in the process of constructing a product
- Failure to Provide Adequate Warnings or Instructions:
    - Failures to provide adequate warnings or instructions regarding the product's proper use
    - Breach of warranty:
    - If condition constitutes a breach of warranty, can form the basis of tort liability even to a party not in privity
    - Target Defendant: Manufacturers; Product Sellers

4. CHINA

4.1. Introduction

The December 2015 “Interim Provisions on Light and Small Unmanned Aircraft Operations (轻小无人机运行试行规定)” issued by China’s civil flight regulatory agency, the Civil Aviation Administration of China (CAAC: 中国民航总局), regulate the operation of unmanned aircraft systems (UAS) with a maximum empty weight of 116 kilograms or less, or a maximum take-off gross weight of 150 kilograms or less, and a calibrated air speed of no greater than 100 kilometers per hour.

UAS weighing 1.5 kilograms or less are generally not required to follow the Provisions. The People’s Republic of China (中国) has not passed any legislation specifically regulating drones or unmanned aircraft systems (UAS).

Civil aviation and flight activities are primarily regulated by the PRC Civil Aviation Law, the PRC General Flight Rules, and the Regulations on General Aviation Flight Control. They have not, however, expressly extended their application to the flight of UAS. China’s civil flight regulatory agency, the CAAC (中国民航总局), has issued advisory circulars setting up guidelines for the flight of UAS. These interim measures are expected to be updated as the UAS industry and regulatory framework develop.[6]

The CAAC is considering new rules on commercial operations of UAS and issued a draft of the rules to solicit public opinion in December 2015 (Wei et al., 2016).

In addition, the Ministry of Industry and Information Technology (MIIT) is reportedly planning new UAS regulations.[7] The UAS Operation Provisions set forth an online, real-time supervision system comprising the “electric fence,” a system consisting of hardware and software that stops aircraft from entering certain areas, and the “UAS Cloud,” a dynamic database management system that monitors flight data, which has an alarm function for UAS connected to it that is activated when these UAS fly into the electric fence. Airport obstacle control surfaces, as well “prohibited areas, restricted areas, and danger zones” provided by other laws and regulations, are restricted areas prescribed by the UAS Operation Provisions.

UAS connected to the UAS Cloud must follow the restrictions shown in the system, while those not connected to the UAS Cloud must consult with relevant authorities about the restricted areas. UAS flying within visual line of sight (VLOS) must be operated in the daytime. Such a requirement does not apply to UAS flying beyond visual line of sight (BVLOS), but a certain regulatory framework for addressing emergencies applies to BVLOS flights. Both UAS flying within VLOS and BVLOS must give way to manned aircraft.

4.2. Drone Laws and Regulations in China

Drone use in China is allowed but getting a permit from the CAAC (中国民航总局) Flight Standards Division is probably very difficult. Nonetheless, here you are the conditions under which you can fly the little helper:

- Regulations in China divide drones into 7 classes according to weight, all of which require a permit
- Any drone weighing over 116kg requires a pilot license and UAV certification for operation
- Not permitted to fly drones near airports or where aircraft are operating
- Be careful when flying over people or built-up areas restrictions are much looser in rural or less populated areas,
- But much stricter in cities like Beijing or Shanghai refrain from flying your drone at Beihai Park, or near the Forbidden City and other major monuments (don't fly it within the first or second ring of the city).
- All drones under 7Kg are permitted to be flown in China.
- If your drone weighs 7Kg -116Kg a license from CAAC is required
- Any drone weighing over 116Kg requires a pilot's license and UAV certification for operation
- Drone flights in controlled areas require approval in advance.
- Approval from CAAC is needed for all commercial drone flights.
- Avoid flying near airports and flight paths and exercise caution when flying over built up areas or over people.

Hire UAV Pro has m any drone businesses and operators ready to assist in China listed on our website. You can also register your business at Hireuavpro.com to be amongst one of the most trafficked websites in the world for drone pilots and businesses. For any further questions, always refer to us by contacting hireuavpro.com directly.

4.3. CAAC (中国民航总局) Flight Standards Division (飞行标准司)

I would like to introduce the Flight Standard Division in China as the followings.

1. Civil Unmanned Aircraft Pilot System Provisional Regulations: 18 November 2013.
2. The Light and Small UAV Operation Regulation: 29 December 2015.
3. Civil UAV Driver Management Regulations, Issue Date: 11 May 2016

4. Civil Unmanned Aircraft Systems Air Traffic Management Office 21 September 2016
5. Civil unmanned aircraft real name registration management regulation: 16 May 2017

4.4. UAS Operation Provisions in China

The UAS Operation Provisions in China are also applicable to "plant protection UAS" used for agricultural, landscaping, or forest protection purposes with a maximum take-off gross weight of 5,700 kilograms or less and flying no higher than 15 meters above the surface, and unmanned airships with an inflatable volume of 4,600 cubic meters or less (Wei et al., 2016).[8]

1. UAS Categories in China

The UAS Operation Provisions in China divide UAS and unmanned airships subject to its regulation into seven categories, mainly based on weight and use, as follows:

- **Category I:** UAS weighing 1.5 kilograms or less.
- **Category II:** UAS with an empty weight between 1.5 kilograms and 4 kilograms or with a take-off gross weight between 1.5 kilograms and 7 kilograms.
- **Category III:** UAS with an empty weight between 4 kilograms and 15 kilograms or with a take-off gross weight between 7 kilograms and 25 kilograms.
- **Category IV:** UAS with an empty weight between 15 kilograms and 116 kilograms or with a takeoff gross weight between 25 kilograms and 150 kilograms.
- **Category V:** Plant protection UAS.
- **Category VI:** Unmanned airships.
- **Category VII:** Category I and II UAS that can operate 100 meters beyond visual line of sight.

Category I UAS are required to be operated safely and to avoid causing injury to others but are not otherwise subject to the UAS Operation Provisions. Nor do the Provisions apply to model aircraft and indoor flights, except under certain conditions specified by the Provisions.

2. Electric Fence and UAS Cloud

The UAS Operation Provisions set forth an online, real-time supervision system that has two components: the "electric fence" and the "UAS Cloud." The "electric fence" is a system consisting of hardware and software that stops aircraft from entering certain areas. The UAS Cloud is a dynamic database management system that monitors flight data, including operation information, location, altitude, and speed, in real time. The UAS Cloud has an alarm function for UAS connected to it that is activated when these UAS fly into the electronic fence.

UAS under categories III, IV, VI, and VII must install and use the electric fence and connect to the UAS Cloud. Operators must report at least every second when in densely populated areas and at least every thirty seconds when in non-densely populated areas. UAS under categories II and V are required to install and use the electric fence, connect to the UAS Cloud, and report at least every second if they are operated above the airspace of key areas and in airport clear zones (Wei et al., 2016).[9]

"Key areas" is defined by the Provisions to include military sites, nuclear plants, administrative centers and their neighboring areas, and areas temporarily designated as key areas by local governments. A qualified UAS Cloud provider must be approved by the CAAC for a trial operation, among other requirements specified by the UAS Operation Provisions. A UAS Cloud system developed by the Aircraft Owners and Pilots Association of China, "U-Cloud," has been approved for operation during a two-year period from March 4, 2016, to March 3, 2018.[10]

3. Restricted Areas

Airport obstacle control surfaces are restricted areas prescribed by the UAS Operation Provisions. "Prohibited areas, restricted areas, and danger zones" provided by other laws and regulations are also restricted areas under the Provisions.[11] UAS connected to the UAS Cloud must follow the restrictions shown in the system, while those not connected to the UAS Cloud must consult with relevant authorities about the restricted areas. In 2009, the CAAC issued rules on air traffic control for civil UAS, which subject civil UAS to the relevant provisions of the Civil Aviation Law, the Basic Rules of Flight, the Regulation on Flight Control of General Aviation, and other rules concerning air traffic control issued by the CAAC.[12]

4. Flight Specifications

According to the UAS Operation Provisions, UAS flying within visual line of sight (VLOS) must be operated in the daytime. Such a requirement does not apply to UAS flying beyond visual line of sight (BVLOS), but a certain regulatory framework for addressing emergencies applies to BVLOS flights. Both UAS flying within VLOS and BVLOS must give way to manned aircraft (Wei et al., 2016).[13]

5. Insurance

In compliance with the PRC Civil Aviation Law, the UAS Operation Provisions require UAS operators to buy insurance for UAS covering liability for third parties on the ground, a requirement deemed to be "consistent with best practices" (Wei et al., 2016).[14]

6. Pilots

The UAS Operation Provisions require a pilot-in-command to be appointed who is directly in charge of the operation of the UAS and has the right to make final decisions.[15] Qualification requirements for UAS pilots are prescribed by another CAAC advisory circular issued in 2013, the Interim Provisions on the Administration of Civil Unmanned Aircraft System Pilots.[16]

4.5. Drone Industry in China

According to the reliable sources, China is to be the largest exporter of military drones.

- The best-known Chinese military drones are the Wing Loong (翼龙) family, made by Aviation Industry Corp of China (中国航空工业集团公司), and China Aerospace Science and Technology Corporation (中国航天科技集团公司) (Wikipedia, n.d.).
- CH drones have been sold to military users in more than 10 countries, while the Wing Loong (翼龙) series, which made its maiden flight last year, captured the largest contract ever signed for a Chinese export drone.
- China's deployment of unmanned aerial vehicles is not limited to military purposes.
- In Liaoning Province (辽宁省), the local government is allegedly using the UAV to monitor the border between China and the North Korea and is building two coastal UAV bases to monitor the Provinces' jurisdiction in the Yellow Sea (黄海) and the Bohai Bay (渤海湾) area.
- China will launch the Wing Loong ID in 2018, the latest model of the domestically-developed Wing Loong UAS family, with a series for a new model also underway, its developer AVIC has announced. The Wing Loong ID is developed by Chengdu Aircraft Design & Research Institute (CADI), a subsidiary of the state-owned Aviation Industry Corporation of China

(AVIC) in southwest China's Sichuan Province (四川省). It will conduct it and maiden flight and enter the market in 2018, said Aviation Industry Corporation of China (AVIC:中国航空工业集团有限公司) at the Wing Loong UAS Development Conference (Xinhua, 2018).

5. JAPAN

5.1. Overview

Japan recently amended its Aviation Act and passed a new law to regulate the flights of unmanned aerial vehicles (UAVs). UAVs are prohibited from flying near airports and over densely populated areas and important facilities. Legislation to regulate drones was recently proposed in Japan after an April 2015 incident in which a small drone was discovered on the roof of the Japanese Prime Minister's office building in Tokyo.[17] At that time, there was no regulation in place to prohibit the flying of such devices. The drone operator was indicted and received a suspended sentence of two years' imprisonment for the criminal act of forcible obstruction of business (Murai, 2016). Following this incident, the ruling Liberal Democratic Party (LDP) submitted a bill to the Diet (Japan's Parliament) in June 2015 to regulate flights of unmanned aerial vehicles (UAVs) over certain areas.[18] The second bill, which amended the Aviation Act, was submitted by the Cabinet in July 2015.[19] The Diet passed both of these bills, whose provisions are discussed below.

5.2. Aviation Act and Drones in Japan

The Aviation Act amendment was the first of the two bills to be enacted. The amendment was promulgated on September 11, 2015 and became effective on December 10, 2015.[20] Under the amendment, a UAV operator is prohibited from flying a UAV, absent permission from the Ministry of Land, Infrastructure and Transportation (MLIT), in the following areas:

- Where air traffic is expected, such as airports and their approach areas, and areas above 150 meters
- Densely populated residential areas[21]
- The amendment also sets the conditions for UAV flights:
- UAV flights may be made only between dawn and dusk.
- An operator must monitor the UAV and its surroundings with his/her own eyes at all times.

- In-flight UAVs must keep more than 30 meters' distance from people and objects.
- UAVs must not fly over a place where an event attended by many people is being held.
- UAVs must not carry specified dangerous items, such as explosives and flammable objects.
- UAVs must not drop items while in flight.[22]

These conditions may not apply in the emergency situations or when an operator obtains prior approval from MLIT. A UAV that weighs 200 grams (7 ounces) or less is not subject to the rules in the Aviation Act. The MLIT has requested that UAV operators report accidents, collisions, UAV falls, and near-miss incidents. Eleven cases were reported between the enforcement date of the amendment to the Aviation Act and March 30, 2016.[23] Additional cases were reported to the police instead of the MLIT. [24] Flying a UAV over a prohibited area or violating the conditions of flight is punishable by a fine of up to 500,000 yen (approximately US$4,000).[25]

5.3. Aviation Act on Prohibition of Flying UAVs Over Important Facilities

The Aviation Act on Prohibition of Flying UAVs over Important Facilities and Their Peripheries was promulgated on March 18, 2016, [26]and becomes effective three months from the date of promulgation. [27] The Act prohibits flying UAVs over designated facilities, such as the Diet building, the Prime Minister's office building, buildings of designated government agencies that are involved in crisis management, the Supreme Court building, the Imperial Palace, embassies, and nuclear facilities. These no-fly areas generally extend to within a 300-meter radius of such designated facilities. When a police officer finds a person flying a UAV over a designated area, the police officer can order the person to stop the operation of the UAV. If the person does not follow the order, the police officer can take necessary measures, such as obstructing the flight of and destroying the UAV in order to remove any danger it poses. In addition, the person who did not follow the officer's order is punishable by up to one year of imprisonment or a fine of up to 500,000 yen.

5.4. The Contents of Drone Rules in Japan

- You must fly your drone at least 9 km away from airports
- You must not fly your drone over crowds.
- You must stay away from all power lines.
- In accordance with the road transport law

Figure 3.

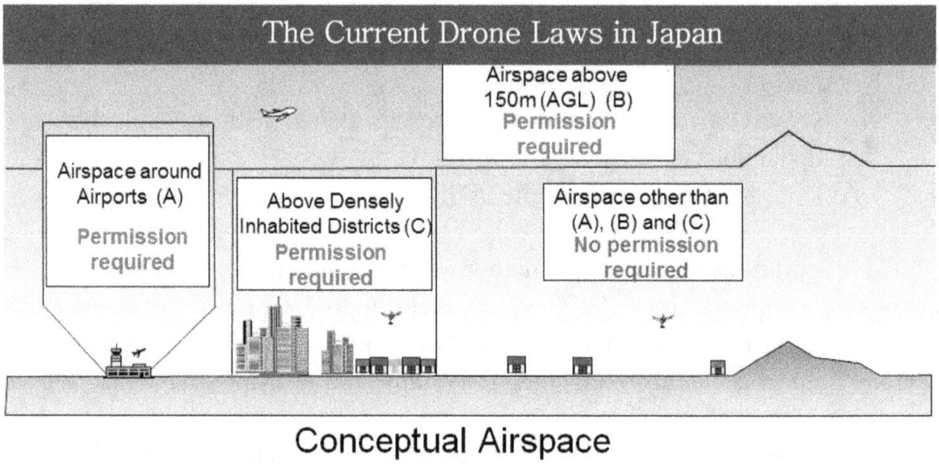

- You are not allowed to fly your drone over any roads
- In accordance with the land property law
- You are not allowed to fly over any property without permission from the owner
- Osaka and Tokyo have banned drone use in all parks within the city limits
- There are currently several proposed drone laws within the Japanese government but currently the only laws that apply are those listed above

5.5. Drone New Rules in Japan

The new rules came into force on Dec. 10, 2015 (TTILIT, n.d.). The details of the rules are as follows:

1. Definition

The term "UA/Drone" means any airplane, rotorcraft, glider or airship which cannot accommodate any person on board and can be remotely or automatically piloted (Excluding those lighter than 200g.

The weight of a UA/Drone includes that of its battery).

2. Prohibited Airspace for Flight

Any person who intends to operate a UA/Drone in the following airspaces is required to obtain permission from the Minister of Land, Infrastructure, Transport and Tourism.

1. Airspace around airports. (air space above approach surface, horizontal surface, transitional surface, extended approach surface, conical surface and outer horizontal surface.)
2. Above Densely Inhabited Districts (DID), which are defined and published by the Ministry of Internal Affairs and Communications.
3. Above 150 meters (492 ft.) within airways or 250 meters elsewhere and within 9 km of airports
4. (Civil Aeronautics Act No. 118 of 2006 and Ordinance for Enforcement of the Civil Aeronautics Act).
5. In all of Metropolitan Tokyo's 81 public parks and gardens.
6. (Tokyo Metropolitan Government ordinance) (violation may result in a fine of up to 50,000 yen: about $410Other prefectures that have passed similar legislation to Tokyo as follows.
7. At least 17 prefectural governments including Tokyo and five municipal governments already have introduced drone regulations.
8. It is illegal to fly drones in these following areas:
    a. In highly populated urban areas, which includes most Japanese cities and all of Tokyo's 23 wards for drones weighing above 200g (that is pretty much any drone with a camera).
    b. Over the Prime Minister's Office, or within 300m of the PM.
    c. Over the Imperial Palace, or within 300m of the Imperial Palace.
    d. Over other key facilities including nuclear power plants.
    e. Osaka and Tokyo have banned drone use in all parks within the city limits.
    f. A new law has been passed recently that allows law enforcement officers the right to destroy drones found in violation of the law so be careful!

3. Exception

Requirements stated in "Airspace in which Flights are Prohibited" and "Operational Limitations" are not applied to flights for search and rescue operations by public organizations in case of accidents and disasters.

4. Permission and Approval

You are required to submit an application in Japanese for the permission or approval to the Ministry of Land Infrastructure, Transport and Tourism at least 10 days (excluding Saturdays, Sundays, and holidays) before you fly a UA/Drone. For further information, please contact UA/Drone Counseling Service.

6. AUSTRALIA

6.1. Introduction

Australia has regulated unmanned aircraft since 2002. The relevant regulations are being substantially revised in 2016, including new rules related to using remotely piloted aircraft (RPA) for non-recreational purposes that come into force in September 2016. The new rules provide for commercial operations of very small RPA (weighing less than 2 kilograms/4.4 pounds) to be conducted without the need for a remote pilot license or operator's certificate, provided that these are operated under the standard conditions established in the new regulations.

Small RPA (2–25 kilograms/4.4–55 pounds) will also be able to operate over a person's own land for certain purposes and under the standard conditions without the need for certification and a license, while the use of medium RPA (25–150 kilograms/55–330.7 pounds) for the same purposes and under the standard conditions will only require a remote pilot's license. Operators of large RPA, as well as smaller RPA for other non-recreational purposes, will still be required to obtain a remote pilot license and operator's certificate. Large RPA must also have airworthiness certification.

To be eligible for a remote pilot license, operators must obtain certain qualifications, complete certain training, and have a minimum number of hours of experience flying RPA. Entities wishing to obtain an operator's certificate to operate large RPA or smaller (non-excluded) RPA for non-recreational purposes must have the facilities, procedures, and personnel needed to operate the RPA safely. There is currently some uncertainty with respect to the privacy rules applicable to the use of drone technology for collecting or recording information.

The Australian Civil Aviation Safety Authority (CASA) states that "Australia was the first country in the world to regulate remotely piloted aircraft, with the first operational regulation for unmanned aircraft in 2002" (Australian Government, n.d.). The current rules related to remotely piloted aircraft (RPA) and model aircraft are contained in Part 101 of the Civil Aviation Safety Regulations 1998 (Cth) (CASR) (Civil Aviation Act 1988, 1988).

6.2 Review of Regulations and Guidance Material Related to Unmanned Aircraft Systems (UAS)

The objective of this project is to "provide an up to date regulation and more comprehensive guidance to industry on the regulatory requirements and approval processes for the commercial operation of RPAS [remotely piloted aircraft systems] in Australia." On March 24, 2016, an amending regulation, the Civil Aviation Legislation Amendment (Part 101) Regulation 2016 (Cth) (2016 Regulation), was promulgated. The amendments will come into force on September 29, 2016. CASA states that the amendments "reduce the cost and legal requirements for lower-risk remotely piloted aircraft (RPA) operations. More complex operational matters will be dealt with in a new manual of standards to be developed with industry, providing greater flexibility and responsiveness in this rapidly evolving area" (CASA, 2016). This report outlines the rules contained in Part 101, as amended by the 2016 Regulation.

6.3 Civil Aviation Safety Regulations (CASR): Part 101

There are currently three subparts in Part 101 of the CASR that specifically regulate different sizes and uses of remotely piloted aircraft (RPA) and model aircraft: Subpart 101. C contains provisions that are applicable to unmanned aircraft generally; Subpart 101.F, as amended by the 2016 Regulation, applies to the operation of "large" RPA and of "very small," "small," and "medium" RPA for purposes other than sport or recreation; and Subpart 101.G relates to model aircraft, which are only used for recreational purposes.

Subpart 101.A also contains some preliminary provisions relevant to operating RPA, and Subpart 101.B includes a general prohibition on the unsafe operation of unmanned aircraft.

The 2016 Regulation will insert a new Subpart 101. AB that provides for the general authorization for people supporting the operation of model aircraft and RPA. As noted above, 2016 Regulation also provides for CASA to issue a "Manual of Standards" prescribing matters required or permitted by the regulations, or that are otherwise "necessary or convenient" to be prescribed in terms of giving effect to Part 101. Other changes include the following:

- Changing the terminology relating to unmanned aircraft from "UAV" (unmanned aerial vehicle) to RPA, to align with that used by the International Civil Aviation Organization.
- Creating new weight classifications for RPA, being "very small" (less than 2 kilograms/4.4 pounds), "small" (2–25 kilograms/4.4–55 pounds),

"medium" (25–150 kilograms/55–330.7 pounds), and "large" (more than 150 kilograms/330.7 pounds).[28]

- Introducing the concept of "excluded RPA," which relates to RPA operations considered to be of lower risk, as determined by RPA category and operational use, and that will consequently have reduced regulatory requirements.
- Establishing a set of standard RPA operating conditions, which must be complied with in order for certain operations to be considered excluded RPA operations.
- Establishing a new system that allows very small RPA to be operated for commercial purposes without the need for prior certification or licensing, provided certain conditions are met.

The amendments prohibit autonomous flights "until such time as suitable regulations can be developed by CASA." However, the explanatory statement accompanying the 2016 Regulation states that "there is scope for autonomous flight to be approved by CASA on a case-by-case basis in the meantime."[29]

6.4 Operating Large RPA and Operating Other RPA for Non-Recreational Purposes: Subpart 101.F

As noted above, Subpart 101.F (as amended) applies to the operation of very small, small, and medium RPA "other than for the purpose of sport or recreation," and to the operation of any large RPA. In many circumstances, the person actually operating such RPA must hold a "remote pilot license" and the relevant entity must hold a certificate authorizing the operation (currently referred to as a UAV operator's certificate or UOC) (Australian Government, n.d.). Under the 2016 Regulation, these requirements will not apply if the operation involves an "excluded RPA." In general, RPA must not be operated within 30 meters of a person "not directly associated with the operation" of the RPA. The 2016 Regulation creates new exceptions to this rule, stating that it does not apply if the person is standing behind the RPA while it is taking off, or to very small, small, and medium RPA where the person has consented and the RPA is operated no closer than 15 meters to him or her.[30] The amendments also provide for the Manual of Standards to prescribe requirements relating to the operation of RPA in certain areas.[31]

6.5 Flying Drones or Model Aircraft Recreationally

Our recreational drone safety rules are designed to protect other people in the air and on the ground. You must not fly your drone in a way that creates a hazard to another aircraft, person or property, so follow our rules every time you fly. These rules do not apply to all drone flyers. If you hold a remote pilot licence (RePL) and operate according to a remotely piloted aircraft operator certificate (ReOC) or have an authorization from the Australian CASA, you will be exempt.

1. The Rules

- You must not fly your drone higher than 120 meters (400 ft) above the ground.
- You must not fly your drone over or near an area affecting public safety or where emergency operations are underway (without prior approval). This could include situations such as a car crash, police operations, a fire and associated firefighting efforts, and search and rescue operations.
- You must not fly your drone within 30 meters of people, unless the other person is part of controlling or navigating the drone.
- You must fly only one drone at a time.
- If your drone weighs more than 100 grams:
- You must keep your drone at least 5.5km away from controlled aerodromes (usually those with a control tower)
    - You may fly within 5.5km of a non-controlled aerodrome or helicopter landing site (HLS) only if manned aircraft are not operating to or from the aerodrome. If you become aware of manned aircraft operating to or from the aerodrome/ HLS, you must manoeuver away from the aircraft and land as soon as safely possible. This includes:
    - Not operating your drone within the airfield boundary (*without approval)
    - Not operating your drone in the approach and departure paths of the aerodrome (*without approval)
- You must only fly during the day and keep your drone within visual line-of sight.
    - This means being able to orientate, navigate and see the aircraft with your own eyes at all times (rather than through a device; for example, through goggles or on a video screen).
- You must not fly over or above people. This could include festivals, sporting ovals, populated beaches, parks, busy roads and footpaths.
- You must not operate your drone in a way that creates a hazard to another aircraft, person, or proper
- You must not operate your drone in prohibited or restricted are

2. Getting Licensed and Certified to Fly a Drone

If you want to fly a drone commercially, you might need to be licensed and/or certified by us. If you intend to fly a drone commercially that weighs between 100g and 2kg, you can fly your drone in our ‘excluded’ category. Find out more about the excluded category and the standard operating conditions that apply. If you want to fly commercially outside of these standard conditions or your drone weighs more than 2kg, you will need to be licensed and/or certified to fly. A remote pilot license (RePL) is your individual permission to fly. If you hold a RePL, you will need to be employed by someone who holds a certificate to fly. These operators hold a remotely piloted aircraft (RPA) operators certificate, or ReOC (Australian Government, n.d.).

7. INDIA

7.1. India Drone Laws and Regulations

Drones in India have been banned since October 2014, regardless of their use. India is currently in treated, civil operation of drones will require approval from the Air navigation service provider, defense, ministry of home affairs, and other concerned security agencies within India (Australian Government, n.d.).

1. Drone Laws India 2018

Aviation ministry of India has not yet confirmed or updated any new rules for the year 2018. Since the year has just started and the elections are coming soon, the focus of Government is more on the serious matters. However in the past tweet by Ashok Gajapathi Raju Government is keen on new rules but there is no official announcement yet on this. Every drone lover is waiting for the updates but we think there is little more wait time (StartUp World, n.d.).

2. Drone Draft Regulations India 2017

2017 drone regulations draft, India is announced by The Ministry of Civil Aviation on 2017. During the press briefing, Aviation Minister P. Ashok Gajapathi Raju stated the drone industry would help the country growth in various sectors (Viroo, 2017). In April 2016, a draft policy for the operation of drones was released but it is yet to see the light of day. The policy framed quite a few guidelines for flying of drones keeping in mind the recreational and R&D scope of them apart from them being used for surveillance and commercials purposes.

3. How to Apply for a Drone License

The first step is to register your drone with a unique identification number or a UIN. For this, you will need to have an address proof, a permit from the police and the telecom department. Once you have submitted the docs, the UIN will be generated for you which then needs to be installed on your device before you fly it. As said, it can be a lengthy, official process before you can actually fly your drone. You need to apply for a permission from a civil or defense Air Navigation System (ANS) provider. If you are flying your drone over any property, you need to get permissions from the property/land owner. You also need to get a security clearance from the Bureau of Civil Aviation Security of India. Be it a recreation or commercial drone, you need all the permissions. The DGCA guidelines also state that the owners need to have insurance for their drones with the liability that they might have to deal with for any damage. You can go ahead and check out the application for permission for aerial photography to understand what all needs to be done to obtain permissions.

4. The Main Points Proposed in the Regulation

- There are 5 types of drones: nano, micro, mini, small and large.
- These range from less than 250 grams in weight to over 150 kg.
- Drones under 250 g (nano) won't need security clearance.
- Micro category (250 g to 2 kg) will get approvals in 2 days.
- Whenever you operate a drone, you will need different approval. Apart from nano drones, all other categories will need an air defense clearance so that aviation as well as security authorities are aware of the flight path.
- There will be no-drone zones such as above operational aerodromes and within 5 km of Vijay Chowk in Delhi, within 500 meters from strategic locations, from mobile platforms such as car, ship or air craft, over eco-sensitive zones like national parks and wildlife sanctuaries (unless approved by Environment Ministry).
- Drones less than 2 kg and operating under 200 feet of height, once registered, can be flown without nods.
- Drones can be used for photography, medical uses, ad film making and so on. E-commerce companies should be able to use drones as well.
- Air-rickshaws or passenger drones can also be considered under this policy

7.2. Use of Drones in India

Currently, people are using drones for aerial photography and for those gorgeous wedding shots. Drones are also used for getting aerial shots in movies and for making documentaries. But right now in India, a few government agencies, law enforcement agencies and the defense forces are legally allowed to use a drone. Apart from these, drones can also be quite useful for various other purposes. For examples, they can be used for remote sensing during natural calamities and disasters to pinpoint danger. Drones can also be used in agriculture; they can help to identify moisture content and nutrient soil availability; thereby passing on useful information which can be helpful to farmers.

1. Drone Permissions in India

It is not easy to get permissions for flying drones in India. There is a lengthy, official process which drone users have to flow. Also, importing drones into India is not allowed, unless you have official permissions on government letterheads. The Customs department has drones under "Prohibited Goods", so don't try to get them from outside India. Although there is no ban on the sale of drones in India, there are restrictions on their usage.

2. Bringing Drone to India

Bringing a drone to India is not that easy unlike earlier years. If you are lucky and clearance happens without check and the drone is with you otherwise customs will hold the drone, give you a copy of withholding and will give it back to you when you are flying back from India. But yes, there is a way to bring drone and bringing is not illegal though. It just needs approvals and the process of approval is lengthy. If you have anyone in custom or Airforce who can fasten the legal process for you then it will be easy to bring drone with you.

3. Top Drone Videos From India

There are many professional drone photographers and filmmakers in India who fly drones to make videos abiding by all rules and regulations. They follow all the guidelines of flying drones and make professional shoots for a hobby as well as business. You can check few of the samples of professional drone photographer from Pixel Do, Mumbai, India

4. Flying Drones in India

Hire UAV Pro has many drone businesses and operators ready to assist in India listed on our website. You can also register your business at Hireuavpro.com to be amongst one of the most trafficked websites in the world for drone pilots and businesses. For any further questions, always refer to us by contacting hireuavpro.com directly.

8. KOREA

8.1. New Korean Drone Laws

The South Korean government has unveiled a new set of safety standards for drone flying following the recent revision of a related law, the Ministry of Land, Infrastructure and Transport (MOLIT) said on November 14 (TUE), 2017, South Korea's Yonhap news agency reported. The amended aviation safety law came into force last week, allowing drones to fly at night or beyond the pilot's visual line of sight once safety requirements are met. Under the standards, drones should pass a set of safety tests and requirements to be flown at night or out of their pilot's direct line of sight. To prepare against emergencies such as loss of communication or malfunction, flyers should install a fail-safe device on their small unmanned aerial vehicles that enables them to fall safely.

Drones should also be equipped with an anti-collision system and a global positioning system transmitter that helps track their locations in case of crashes. In order to fly drones at night, flyers are required to buy insurance against possible accidents, place an "observer" and install an anti-collision light that can be visible up to 5 kilometers away. Drones should also be equipped with an infrared camera and other first-person view devices. In addition, the standards call for pilots and observers to receive education about emergencies and carry emergency manuals. The revised law also allows drones to deliver packages long distance when users meet safety requirements, as part of efforts to support the fast-growing drone industry (Bernama, 2017). Now the development of Drone enable ordinary people with little or no expertise to access the sky.

8.2. Applicable Laws

With the explosion of online shopping and home shopping in Korea traditional methods delivery products is being constantly challenged. Currently drones are regulated as an "ultra-light plane" under the Aviation Act. With effect as of 30 March 2017, the Aviation Act will be abolished and replaced with the Aviation Security Act (航空安全法), the Aviation Business Act (航空事業法) and the Airport Facility Act (空港施設法).

Relevantly, ultra-light planes will, in the main be regulated by the Aviation Security Act and the Aviation Business Act. Accordingly, we will briefly examine the provision on the use of drones for business purpose under such legislation. The following summarizes the flow of application that may need to be made, to operate an ultra-light plane business (Shin & Kim, 2017).

8.3. Drone Laws in the South Korea

Drone use is allowed in South Korea, but there are several drone laws that need to be followed when flying in the country. Operators must ensure that they follow the following drone laws when flying in South Korea. You cannot fly higher than 150 meters (492 feet). You cannot fly within 5.5km of airfields or in areas where aircraft are operating

You must fly during daylight hours and only fly in good weather conditions Avoid flying over people, or crowds and respect others privacy when flying your drone

- You cannot fly near Seoul Plaza, military installations, government facilities, power plants, or areas of facilities related to national security
- You cannot fly when there is low visibility or yellow dust
- Do not fly your drone beyond line of sight (UAV Systems, n.d.)

8.4. Registration, Reporting, Flight Approval, Operation License

1. Registration of an Ultra-Light Plane Business

A person intending to run an ultra-light plan business must apply for registration with the Minister of Land, Infrastructure and Transport (国土交通部长官: Article 48 of the Aviation Business Act). An ultra-light plane business means: 'the business of commercially performing tasks designated by the Minister of Land, Infrastructure and Transport for using unmanned aerial vehicles to meet other's demand (Article 2 of the Aviation Business Act).

- **Safety:** The operation of the business must not threaten public safety or national security;
- **Minimum Capital**: For corporate applications, their capital must be at least KRW 30 million. There is no minimum capital requirement where only with the unmanned aerial vehicles with the maximum of takeoff weight 25kgs or less are used; and
- **Insurance:** Personal injury insurance coverage of at least KRW 150 million and property insurance with coverage at least KRW 20million.

There are issues arising from a prudential perspective for a corporate applicant that has a large-scale operation using drones. Under Article of the Aviation Business Act, ultra-light plane business must be limited to 'commercial' business to meet the demand of 'others'. Using drones wholly for personal and using drone for one's own business would seems to fall outside the ambit of Article 2.

2. Reporting Requirement

A person who owns or has the right an ultra-light plane shall report to the Minister of Land, Infrastructure and Transport for the type, usage, name of the owner etc. of such plane and whether personal can be collected (Article 122 of the Aviation Security Act).

Safety certification must be obtained from the Korea Transportation Safety Authority (韓国交通安全公團) for a drone with the maximum takeoff weight of 25kg or more. No report needs to be filed for a drone weighing 12kg or less used as hobby, however, if such drone is used for a ultra-light plane, it must be reported even if it weighs 12kg or less.

3. Flight Approval

Flight approval is not required for a drone weighing 12kg or less. However, flight approval is required for drones intended to be flown in aerial zones, including controlled areas surrounding airfields, prohibited flight areas and at an altitude of 150meters or higher.

Such zones can be searched by the "Ready to Fly" application developed by the Ministry of Land, Infrastructure and Transport and the Korea Drone Association. The Ministry of Land, Infrastructure and Transport has developed a one-stop approval system where applications for flight approval can be obtained more easily. As of 1 January 2017, this one stop approval system can be used by apply for flight approvals from Ministry of National Defense (国防部) for aerial filming and for drone that fly military aerial zones, such as areas "near" an Airforce airfield (KBLA, n.d.).

8.5. Operation License

A person who obtains flight approval can fly relevant drone. According to the Enforcement Decree of the Aviation Security Act enacted on March 30, 2017, operator will be required for a drone that weighs 12kgs or more that is used for an ultra-light business.

Drones used for personal purposes, including leisure activities, regardless of weight and drone used for business purpose weighing less than 12kg may be flown without such operator certification.

8.6. Legal Issues on the Drone

While markets are currently developing drones for disaster monitoring, safety inspections etc., drones are already been used for various business sectors including air delivery. With the importance placed on the development in drones by the Korean government, it has already announced plans to pair back the restrictions on nighttime flight and rang of visibility.

- Dropping of falling Drones likely to endanger human lives or property;
- Flight in manner likely to endanger human lives or property in the sky above densely populated or crowded areas
- All drones should be brought on carryon luggage if possible. This is because according to the Montreal Convention, airlines are only liable for losses up to ~$1,000 USD.
- Flight under influence of alcohol or drugs.

One issue the requires further consideration is the requirement that a person wanting to run an ultra-light plane business is required only to take out coverage for KRW 150 million for personal insurance injury and KRW 20 million for property damage. This requirement seems low for a business that is operating thousands of drones. We believe that the insurance requirement should be based on a per drone basis, or there should be a laddered coverage based on number of drones used in the business. Another issue that is particularly relevant with the context of drones is privacy. The Korean privacy laws are one of the strictest under the Aviation Security Act around the world.

When a third person on the ground has suffered the personal injury or property damage caused by falling or collision of Drone (UAS) in the air, the perpetrator is liable for compensation of damages to the victims under Article 750 (Tort) of the Korean Civil Code and Article 709 (Tort) of the Japanese Civil Code. However, it is expected that Drone will be able to fly between the South Korea and North Korea, between Korea and China or Japan as well as the neighboring countries in the future.

9. CONCLUSION

The apparent abundance in new flight-stabilizing features, lighter and higher-quality, rugged cameras for airborne purposes have all contributed to the growing popularity of drones worldwide.

However, past issues remain unanswered as technology progresses and additional UAV manufacturers claim new customers and social Flight Clubs enlist fresh enthusiast amateur pilots. The air is still largely unregulated and unmarked, especially to the naked eye, unequipped with height measuring methods, without prior knowledge of any restrictions regarding the filming of surrounding people, and the seriousness of the threat a drone poses as it zooms past or above people. Enhanced availability of better GPS-trackers, quieter copters, and smaller 'footprint' also raise new legal issues and require current and up-to-date regulation, but the vast majority of the world still remains behind on effective drone control.

The dominating problems occur all over the world - regulation (or lack thereof), Privacy issues, and avoiding collision between civilian/military aircrafts and 'toys', and of course, the apparent threat of interference from signal jamming and virtual radio-walls set up that may interrupt civilian air-traffic. While markets are currently developing drones for disaster monitoring, safety inspection, etc., drones are already been used for various business sectors, including delivery. They remain in the test phase in Korea, but no doubt will be implemented in the market with time (IDC, n.d.).

However, the current regulations that require drone operation within the range of visibility and prohibiting night time flight, may prevent such uses as envisaged by the relevant industries in the near future. With the importance placed on the developments in drones by the Korean government, it has already announced plans to pair back the restrictions on night time flight and range of visibility. In order to transport package, products or merchandise etc. Drones will be able to fly over another countries in future. However, it is expected that Drone will be able to fly between the South Korea and North Korea, between Korea and China or Japan as well as the neighboring countries in the near future.

There is no the international legal guidance and regulations for regulating the jurisdiction, limited amount of compensation for the personal or property damage caused by the Drones accidents of third persons on the ground in another country when the persons were injured or dead as well as destroy, damage or loss of goods.

In order to protect victims and develop Drone industry, I would like to propose that it is indeed a great necessary and desirable for us to legislate an unified and global "Draft for the Convention for the Unification of Certain Rules Relating to Drone's Operate and Transport" including the establishment of standards for flying in the dense populated area and near at airport, preventing infringement of personal privacy due to the photo by drone, liability of damage for compensation caused by drone's accidents, liability for transport contract or tort of Drone's operator etc. by ICAO Legal Committee as soon as possible.

However, it is expected that Drone will be able to fly between the South Korea and North Korea, between Korea and China or Japan as well as the neighboring countries in the near future.

There is no the international legal guidance and regulations for regulating the jurisdiction, limited amount of compensation for the personal or property damage caused by the Drones accidents of third persons on the ground in another country when the persons were injured or dead as well as destroy, damage or loss of goods.

In order to protect victims and develop Drone industry, I would like to propose that it is indeed a great necessary and desirable for us to legislate an unified and global "Draft for the Convention for the Unification of Certain Rules Relating to Drone's Operate and Transport" including the establishment of standards for flying in the dense populated area and near at airport, preventing infringement of personal privacy due to the photo by drone, liability of damage for compensation caused by drone's accidents, liability for transport contract or tort of Drone's operator etc. by ICAO Legal Committee as soon as possible.

REFERENCES

Australian Government. (n.d.). Retrieved from: https://www.casa.gov.au/aircraft/standard-page/commercial-unmanned-flight-gaining-your-remotely-piloted-aircraft-pilot

Bernama. (2017). S. Korean gov't unveils new safety rules on drone flight. *New Straits Times*. Retrieved from: https://www.nst.com.my/world/2017/11/303032/skorean-govt-unveils-new-safety-rules-drone-flight

Böckstiegel. (2001). *Zeitschrift Für Luftund Weltraumrecht*. Köln: Institut für Luft- und Weltraumrecht der Köln Universität.

CASA. (2016). Red tape cut for remotely piloted aircraft. *Australian Government: Civil Aviation Safety Authority*. Retrieved from: https://www.casa.gov.au/publications-and-resources/media-release/red-tape-cut-remotely-piloted-aircraft

Cheng. (1998). *The Use of Airspace and Outer Space Cooperation and Competition*. The Hague: Kluwer Law International.

Cheng & Kim. (2000). *The Utilization of the World's Air Space and Free Outer Space in the 21st Century*. The Hague: Kluwer Law International.

Dempsey. (2008). Air Law. Institute of Air and Space Law, McGill University.

Dempsey & Milde. (2005). International Air Carriage Liability: The Montreal Convention of 1999. Institute of Air and Space Law.

Diederiks-Vershoor, I. H. P. (2001). An Introduction to Air Law. Kluwer Law and Taxation Publishers.

Franke. (2015, March 5). *Civilian Drones: Fixing an Image Problem?* ISN Blog.

Franke, U. E. (2015). Civilian drones: Fixing the problems. *THE CSS Blog Network. Center for Security Studies.* Retrieved from: https://isnblog.ethz.ch/security/civilian-drones-fixing-an-image-problem

Haanappel, P. P. C. (2003). *The Law and Policy of Air Space and Outer Space: A Comparative Approach.* The Hague: Kluwer Law International.

Heffernan, D., & Urban, J. (2017). US regulation of commercial uses of drones. *Expert Guides.* Retrieved from: https://www.expertguides.com/articles/us-regulation-of-commercial-uses-of-drones/ARVLNBXG

History of Cabinet Bill No. 75 of 189th Diet Session, HR, Japan.

History of House of Representatives (HR) Bill No. 24 of 189th Diet Session, HR, Japan.

ICAO. (2011). ICAO's circular 328 AN/190: Unmanned Aircraft Systems. *ICAO.* Retrieved from: https://www.icao.int/Meetings/UAS/Documents/Circular%20 328_en.pdf

ICAO. (2017). Drone enable, ICAO's unmanned aircraft systems (UAS) industry symposium. *ICAO.* Retrieved from: https://www.icao.int/Meetings/UAS2017/Pages/default.aspx

ICAO. (2018). Drone enable/2 symposium. *ICAO.* Retrieved from: https://www.icao.int/Meetings/DRONEENABLE2/Pages/default.aspx

ICAO. (n.d.). Unmanned aircraft systems advisory group (UAS-AG). *ICAO.* Retrieved from: https://www.icao.int/safety/UA/Pages/Unmanned-Aircraft-Systems-Advisory-Group-(UAS-AG).aspx

IDC. (n.d.). Legal issues surrounding the use of drones. *IDC Legal Clinic for Start-Ups.* Retrieved from: http://legalstart.idc.ac.il/legal-issues-surrounding-the-use-of-drones/

KBLA. (n.d.). New Korean drone laws. *KBLA.* Retrieved from: http://www.kbla.info/index.php/the-kbla-blog/1079-new-korean-drone-laws

Kim. (2000). *The Innovation of the Warsaw System and the IATA Inter-carrier Agreement. In The Utilization of the World's Air Space and Free Outer in the 21st Century*. Kluwer Law International.

Kim. (2008). Essays for the Study of the International Air and Space Law. Korean Studies Information Co. Ltd.

Lee, S. (1997). *Aviation Accident Law*. New York: II, Matthew Bender & Co., Inc.

Matte. (1981). *Treatise on Air-Aeronautical Law*. Montreal: Institute and Centre of Air and and Space Law, McGill University.

Meola, A. (2017). Drone market shows positive outlook with strong industry growth and trends. *Business Insider.* Retrieved from: https://www.businessinsider.com/drone-industry-analysis-market-trends-growth-forecasts-2017-7

Murai, S. (2016). Man who landed drone on roof of Japanese prime minister's office gets suspended sentence. *Japan Times*. Retrieved from: http://www.japantimes.co.jp/news/2016/02/16/national/crime-legal/man-landed-drone-roof-japanese-prime-ministers-office-gets-suspended-sentence

NCSL. (n.d.). Transportation. *National Conference of State Legislatures (NCSL)* Retrieved from: http://www.ncsl.org/research/transportation/2017-unmanned-aircraft-systems-uas-state-legislation- update.aspx

Operation and Certification of Small Unmanned Aircraft Systems, Final Rule, 81 Fed. Reg. 42,063 (June 28, 2016), codified at 14 C.F.R. Part 107.

Rod, D. (1989). *Aviation Insurance*. London: Butterworths.

Shawcross & Beaumont. (1986). *Air Law*. Montreal: De Daro Publishing.

Shin & Kim. (2017). New Korean drone laws. *Shin & Kim.* Retrieved from: https://www.shinkim.com/eng/media/newsletter/927

Start Up World. (n.d.). Flying a drone in India: rules & regulations – things to know. *Start Up World.* Retrieved from: https://www.startupworld.com/news/technology-news/flying-a-drone-in-india-rules-and-regulations-things-to-know

Tice, B. (1991). Unmanned Aerial Vehicles – The Force Multiplier of the 1990s. *Airpower Journal.*

Tice, B. P. (2009, July 24). Unmanned Aerial Vehicles – The Force Multiplier of the 1990s. *Airpower Journal.*

UAV Systems. (n.d.). South Korea drone laws. *UAV Systems.* Retrieved from: https://www.uavsystemsinternational.com/drone-laws-by-country/south-korea-drone-laws

Viroo. (2017). Buy drones in India! Drone legality in India (2018). DGCA, WPC, customs new rules & regulations. *YouTube.* Retrieved from: https://www.youtube.com/watch?v=UVhRt2m2Soc

Wei, J., Ellet, E. T., Ellman, L., Rizzi, P. R., & Zhou, L. (2016). China launches first operational rules for civil unmanned aircraft. *Hogan Lovells.* Retrieved from: http://www.hlmediacomms.com/2016/01/21/china-launches-first-operational-rules-for-civil-unmanned-aircraft

Wikipedia. (n.d.). China aerospace science and technology corporation. *Wikipedia.* Retrieved from: https://en.wikipedia.org/wiki/China_Aerospace_Science_and_Technology_Corporation

Xinhua. (2018). China to fly new member of Wing Loong UAS family in 2018. *China Daily.* Retrieved from: http://www.chinadaily.com.cn/a/201801/25/WS5a696d6ea3106e7dcc1368ed.html

ENDNOTES

1 When used, UAVs should generally perform missions characterized by the three Ds: dull, dirty, and dangerous.

2 “Drones smuggling porn, drugs to inmates around the world”. 17 April 2017.

3 Registration and Marking Requirements for Small Unmanned Aircraft, 80 Fed. Reg. 78,594 (Dec. 16, 2015) (drone owners must provide the FAA with their names, addresses, e-mail address, any additional information the FAA requests, and a $5 registration fee).

4 This section states that the FAA “may not promulgate any rule or regulation regarding a model aircraft.” Pub. L. No. 112-95, § 336(a), 126 Stat. 11, 77 (2012) (codified at 49 U.S.C. § 40101 note).

5 Taylor v. Huerta, No. 15-1495 at 3 (D.C. filed May 19, 2017).

6 轻小无人机运行规定(试行) [Interim Provisions on the Operation of Light and Small Unmanned Aircraft] (UAS Operation Provisions) (CAAC, Dec. 29, 2015), http://www.caac.gov.cn/XXGK/XXGK/GFXWJ/201601/P020160126526845399237.pdf, archived at https://perma.cc/KJW9-9TQ4; 民用无人驾驶航空器系统驾驶 员管理暂行规定 [Interim Provisions on the Administration of Unmanned Civil Aircraft System Pilots]

(CAAC, Nov. 18, 2013), http://govinfo.caac.gov.cn/000014170/201312/P020131206515715975483.pdf, archived at https://perma.cc/V4PT-9ER6.

7 工信部调研深圳无人机企业, 监管法案或年内出台 [Ministry of Industry and Information Technology (MIIT) Conducting Research on Unmanned Aircraft Enterprises in Shenzhen, Administrative Regulations May Come Out This Year], Sina (Feb. 29, 2016), http://tech.sina.com.cn/it/2016-02-26/docfxpvysx1694773.shtml, archived at https://perma.cc/5PK6-TRUH.

8 UAS Operation Provisions §§ 2.1–2.3.

9 UAS Operation Provisions § 14.1.

10 无人机监管系统" 优云 (U-Cloud)" 正式获批 [UAS Supervisory System U-Cloud Officially Approved], Xinhuanet (Mar. 7, 2016), http://news.xinhuanet.com/science/2016-03/07/c_135163844.htm, archived at https://perma.cc/REV5-MGJM.

11 UAS Operation Provisions § 10.

12 民用无人机空中交通管理办法 [Provisions on Managing Air Traffic of Unmanned Civil Aircraft] (June 26, 2009),http://www.caac.gov.cn/XXGK/XXGK/GFXWJ/201511/t20151102_7975.html, archived at https://perma.cc/UJ4F-N4L7.

13 UAS Operation Provisions §§ 11 & 12.

14 UAS Operation Provisions § 14.2.

15 UAS Operation Provisions § 4; Jun Wei et al., *supra* note 2.

16 民用无人驾驶航空器系统驾驶员管理暂行规定 [Interim Provisions on the Administration of Unmanned Civil Aircraft System Pilots] (CAAC, Nov. 18, 2013), http://govinfo.caac.gov.cn/000014170/201312/P020131206515715975483.pdf, archived at https://perma.cc/V4PT-9ER6.

17 UPDATE: Radiation Detected on Drone Found on Roof of Prime Minister's Office, Asahi Shimbun (Apr. 22, 2015) (on file with author).

18 History of House of Representatives (HR) Bill No. 24 of 189th Diet Session, HR, http://www.shugiin.go.jp/internet/itdb_gian.nsf/html/gian/keika/1DBDDC2.htm (last visited Apr. 5, 2016), *archived at* https://perma.cc/78GZ-JE5A. *See also Japan's Lower House Passes Law Restricting Use of Drones*, RT (July 9, 2015), https://www.rt.com/news/272731-japan-law-drones-ban, *archived at* https://perma.cc/58SH-73GX.

19 History of Cabinet Bill No. 75 of 189th Diet Session, HR, Japan http://www.shugiin.go.jp/internet/itdb_gian.nsf/html/gian/keika/1DBDE56.htm (last visited Apr. 5, 2016), *archived at* https://perma.cc/8J3B-U5C7.

20 Order to Set Enforcement Date of Partial Amendment of Aviation Act, Order No. 371 of 2015 (Oct. 30, 2015), Japan.

21 Aviation Act art. 132, *amended by* Act No. 67 of 2015; Aviation Act Enforcement Ordinance, Ministry of Transport Ordinance No. 56 of 1952, *amended by*

Ministry of Land, Infrastructure and Transport Ordinance No. 79 of 2015, arts. 236 & 236-2.

22 Aviation Act art. 132-2, *inserted by* Act No. 67 of 2015; Aviation Act Enforcement Ordinance art. 236-4.

23 平成27年度無人航空機に係る事故等の一覧(国土交通省に報告のあったもの)[Fiscal Year 2015, Chart of Accidents, etc. of UAVs (Cases Reported to MLIT)], MLIT, http://www.mlit.go.jp/common/001125882.pdf, *archived at* https://perma.cc/TT3C-KYWV

24 ニアミス落下... ドローンのトラブル続々報告 [Near Misses, Falls . . . Continual Reports of Drone Troubles], Yomiuri Online (Mar. 31, 2016),http://www.yomiuri.co.jp/national/20160331-OYT1T0086.html, archived at https://perma.cc/DJQ4-E3AQ.

25 Aviation Act art. 157-4, *inserted by* Act No. 67 of 2015.

26 History of HR Bill 24 of 189th Diet Session, HR, Japan. http://www.shugiin.go.jp/internet/itdb gian.nsf/html/gian/keika/1DBE1E2.htm (last visited Apr. 5, 2016), *archived at* https://perma.cc/LK5U-XYM7.

27 Act on Prohibition of Flying UAVs over Important Facilities and Their Peripheries, Act No. 9 of 2016.

28 *See id.* ; Civil Aviation Legislation Amendment (Part 101) Regulation 2016 – Explanatory Statement, https://www.legislation.gov.au/Details/F2016L00400/Explanatory%20Statement/Text, *archived at* https://perma.cc/GXC5-372F

29 Civil Aviation Legislation Amendment (Part 101) Regulation 2016 – Explanatory Statement, *supra* note 14.

30 2016 Regulation, sch 1, items 24 (inserting new r 101.245(2) & (3)).

31 *Id.* Sch. 1, item 25 (inserting new r 101.247).

Chapter 8
Automated System of Controlling Unmanned Aerial Vehicles Group Flight:
Application of Unmanned Aerial Vehicles Group

Tetiana Shmelova
https://orcid.org/0000-0002-9737-6906
National Aviation University, Ukraine

Dmitriy Bondarev
National Aviation University, Ukraine

ABSTRACT

In this chapter, the authors present a problem of a performance of unmanned aerial vehicles (UAVs) group flights for a solution of different tasks using criteria of efficiency (safety, regularity, efficiency, economy) and criteria of reliability (connectivity, structural redundancy, survivability, and compactness of connections; the relative distance between UAVs; centrality and periphery of UAVs in the group; the level of system centralizing; etc.). It used graph theory for quantitative estimation of effectiveness of UAVs group flight. It presented all types of UAVs connections in the group (a star, ring, tree, with a common tire, mixed, cellular, etc.). The algorithm for finding central drone repeater (CDR) in a group of the UAVs for sending a control signal to other UAVs in the group was obtained. Examples of for determining the central drone and of the optimal topology in a group of the UAVs in flight are presented.

DOI: 10.4018/978-1-5225-7900-7.ch008

BACKGROUND

In recent years we have observed that Unmanned Aerial Vehicles (UAVs) are very popular in all countries. Nowadays drones are used to perform many tasks that were previously difficult to solve. The ubiquitous and effective use of the UAVs is seen in both military and civil aviation, in particular in combating the effects of emergencies and natural disasters, agriculture, aerial photography, and communications retransmission among others.

For example, 100 controlled UAVs with Light-Emitting Diode (LED) elements installed on them simultaneously depict various colorful figures in the air while using multi-colored lights (Figure 1). They («Drone 100») was synchronously running and lighting in the night sky over the airport of the German city of Tornesch to the accompaniment of the orchestra performing the best classical music. Controlled on the ground by a crew using Personal Computers (PCs) with Intel software, the mass of drones lit up the night sky in sync to a live performance of Beethoven's Fifth Symphony and executed a stunning light show resembling a fireworks display. It was a new Guinness World Record for group flight of UAVs (Guinness World Records, 2016). The astonishing footage was shown for the first time during Intel CEO Brian Karzai's keynote on the opening day of CES technology trade show in Las Vegas, USA. A spectacular display of drone technology by Intel Corporation involving 100 small aircraft being launched skywards information has earned a new world record title for the Most Unmanned Aerial Vehicles (UAVs) airborne simultaneously (Guinness World Records, 2016). The record of the show was set in collaboration with ARS ELECTRONICA FUTURELAB to push the limits of the UAV industry and to showcase what UAVs can be used for.

The authors offer to use the opportunity to control a group of UAVs from a central UAV, which is a repeater of communication from an unmanned station. The central retransmission drone is determined using the method of organizing control in Local Computer Networks (Olefir, 2007). We are considering new methods to further enhance the performance of aviation chemical works through the integration of a group of manned and unmanned aircraft, such as airplanes, helicopters, and UAVs (Figure 2).

Remotely Piloted Aircraft Systems (RPAS) are a new component of the aviation system, one which International Civil Aviation Organization (ICAO) states that the aerospace industry is working to understand, define and ultimately integrate. These systems are based on cutting-edge developments in aerospace technologies, offering advancements which may open new and improved civil/commercial applications as well as improvements to the safety and efficiency of all civil aviation. It is obvious that the effectiveness of UAV group flights in many tasks such as monitoring forest fires, search and rescue operations, agricultural application in the processing of

Figure 1. The group flight of UAV on the show (Guinness World Records, 2016)

Figure 2. Performance of aviation chemical works by a group of manned and unmanned aircraft

crops, retransmission of communications and movement of goods is much higher than with single UAV flights.

Unmanned aircraft have several advantages, namely low operating cost, good concealment and flexibility, simplicity and availability of technology compared to manned aircraft and UAV can be used in cases where the usage of manned aircraft is deemed to be impractical, expensive and/or risky (Austin, 2010; Gulevich, 2012). The main advantage of the use of UAVs is the fulfillment of tasks related to the risk for a person and efficiency in solving national economic problems. In this sense, the use of a group of UAVs (UAV group) is even more appropriate: for retransmission of communications in those places where it is impossible to install antenna coverage through complex relief, in agriculture (e.g., group spraying fields), in aerial photography (e.g., group survey of large areas, monitoring of forest fires, and patrolling of territories) and the movement of cargo. Understandably, there is the use of a UAV group for military purposes. Therefore, detailed analysis, classification, purpose, design, and evolution of RPAS, both in the military and civilian segments, require on-going research. According to Niciforov (2010), the international classification of UAVs has been given as well as conducting the analysis of the UAVs of foreign production in particular when applied in the forest sector.

An American aviation expert and a retired colonel, John Warden, predicts that by the year 2025 about 90% of the planes will be unmanned, and only 10% will be piloted, and pilots will be a "golden reserve" for the most important and difficult tasks (Amelin, 2013; Chekunov, 2010; Ganin, 1999; Ignatiev, 2010). A similar situation is observed in connection with the development of UAVs for civilian use. This is due to a number of important benefits, first of all, the absence of a crew on board an aircraft, and thus eliminating the risk of death. The ability to perform maneuvers with large overloads that exceed the physical capabilities of the pilot and the ability to operate a large length and range in the absence of a crew tiredness factor is another important benefits. And, finally, the relatively small cost of manufacturing, operating and maintaining UAVs is another benefit that has attracted the development of UAVs for civilian use (Kreps & Zenko, 2014). The majority of research on the group's actions of the UAVs are guided by the main monograph (Fradkov, 2013) which uses the approaches of the classical theory of management to consider the management of a group. This is organized in the form of "order" which involves a separate representation of the movement of a particular object of the group mathematical model for the lateral and longitudinal movement of the center of mass. In this case, there is a task of developing some "optimal" route for the UAV group. A number of research papers are devoted to the decision of the task of planning a route when one UAV and/or the entire group of UAVs are operated (Kucherov & Kozub, 2015, 2016, 2017; Marsh, 2005). The additional useful properties of the UAV group as compared to the use of one UAV are noted.

According to Montgomery,(1998), the operation of an inhomogeneous group consisting of UAV helicopter and aircraft types requires a complex nature of information exchange through the use of flight control channels, operator channel, and the interaction of autonomous elements of the system with each other. At the same time, the complexity of tasks in the case of the management of a UAV group is noted, which consists in the inappropriateness of the application of the classical theory of control (Marsh, Calbert, Gossink & Kwok, 2005).

Therefore, it is proposed to use the theory of graphs to simulate an increase in the effectiveness of group flights of unmanned aerial vehicles. The theory of graphs in information technology (e.g., connecting computers to a network), using various types of topologies is widely used for object modeling. To,do this, as elaborated above, the tasks that need to be completed are: A comparative analysis of the effectiveness of the use of group flights of the UAV for the intended purpose; Categorization and coding of UAVs according to existing classifications; and, Modelling the UAV group flight for aerial photography using graph theory and determining the topology of a connected UAV system.

APPLICATION OF UNMANNED AERIAL VEHICLES GROUP FLIGHT

An aircraft is defined as any machine that can derive support in the atmosphere from the reactions of the air other than the reactions of the air against the earth's surface. An aircraft which is intended to be operated with no pilot on board is classified as unmanned. An unmanned aircraft which is piloted from a remote pilot station is a Remotely Piloted Aircraft (RPA). The remote pilot station (RPS) is defined as the component of the remotely piloted aircraft system containing the equipment used to pilot the remotely piloted aircraft (ICAO, 2005; 2015). RPAS can go faster and generally further as long as the vehicle allows for it (no human fatigue in the plane). They can also fly longer hours and even if the plane crashes, the pilot will be safe.

The main advantage of using UAVs is when completing certain tasks that involve risk to humans and efficiency in solving economic problems. In addition, they can be used for border patrol security using the software to fly the planes. The goal of ICAO in addressing unmanned aviation is to provide the fundamental international regulatory framework through Standards and Recommended Practices (SARPs), with supporting Procedures for Air Navigation Services and guidance material. RPAS offers a less stressful environment; it is used for better Decision Making (DM). It presents a safer environment.

As a general principle, the RPS functions in the same manner as the cockpit/ flight deck of a manned aircraft and should, therefore, offer the remote pilot an equivalent capability to command/manage the flight. RPA are piloted from RPS utilizing a command and control (C2) link. Together with other components such as launch and recovery equipment, if utilized, the RPA, RPS and C2 link comprises an RPAS. An RPA can be piloted from one of many RPS during a flight; however, only one RPAS should be in control of the RPA at a given moment in time (ICAO, 2015). Group of RPA may be piloted from a group of many RPS during a flight or an organization of RPA group as informational net (Figure 3).

In the long run, the use of RPAS will be cheaper than paying for the personnel to do the task. They are able to fly into the zones where it would be dangerous for pilots in manned aircraft, RPAS are also capable of flying for long periods of time, can enter the environments which are dangerous to the human life and can almost, if not totally, eliminate the exposure risk of the aircraft operator.

To control a group of UAVs, the authors suggest choosing and using a Drone - Repeater to connection to the operator on the ground and control the remaining UAVs (Figure 4). To select the effective Drone – Repeater and optimal configuration network of UAV's group, Olefir (2006) offered to use the same method of server selection in computer networks.

Figure 3. Control systems by one or a group of the UAV

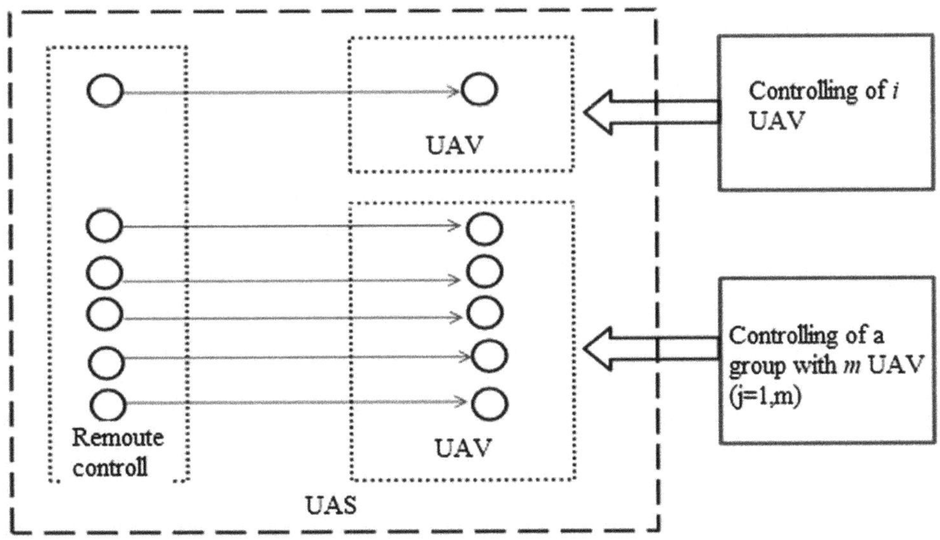

Figure 4. Control system of UAV's group from RPAS using Drone - repeater

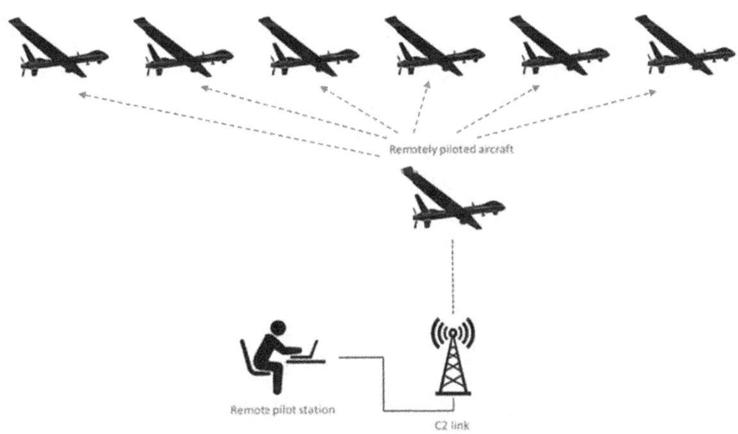

Although there are several advantages of using UAVs as previously mentioned above there are a few disadvantages. These include the limited capacity due to the small sized UAVs that can be compensated for the use of group flight. In addition, the group of small and light UAVs that effectively interact with each other appear to be more useful qualities such as speed task and increasing the density of surface area when spraying crops. Furthermore, the use of the group of UAVs tends to improve the overall performance in the completion of tasks such as photo/video monitoring, agricultural work, retransmission of communications, and cargo movement as well as in areas where a group of UAVs quickly and efficiently perform the task.

Obviously, the effectiveness of group flying UAVs in monitoring forest fires, search and rescue operations, agricultural crops processing, relay communications and the movement of goods is much higher than in single UAV flights. In this sense, the use of multiple UAVs with a UAV being a repeater is more appropriate so that proper communications are established and relayed to those places where the antenna coverage cannot be established because of certain difficulties such as challenging terrain, agriculture (group of spraying fields), aerial photography (group survey large areas, monitoring of forest fires, patrol areas, etc.) and moving cargos.

A group of UAVs performs different tasks. In order to manage a group of UAVs, it is expedient to apply graph theory as mathematical tools for modeling. Imagine there was a group from n – UAVs. It is the proposal to represent the configuration of UAVs group as a graph with n tops (UAVs). They are connected by m arcs between each other to provide information data channels for UAVs' control. It is necessary to calculate the efficiency of UAV structure using methods of graph theory. Considering a potential problem during a UAVs' group flight conducting aerial photography is one example. It is proposed to build a scheme with a group of UAVs and the objectives of which are to maximize the overlap areas for map information. Therefore, the main problem is the lack of optimization of UAVs group operations that depend on the task purpose.

As previously mentioned, the advantages of UAVs are to perform the tasks associated with the risk for man and effectiveness in solving economic problems. In this sense and to reiterate this point, the use of UAVs is more appropriate when performing certain tasks such as providing relay communications in certain places where the antenna coverage cannot be established because of difficult, and often challenging, terrain, agriculture (e.g., group of spraying fields), photo/video monitoring (e.g., group survey of large areas, monitoring of forest fires, patrol areas) and transportation of cargos. Obviously, the use of UAVs has not been limited to civil purposes. According to the European Organization for the Safety of Air Navigation [Eurocontrol] (2004; 2015), UAVs had been used for military purposes since 1961. While comparing usage of UAVs' group with one UAV we have more useful properties: faster coverage of area fragment and consequently more effective at photo/video monitoring, relay communications, and agricultural operations.

However, despite a number of advantages, there are some drawbacks. These include the main problem associated with the usage of airspace allocation of the frequency range for the management of UAVs and transmission of information from the airborne units to the ground. Also, a lack of optimization of the structure of the UAV is another example of such drawback. Therefore, the foregoing has been optimized by graph theory. In the investigated ununiformed group consisting of UAV helicopter and aircraft types, it is a complex information exchange through the use of flight control channels, carrier channel together with interaction autonomous system elements. It is noted that the complexity of problems in the case of management of group UAVs is the inappropriate use of classical control theory. To manage a group of UAVs, it is expedient to apply graph theory as mathematical tools for modelling and optimization of complex structure involving and combining knowledge from various disciplines such as: mathematics to solve complex equations; physics for constructing electronic circuits; construction for the most rational distribution facilities and the construction of roads; biology to solve problems of genetics; economics for finding solutions minimal cost; and, information technology to determining the effectiveness topology of local and global computer networks and other applications.

The Topology of an Aircraft Group

For a UAVs' group flight it is advisable to apply graph theory. The group structure may have a different configuration, location and connections between network nodes but the most common structure is: fully connected, a star, ring, and tree with a common tire, mixed, and cellular. From the UAVs' topology, it is performed by the flight of an aircraft group that depends on the effectiveness of the task purpose. For the management of the UAVs, a system for managing one or a group of UAVs is proposed, depending on the purpose of UAVs and their assigned tasks.

Taking into account the limited and dependence of the use of UAV group on its intended purpose, it is tasked to analyze the network topology indicators for the implementation of the group flight. The analysis of existing topologies of information networks and their capabilities (Table 1) shows the benefits of a hybrid over classical variants such as "tire", "circle" and "star" (Kuzin, 2011; Olefir, 2006).

For example, a star-circle topology of a group of UAVs is shown in Figure 5. UAV flights use the general criteria in Table 1. In addition, when performing group flights of the UAVs it is expedient to apply specific criteria for the reliability of the group structure: connectivity, structural redundancy, uneven distribution of connections, structural compactness, a degree of centralization in the system, and survivability.

The algorithm of finding central drone in a group of UAVs in flight for relaying control signals from other UAVs was obtained and presented below.

The Algorithm of finding Central Drone Repeater (CDR) in a group of the UAVs in flight

1. Build graph *G (n, m)* for UAVs group
2. Build an adjacency matrix $A = \left\| a_{ii} \right\|$ of graph *G (n; m) of* UAVs group:

$$a_{ij} = \begin{cases} 1, i \leftrightarrow i \\ 0, i \not\subset i \end{cases}$$

Table 1. Analysis of network topologies

Network	Deployment	Reliability	Delivery of packages	Access scheme
Tire 	Simple deployment, (up to 10 nodes)	Low-reliability	Simultaneous delivery	Competitive access
Circle 	Simple deployment	Low-reliability	Simultaneous delivery	Mark up access
Star 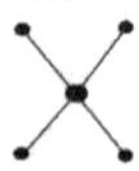	The need for a hub when deployed	Critical element, Hub	Delivery delayed	Address Access
Star 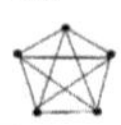	Need for additional equipment	High-reliability	Address delivery	Address Access
Tire-star 	Need for additional equipment	High-reliability	Need to distribute traffic	Tire Competitiveness, Star Targeting
Star-circle 	Need for additional equipment	Dependence on the hub	Need to distribute traffic	Equal access at the expense of markers
Hybrid-pantry 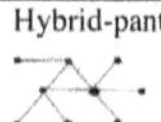	Need for additional equipment	Below just a pantry	Need to distribute traffic	Address Access

Figure 5. Star-circle topology of a group of the UAVs

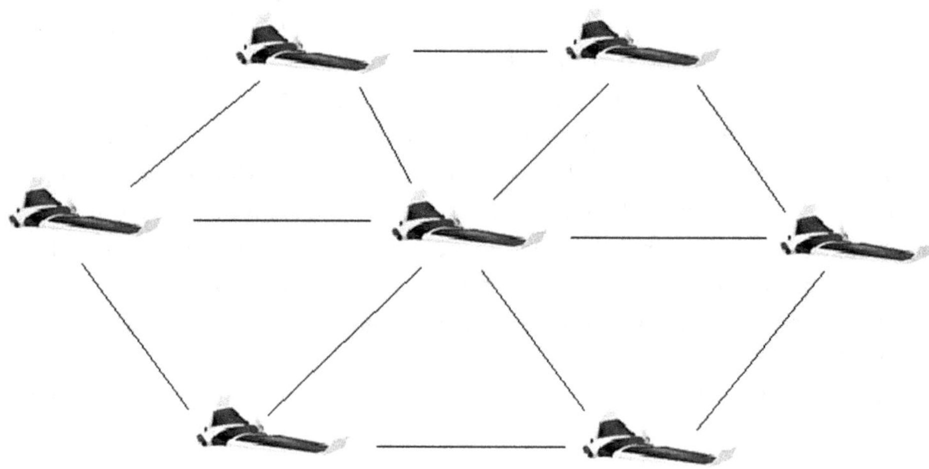

Table 2. The adjacency matrix $A = \|a_{ii}\|$ *for UAVs group*

Tops of the graph $i = \overline{1, n}$	**Tops of the graph** *G (n; m),*					
G (n; m),	U_1	U_2	U_i	....	U_n	$\sum_i U_i$
U_1	1 (0)	1 (0)	1 (0)	1 (0)	1 (0)	
U_2	1 (0)	1 (0)	1 (0)	1 (0)	1 (0)	
U_i	1 (0)	1 (0)	1 (0)	1 (0)	1 (0)	
...	1 (0)	1 (0)	1 (0)	1 (0)	1 (0)	
U_n	1 (0)	1 (0)	1 (0)	1 (0)	1 (0)	
$\sum_i \sum_l a_{ij}$						

Table 3. The matrix of introspection $\rho = \|\rho_{ij}\|$ *of graph G (n; m) for UAVs group*

Tops $i = \overline{1, n}$	**Ribs of the graph** *G(n; m),* $j = \overline{1, m}$						$\sum \rho_{ij}$	$\sum \rho^2$
U_i	1	2	...	*j*	...	*m*		
U_1	1(0)	1(0)	1(0)	1(0)	1(0)	1(0)		
U_2	1(0)	1(0)	1(0)	1(0)	1(0)	1(0)		
...	1(0)	1(0)	1(0)	1(0)	1(0)	1(0)		
U_i	1(0)	1(0)	1(0)	1(0)	1(0)	1(0)		
...	1(0)	1(0)	1(0)	1(0)	1(0)	1(0)		
U_n	1(0)	1(0)	1(0)	1(0)	1(0)	1(0)		
$\sum$								

3. Build the matrix of introspection $\rho = \|\rho_{ij}\|$ of UAVs group graph *G (n; m):*

$$a_{ij} = \begin{cases} 1, i \leftrightarrow j \\ 0, i \not\subset j \end{cases} \text{(Table 3)}$$

Table 4. The matrix of distance matrix $D = \left\| d_{ij} \right\|$ of graph G (n; m) for UAVs group

		Tops of graph *G(n; m)*, $i = \overline{1, n}$						$\sum d_{ij}$
		1	2	…	*i*	…	*n*	
Tops of graph *G(n; m)*, $i = \overline{1, n}$	1	1	d_{ij}	d_{ij}	d_{ij}	d_{ij}	d_{ij}	
	2	d_{ij}	1	d_{ij}	d_{ij}	d_{ij}	d_{ij}	
	…	d_{ij}	d_{ij}	1	d_{ij}	d_{ij}	d_{ij}	
	i	d_{ij}	d_{ij}	d_{ij}	1	d_{ij}	d_{ij}	
	…	d_{ij}	d_{ij}	d_{ij}	d_{ij}	1	d_{ij}	
	n	d_{ij}	d_{ij}	d_{ij}	d_{ij}	d_{ij}	1	
$\sum\sum d_{ij}$								

4. Build distance matrix $D = \left\| d_{ij} \right\|$ of UAVs group graph *G (n; m)* as the minimum distance d_{ij} between the nodes is *i* and *j* (Table 4)

5. Determine the connectivity of UAVs group graph *G (n; m)*

$$L = \frac{1}{2}\sum\sum a_{ij} \geq n - 1, i \neq j$$

$$L_{\min} = n - 1$$

6. Determine the structural redundancy of R of UAVs group graph *G (n; m)*

$$R = \frac{1}{2}\left[\sum\sum a_{ij}\right]\frac{1}{n-1} - 1$$

7. Uneven distribution of relationships of UAVs group graph *G (n; m)*

$$\varepsilon^2 = \sum_{i=1}^{n}\left(\rho_i - \bar{\rho}\right)^2 = \sum \rho_i^2 - 4\frac{m^2}{n},$$

where $\rho_i = \sum_{i=1}^{n} \rho_{ij}$

8. 8. Structural compactness D of UAVs group graph $G(n; m)$ shows the proximity of the parameters (D_i) to each other through the minimum chain length d_{ij} - the smallest distance between i and j. The common proximity of the elements in the network. Structural compactness is characterized by the indicator - the diameter of the structure: $d = max\ d_{ij}$. The values of D_{rel} and D integrally characterize the inertia of the processes in the system, with equal values of R and ε^2 their increase reflects the growth of the number of bonds that disconnect. This situation helps to reduce the reliability of the system as a whole.

$$D = \sum_i \sum_j d_{ij}$$

The relative distance:

$$D_{rel} = \frac{D}{D_{min}} - 1,$$

where $D_{min} = n(n-1)$.

9. The degree of centralization in the system δ is determined using the centrality index:

$$\delta = (n-1)(2Z_{max} - n)\frac{1}{Z_{max}(n-2)},$$

where Z_{max} maximum value of the indicator Z:

$$Z_i = \frac{D}{2}\left(\sum d_{ij}\right)^{-1},\ i = 1, n,\ i \neq j$$

10. Determine the centrality of the nodes C_i of the graph G $(n; m)$. The CDR in a group of the UAVs in flight has $C_i = C_{max}$ and the relative periphery of the node:

$$C_i = \frac{\sum \sum d_{ij}}{\sum d_{ij}}$$

11. Determine the periphery P_i of the graph *G (n; m)*. The CDR in a group of the UAVs in flight has Pi = 0. The relative periphery of the node:

$$P_i = C_{max} - C_i$$

12. Determine the survivability of the network, S, CRD in a group of the UAVs in flight has S > 0:

$$S = \frac{\sum a_i - 2(n-1)}{2(n-1)}$$

13. Determine the moment of the network, *M* - characterizes the controllability of the CRD:

$$M = \frac{\sum_{i=1}^{n} a_i}{a_l^2} \sum_{i=1}^{n} (a_l - a_i)$$

14. Data table of parameters CRD in a group of UAVs.

In order to quantify the reliability of the UAV group flight performance, it is necessary to submit a group flight in the form of a graph for the above-mentioned criteria. Let's consider the flight of a group of five UAVs that perform an aerial search of the fragment of the terrain (Figure 6) with the UAVs group presented in the form of a non-oriented graph *G (n; m)*, which has *n* nodes (UAVs) and *m* arcs (connection). Full-fledged topology will characterize the effectiveness of the group's task of the UAVs.

For example: for making aerial photography of an area fragment with the help of 5 UAV's let us imagine group flight scheme as an undirected graph *G (n; m)*, which has *n* nodes (UAV's) and *m* arcs (connections), that is shown in Figure 7. So, the fully connected topology will analyze the effectiveness of the task group UAV. It is a fully connected topology network where each node is directly connected to the remaining nodes. (Bondarev, Jafarzadeh & Kozub, 2014; Shmelova, Bondarev & Kucherov, 2015;).

Figure 6. Presentation a group the UAV's in a fully connected network proceeding aerial photography, n=5; m=10

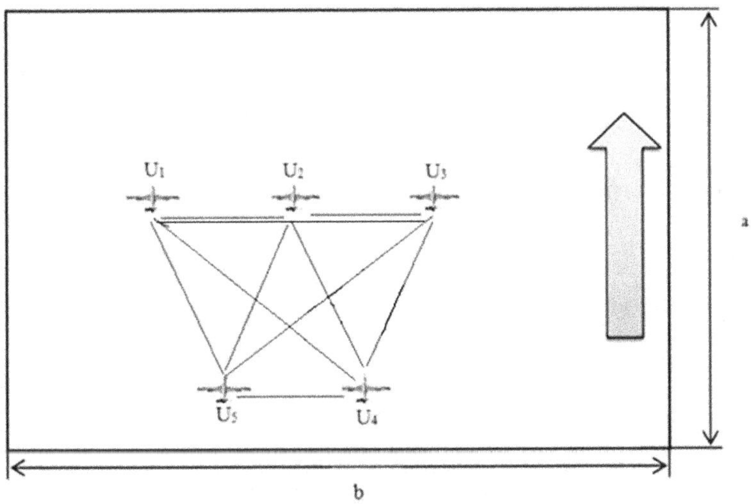

Figure 7. Representation of a UAVs group in flight for aerial photography, n=5; m=10

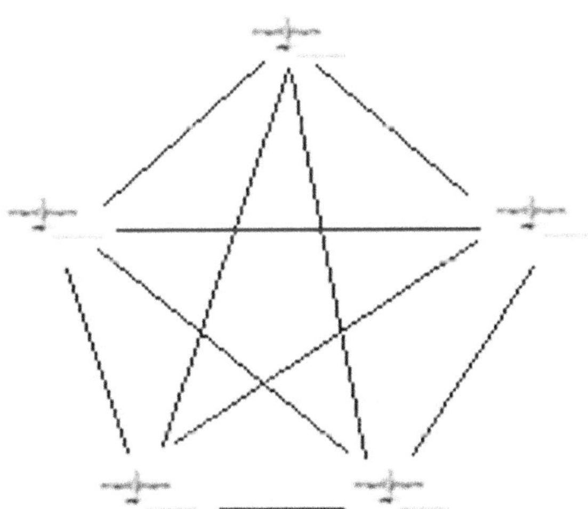

We may represent UAV groups as a graph network and obtain parameters of the reliability of the network, such as connectivity, structural redundancy, uneven distribution relationships, structural compactness, the degree of centralization in the system, and survivability. Undirected graph *G (n; m)* is considered to be connected if from any node (UAV's) there is the way to other nodes (the path may consist of any number of ribs). In our case, the graph is connected because all UAV's are connected between each other.

Applications of Algorithm of finding Central Drone Repeater in a group of the UAVs in flight for determining the central drone and optimal topology of configuration in a group of the UAVs in flight in examples are obtained.

Illustrative Examples

Example 1: To select an optimal structure for goal task - UAVs group in flight for aerial photography.

We are building graph G (n; m) (Figures 8) for UAVs group according to UAVs group in flight (Figures 7).

We represent a group of UAVs using adjacency, introspection and distance matrixes; let us calculate criteria of the topology of the structures of the UAV's group (Tables 5, 6).

Connectivity meets the following conditions:

$$L \geq L_{min,}$$

$$L = \frac{1}{2} \sum_{i=1}^{n} \sum_{i=1}^{n} U_{ii} \geq n - 1 = 10$$

$$L_{min} = n - 1 = 4$$

where: n is number of UAV's in a group; L_{min} is the minimum number of required connections of UAV's group; U_{ij} are the tops of graph *G (n; m)*, $i = \overline{1, n}$

So, if inequality observed $L \geq L_{min}$, then the scheme is connected.

Determination of *structural redundancy R,* it is exceeding the total number of connections over the minimum necessary.

$$R = \frac{1}{2} \left[\sum_{i=1}^{n} \sum_{i=1}^{n} U_{ii} \right] \frac{1}{n-1} - 1 = 1{,}5$$

where: U_{ii} are the tops of graph *G (n; m)*, $i = \overline{1, n}$; $j = \overline{1, n}$; n is the number of UAVs in a group;

Figure 8. Representation of a UAV group in the form of a non-oriented graph G (5; 10) of five elements of a full fledged network

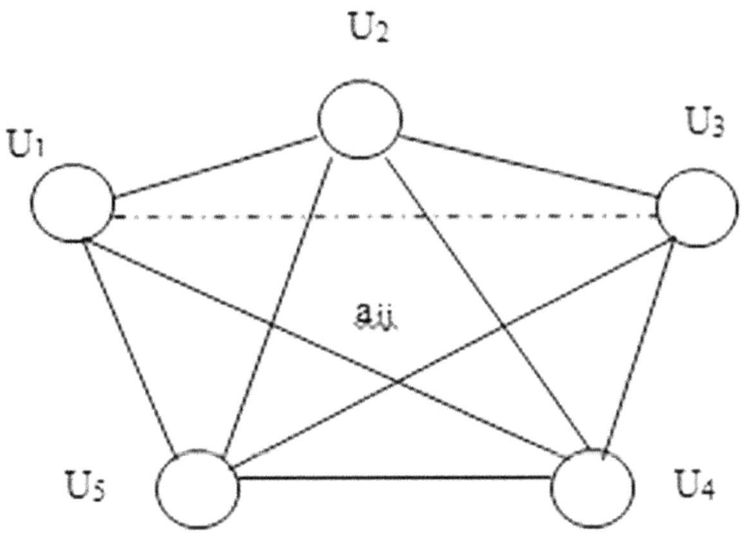

Table 5. Criteria of reliability UAV's group structure

Nº	Criteria of reliability	Symbol
1.	Graph Connectivity	L
2.	Structural redundancy	R
3.	The uneven distribution of connections	ε^2
4.	Structural compactness	D
5.	Relative distance	D_{rel}
6.	Level of system centralizing	δ
7.	Centrality	CDR
8.	Periphery	P
9.	Survivability	S
10.	Moment of the network	M

If R > 0 it is maximum redundancy, R = 0 - minimal redundancy, R < 0 -

Table 6. The matrix of adjacency of fully connected UAV's group $A = \|Uij\|$

Tops of graph	Tops of graph G (n; m),					
G (n; m)	U_1	U_2	U_3	U_4	U_5	$\sum_j U_{ij}$
U_1	1	1	1	1	1	5
U_2	1	1	1	1	1	5
U_3	1	1	1	1	1	5
U_4	1	1	1	1	1	5
U_5	1	1	1	1	1	5

incoherent system. For a group of the UAV's as fully connected structure, we have R = 1, 5 > 0, so it is maximum redundancy

Uneven distribution characterizes connections ε^2 non-opportunities structure that has *m* connections and *n* tops to achieve maximum connectivity of our structure (Table 7):

$$\varepsilon^2 = \sum_{i=1}^{n} \left(\rho_i - \bar{\rho}\right)^2 = \sum_{i=1}^{n} \rho_i^2 - 4\frac{m^2}{n} = 0,$$

Table 7. The matrix of an incidence of fully connected UAV's group $\rho = \|\rho_{ij}\|$

Tops $i = \overline{1, n}$	Ribs of the graph G(n; m), $j = \overline{1, m}$											
U_i	1	2	3	4	5	6	7	8	9	10	$\sum_{j=1}^{n} \rho$	$\sum_{j=1}^{n} \rho^2$
U_1	1	1	0	0	0	1	1	0	0	0	4	16
U_2	0	1	1	0	0	0	0	1	1	0	4	16
U_3	0	0	1	1	0	0	1	0	0	1	4	16
U_4	0	0	0	1	1	1	0	0	1	0	4	16
U_5	0	0	0	0	1	1	1	0	0	0	3	9

where: $\rho = \|\rho_{ij}\|$ is the element of the matrix of an incidence; m is the number of the ribs of the UAV's fully connected group structure; *n* is number of the tops of the UAV's fully connected group structure

If $\varepsilon^2 = 0$, so the fully connected structure of UAV's group has even distribution.

Structural compactness parameter indicates the proximity to each other via a minimum chain length d_{ij}.

$$D = \sum_{i=1}^{n} \sum_{j=1}^{m} d_{ij} = 30 \tag{4}$$

where: d_{ij} is the minimum distance between tops in graph G (n; m) of the UAV's group structure.

The structural compactness index characterizes the diameter of the structure: d is the maximum value of d_{ij}.

Values d and D_{rel} integrally characterize the inertia in the system, at equal values of R and their increase reflects the growing number of connections (Table 8). Putting distance matrix whose elements d_{ij} is defined as the minimum distance between nodes i and j. A total intimacy of the elements in the UAVs group equals to 30: it means if the meaning of D is more the system will show more compactness. The relative rate from:

Table 8. Distance matrix $D = \|d_{ii}\|$

Tops of the graph:	**Tops of the graph G (n; m),**					
G (n; m),	U_1	U_2	U_3	U_4	U_5	$\sum_{j=1}^{n} d_{ij}$
U_1	0	1	2	2	1	6
U_2	1	0	1	2	2	6
U_3	2	1	0	1	2	6
U_4	2	2	1	0	1	6
U_5	1	2	2	1	0	6
$\sum_{i=1}^{n} \sum_{i=1}^{n} d_{ij}$						30

$$D_{rel} = \frac{D}{D_{\min}} - 1 = \frac{30}{20} - 1 = 0,5$$

$$D_{\min} = n(n-1)=20$$

Centralizing level of the system is determined by the index of centrality, calculated for a group of the formula:

$$\delta = \left(n-1\right)\left(2Z_{\max} - n\right)\frac{1}{Z_{\max}\left(n-2\right)} = 0$$

$$Z_{i} = \frac{D}{2}\left(\sum_{i=1}^{n} d_{ii}\right)^{-1} = 5,\ i = \overline{1,n}\ ,\ i \neq j\ ,$$

where:

Z_{max} is the maximum rate of Z_i; n is number of UAV's in a group;
Z_i is the index of the centrality of the system;
D is the structural compactness of the UAV's fully connected group structure.

To assess the degree of irregularity (C_i) load elements of UAVs group structures and the degree of centralization system uses the notion of the centrality of its individual elements C_i, calculated using the formula:

$$C_{\max} = \frac{\sum_{i=1}^{n}\sum_{i=1}^{n} d_{ii}}{2\sum_{i=1}^{n} d_{ii}} = 5$$

where:

d_{ii} is the minimum distance between tops of structure U.

Maximum centrality may be for any of the knots (UAV's), as in star topology. The relative periphery of a knot:

$$P_i = C_{\max} - C_i = 0$$

The durability of the group is defined by the number of states, in which the network keeps working.

Durability can be regarded as the most objective and adequate indicator which enables the best to evaluate all aspects of structural and functional reliability of networks, which is in the environment that is constantly changing and subject to permanent modernization in order to improve the quality of its operation.

$$C = \frac{\sum_{i=1}^{n} U_i - 2(n-1)}{2(n-1)} = 2,125$$

If $C > 0$, the loss of at least one communication of structure remains operational.

In case of a UAVs group in flight for aerial photography, we have high effectivity for star topology

With the help of graph theory, we can determine the effectiveness of different types of configuration of the group structures of UAVs (e.g., full connecting, star-shaped, ring, tree, general tire, mixed, and cell) and select an optimal structure for a goal task. The type of lining (i.e., the structure of the UAV group), on which the flight of the group of aircraft depends on the effectiveness of the task.

The quantitative values of the effectiveness of group flights for different types of connections in the UAV group were calculated (Table 9, Figure 9).

Table 9 shows the designation of reliability parameters of the group structure for the UAV flights. The reliability of the group flight of the UAVs is determined by referring to the criteria of the theory of graphs. In order to quantify the reliability of the UAV group flight performance and to satisfy the above criteria, it is necessary to submit a group flight in the form of a graph. Let's consider a group of five UAVs that perform an aerial photo of the fragment of the terrain.

Example 2: To select CRD in an arbitrary network of UAV's group flight with controlling from one RPAS (Figure 10) using Algorithm of finding CDR in a group of the UAVs in flight.

1. The network of UAV's group fight in figure 10 as *G (7; 8)*.
2. Build adjacency matrix $A = \|a_{ii}\|$ of graph *G (7; 8) of* UAVs group (Table 11)

Table 9. Performance criteria of group flights for different topologies of UAVs groups

Scheme	Connectivity	Marks							
Types:	(connective or not)	R	ε^2	D	D_{rel}	d	δ	P	M
Fully connected	+	1,5	0	30	0,5	2	0	0	2,125
Cellular	+	0,4	47,3	68	1,26	3	0,23	4,63	0,4
Ring	+	2,5	105	50	1,5	4	0	0	1,5
Star	-	0	13,3	50	0,6	2	1	22,5	0
Tree	-	0	5,42	96	1,28	4	0,0325	4,38	0
Common tire	+	2,5	-23,8	80	3	6	0,33	3,3	1,5
Mixed	+	1,26	-9,14	755	3,14	9	0,67	101,98	0,26

Figure 9. Graphically representation of group durability

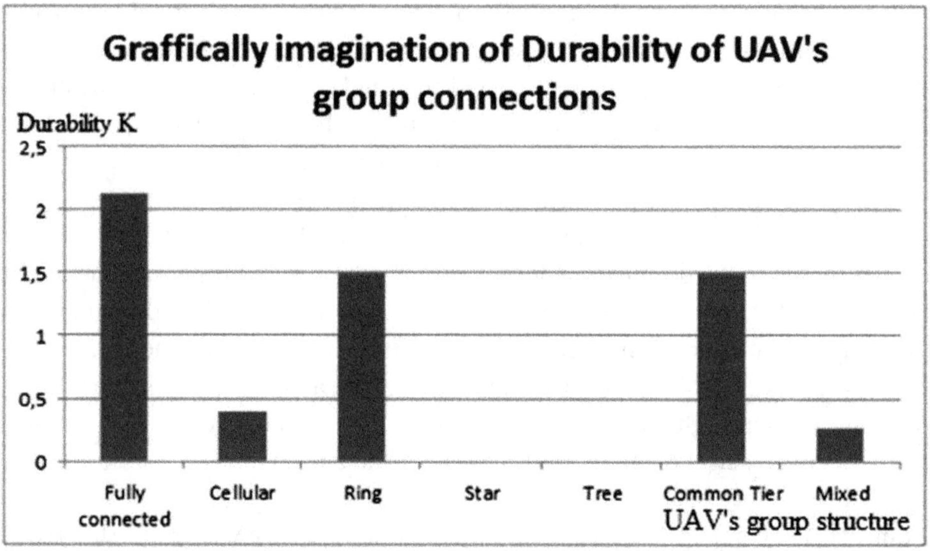

3. Build the matrix of introspection $\rho = \left\| \rho_{ij} \right\|$ of UAVs group graph *G(7; 8)* (Table 12)
4. Build distance matrix $D = \left\| d_{ij} \right\|$ of UAVs group graph *G (n; m)* as the minimum distance d_{ij} between the nodes *i* and *j* (Table 13)

5. Determine the connectivity of UAVs group graph *G (n; m):*

Figure 10. The arbitrary network of UAV's group fight

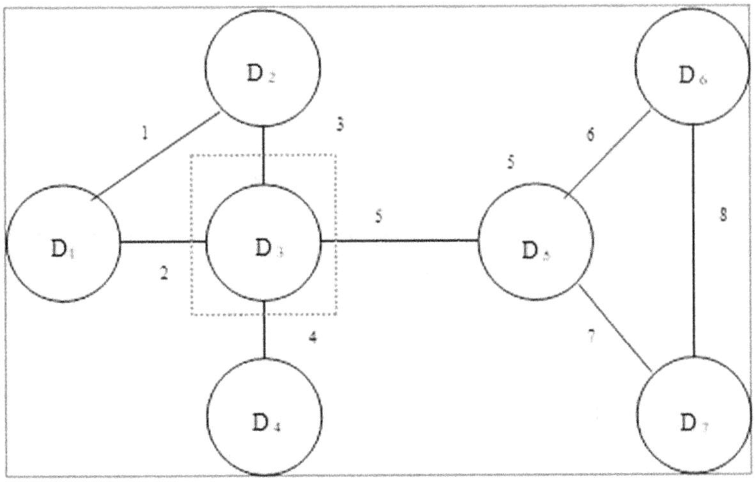

Table 11. The adjacency matrix $A = \left\| a_{ii} \right\|$ for UAVs group

Tops of the graph $i = \overline{1,7}$	**Tops of the graph** ***G (7; 8)***,							
	1	2	3	4	5	6	7	$\sum_i U_i$
1	0	1	1	0	0	0	0	2
2	1	0	1	1	1	0	0	4
3	1	1	0	0	0	0	0	2
4	0	0	1	0	0	0	0	1
5	0	0	1	0	0	1	1	3
6	0	0	0	0	1	0	1	2
7	0	0	0	0	1	1	0	2
$\sum_i \sum_l a_{ij}$								16

$$L = \frac{1}{2} \sum \sum a_{ij} = 8 \geq L_{\min} = 7 - 1 = 6,$$

We have a *connected* graph

Table 12. The matrix of introspection $\rho = \left\| \rho_{ij} \right\|$ of graph G (n; m) for UAVs group

Tops $i = \overline{1,7}$	Ribs of the graph G(7; 8), $j = \overline{1,8}$								$\sum \rho_{ij}$	$\sum \rho^2$
	1	2	3	4	5	6	7	8		
1	1	1	0	0	0	0	0	0	2	4
2	0	1	1	0	0	0	0	0	2	4
3	0	1	1	1	1	0	0	0	4	16
4	0	0	0	1	0	0	0	0	1	1
5	0	0	0	0	1	1	1	0	3	9
6	0	0	0	0	0	1	0	1	2	4
7	0	0	0	0	0	0	1	1	2	4
$\sum$	1	1	0	0	0	0	0	0	2	42

Table 13. The matrix of distance matrix $D = \left\| d_{ij} \right\|$ of graph G (7; 8) for UAVs group

Tops of graph	Tops of graph							$\sum d_{ij}$
	1	2	3	4	5	6	7	
1	0	1	1	2	2	3	3	12
2	1	0	1	2	2	3	3	12
3	1	1	0	1	1	2	2	8
4	2	2	1	0	2	3	3	13
5	2	2	1	2	0	1	1	9
6	3	3	3	3	1	0	1	13
7	3	3	2	3	1	1	0	13
$\sum\sum d_{ij}$								80

6. Determine the structural redundancy of R of UAVs group graph *G (7; 8)*:

$$R = \frac{1}{2}\left[\sum\sum a_{ij}\right]\frac{1}{n-1} - 1 = 0,3 > 0$$

The network has the maximum redundancy. Such a system is reliable in operation.

7. Uneven distribution of relationships of UAVs group graph *G (7; 8):*

$$\varepsilon^2 = \sum_{i=1}^{n} \left(\rho_i - \bar{\rho}\right)^2 = \sum \rho_i^2 - 4\frac{m^2}{n} = 5,43\,,$$

8. Structural compactness *D* of UAVs group graph *G (n; m)* and the relative distance D_{rel}:

$$D_{rel} = \frac{D}{D_{min}} - 1 = \frac{80}{42} - 1 = 0,9$$

$$D = \sum_i \sum_j d_{ij} = 80$$

$$D_{min} = n(n\text{-}1) = 7(7\text{-}1) - 42$$

9. Degree of centralization in the system δ is determined using the centrality index:

$$\delta = \left(n-1\right)\left(2Z_{max} - n\right)\frac{1}{Z_{max}\left(n-2\right)} = 0,7\,,$$

where Z_{max} maximum value of the indicator *Z*:

$$Z_1 = Z_2 = \frac{80}{2}\left(12\right)^{-1} = \frac{80}{24} = \frac{10}{3} = 3,3\,;\; Z_3 = \frac{80}{2}\left(8\right)^{-1} = \frac{80}{2\cdot 8} = 5\,;$$

$$Z_4 = Z_6 = Z_7 = \frac{80}{2}\left(13\right)^{-1} = \frac{80}{2\cdot 13} = \frac{40}{13} = 3,01 \qquad ;$$

$$Z_5 = \frac{80}{2}\left(9\right)^{-1} = \frac{80}{2\cdot 9} = \frac{40}{9} = 4,4\,;\; Z_{max} = 5.$$

10. Determine the centrality of the nodes C_i of the graph *G (7; 8)*:

$$C_1 = C_2 = \frac{80}{12} = 6,6\,;\; C_3 = \frac{80}{8} = 10\,;$$

$$C_4 = C_6 = C_7 = \frac{80}{13} = 6,1;\; C_5 = \frac{80}{9} = 8,9$$

The CDR in a group of the UAVs in flight has $C_3 = C_{max} = 10$.

11. Determine the periphery P_i of the graph *G (n; m)*:

$C_{max} = C_3 = 10;$

$P_1 = P_2 = 10 - C_1 = 10 - 6{,}6 = 3{,}4;$

$P_3 = 0;$

$P_4 = P_6 = P_7 = 10 - 6{,}1 = 3{,}9;$

$P_5 = 10 - 8{,}9 = 1{,}1$

$P = \Sigma P_i = 3{,}4 \cdot 2 + 3,9 \cdot 3 + 1,1 = 18,7$

$P_i = C_{max} - C_i$

The CDR in group of the UAVs in flight has $P_3 = 0$. Relative periphery of the node

12. Determine the survivability of the network S, CRD in group of the UAVs in flight has S > 0:

$$S = \frac{\sum a_i - 2\left(n-1\right)}{2\left(n-1\right)} = 0,33$$

13. Determine the moment of the network, *M* - characterizes the controllability of the CRD:

$$M = \frac{\sum_{i=1}^{n} a_i}{a_l^2} \sum_{i=1}^{n} \left(a_l - a_i\right) = 12$$

$a_l = 4$ - number of links of the central node

14. Data table of parameters CRD in a group of UAVs (Table 14)

Figure 11 presents the communication system with the central UAV (CDR) for control of UAVs group graph *G*(7; 8).

The group use of UAVs is widely adopted for the solution of national economic and military tasks throughout the world. For instance, for the purpose of preventing fires and floods, the control of the state of large forestry is established in order to take appropriate measures for the timely fire suppression and rescue of the victims. Other tasks include, but not limited to, monitoring of road safety, aircraft, search and rescue operations, and aerial surveying. The application of group action of remotely-manned devices allows timely DM and shortening the time to perform a task or executing it with higher quality. In the course of the application of aircraft, there are the following types of group construction: wedge, bearing, ring, and mixed structure. These structures are characterized by certain geometric parameters of distance, intervals, and excesses that are supported by the installed equipment for the purpose of ensuring flight safety. The option of building a group flight UAV is determined by the specific task assigned to the group. An important component

Table 14. Criteria of reliability UAV's group structure for example 2

N°	Criteria of reliability	Symbol	Results
1.	Graph Connectivity	L	+
2.	Structural redundancy	R	0,3
3.	The uneven distribution of connections	ε^2	5,43
4.	Structural compactness	D	80
5.	Relative distance	D_{rel}	0,9
6.	Level of system centralizing	δ	0,7
7.	Centrality	CDR	3
8.	Periphery	P	18,7
9.	Survivability	S	0,33
10.	Moment of the network	M	12

Figure 11. Communication system with the central UAV (CDR) for control of UAVs group graph

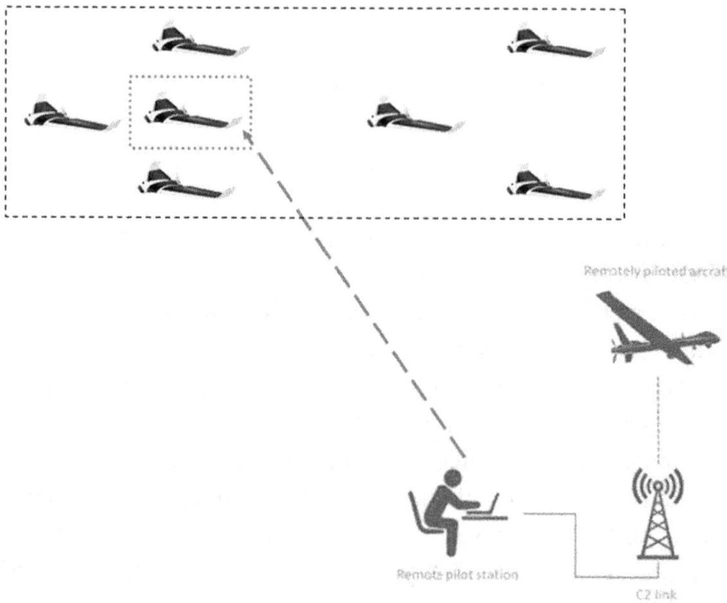

of UAV group flight operations is the availability of integrated and computing equipment, which in turn suggests that the UAV group is part of an informal network within which elements exchange information with each other.

To control a group of UAVs, a flight computer program "Aggregation of heterogeneous information flows" was developed (Shmelova & Stratij, 2015). To coordinate interaction and exchange of information between remote pilots, the database of local RPS NoSQL was also developed. During the development of a database of local RPS and UAV users, RPAS components analysis, UAV, RPS, and *C*2 appeared (Shmelova & Stratij, 2015), taking into account of the UAVs operating procedure that includes the purpose of the flight, flight rules, flight areas, functional level C2 lines and other standards (Figure 12).

The structure of the information system is determined by its topology. The most well-known topologies are full-fledged, star, ring, tree, common tire, cell, and mixed. An analysis of these topologies in terms of reliability and efficiency can be found in the literature on computer topics. It should be noted that the analysis of topology traditionally takes into account the static nature of the structure, which is not always justified in terms of moving media and performance of individual elements of the group. The work of the wireless network is accompanied by malfunctions and the failures of nodes. Therefore, the actual task is to optimize the structure according to

Figure 12. Fragment of NoSQL database of local RPS UAVs users

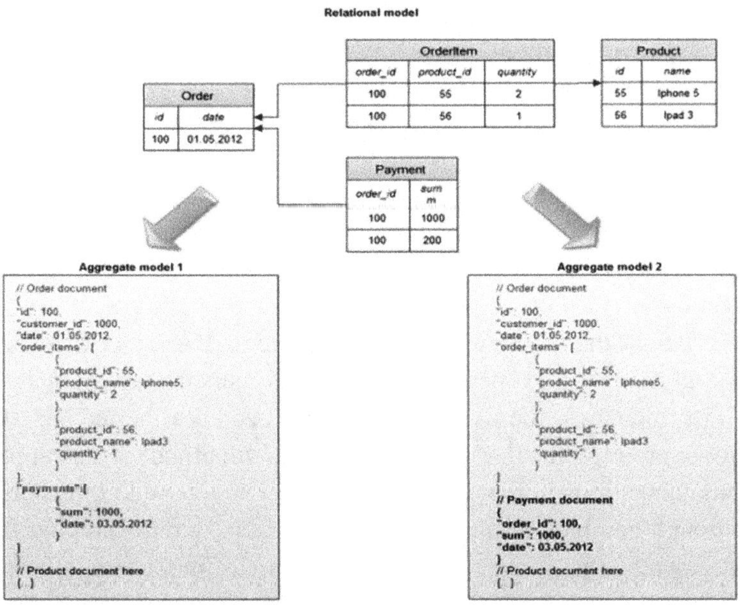

certain criteria (Figure 13). Using graph theory, we can determine the effectiveness of group structures in different types of UAVs' in formation. The type of formation (structure of UAV's group), which is performed by the flight of an aircraft's group depends on the effectiveness of their task purpose. It is intended to apply for flight group performance criteria to all types of connections. Therefore, UAVs have a number of advantages, namely: low cost of operation, low radar and optical visibility, stability and flexibility, and simple and accessible technology for their creation.

Unmanned media can even be used in cases where the use of manned aircraft is impractical, expensive or risky. By 2008, 91% of all UAVs were built in the US. However, currently, almost 40% is produced in other countries, including Ukraine.

Figure 13. Basic criteria for the flight of aircraft

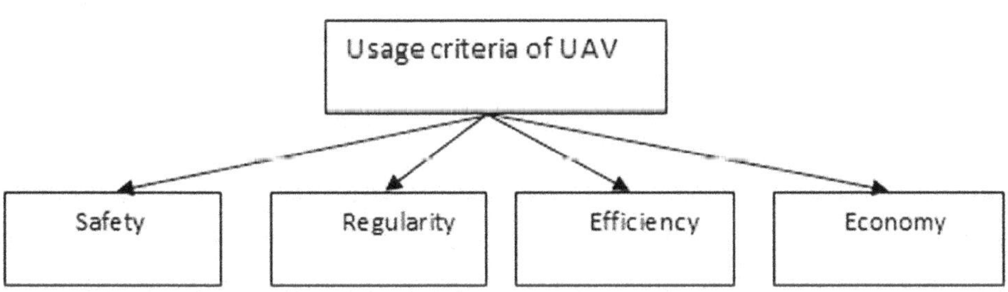

Initially UAVs were used mainly for military purposes: military intelligence, surveillance, detection, recognition and support of objects (purposes); provision of two-way radio relay communication; between radio and radio engineering and radio-electronic intelligence, electronic warfare; detection of the use of chemical, biological and nuclear weapons; delivery of goods; participation in information operations; solving search and rescue tasks; direct aviation support; participation in an airstrike operation; and monitoring the state of the environment. Ignatiev (2010) added that among all of the multipurpose UAVs were used for the military purpose the primary purposes were: reconnaissance - 100% and shock (for the fire by land and air targets) - 39%.

In civil aviation, the main advantage of the use of UAVs is the fulfillment of tasks related to the risk for a person and efficiency in solving national economic problems. In some cases, the use of one aircraft becomes ineffective - then it is expedient to use group (collective) of UAVs, both in civilian and military applications: to relay communications in those places where it is impossible to install antenna coverage through complex terrain; in agriculture (e.g., group spraying fields); in aerial photography (e.g., group shooting of large areas, monitoring of forest fires, and patrolling of territories); and the transportation of cargos. Today, according to UVS-International (the leading international association of unmanned systems), UAVs are produced in 52 countries around the world. Dozens of large companies and small firms compete in this market. An expanded, though not complete, list of manufacturers and models are available from the link in the annual report of UVS-International 2009/2010 UAS Yearbook. Despite the fact that the demands of the military departments on the UAV are large and diverse, not all manufacturers can expect to receive defense orders. As a result, many companies with development in the field of UAVs tend to pay attention to the prospective use of UAVs in the civilian and commercial spheres. In turn, interested state agencies and special services, whose functions are related to the protection, control and monitoring of objects, the elimination of fires, enterprises of the fuel and energy complex as well as firms whose business is associated with the receipt of spatial data, also express their strong interest in the effective application and use of UAVs.

In the area of technological advancements in the UAV sector, the database of local RPS NoSQL was developed to better coordinate interaction and exchange of information between remote pilots. While developing a database of local RPS and UAV users, RPAS components analysis, UAV, RPS, and C2, made their appearance (Shmelova & Stratij, 2015). For optimization, a solution to problems was developed to provide models of determining the optimal landing site in case of an extraordinary situation, search for optimal flight routes UAV with the module «ASS[CCC)SIST» (Acknowledge, Separate, Synergetic [Coordinated, Cooperation, Consolidation) Silence, Inform, Support, Time) for each type of UAV. The investigation into the processes of modeling the DM by UAVs' operators in normal and unusual situations enabled to build the following models: DM in Certainty, DM in Risk and DM in Uncertainty.

During the flight, UAVs may be controlled by RPS. At any given time ti *k*-UAV must be piloted by only one *j*-th RPS, if necessary, at time t_i+1 to be transmitted to the control (*j* + 1)-th RPS (Figure 14). This transfer of flight control of the *j*-th RPS to (*j* + 1)-th RPS is safe and effective and this is provided through the local DSS operators of UAVs.

The UAV flights use the above criteria and distributed RPS. In addition, when performing group flights of the UAVs it is expedient to apply specific criteria for the reliability of the group structure: connectivity, structural redundancy, uneven distribution of connections, structural compactness, the degree of centralization in the system, and survivability.

FUTURE RESEARCH DIRECTIONS

Further research should be directed to the solution of a problem in control systems of drones for different goal tasks. The control systems of drones must use various new modern information technologies to increase the efficiency of the system operation, such as Data Analysis and DM, Big Data, Data Mining, Machine Learning, Intelligent Data Analysis, Artificial Intelligence System, and Knowledge Discovery.

Figure 14. The structure of distributed RPS Mission Control UAVs

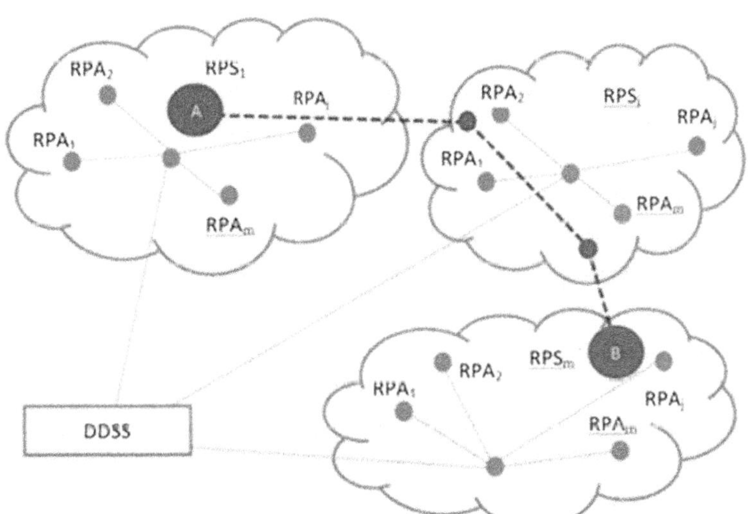

Also, the development of uniform standards for the management of various types of UAVs that perform single and group flights and the systems for the use of drones in various areas of industry, education, and social spheres. It is also necessary to study the behavior of RPAS operators in normal and extreme situations, prerequisites of emergency situations and preventing catastrophic situations. It is necessary to develop modern DSSs of H-O of ANS (pilots, ATCs, UAV operators) in-flight emergencies and in other situations too.

What's Next?

On next steps, the authors are planning to consider Multi-Criteria Decision Analysis (MCDA) through multicriteria assessment of the performance of aviation chemical works by a group of manned and unmanned aircraft such as airplanes, helicopters, UAVs. MCDA is a discipline aimed at supporting decision makers who are faced with making numerous and conflicting evaluations (ICAO, 2009). MCDA aims at highlighting these conflicts and deriving how to arrive at a compromise in a transparent process. The authors are also planning on creating Intelligence Further research that is directed to the solution (e.g., software creation) of practical problems of actions of UAV operators during emergency situations. Models of FE development and of DM by UAVs in FE will allow predicting the H-O's actions with the aid of the informational-analytic and diagnostics complex to have a better understanding of H-O behaviors in the extreme situation. In order to predict and prevent the development of the catastrophe with the use of Intelligent DSSs, It is necessary to analyze all factors influencing the operators within these systems.

CONCLUSION

The problems of using drones for different purposes in civil aviation are considered, and the analyses of the advantages of group flights of UAVs are presented. To control a group of drones, the authors suggest choosing and using a Drone, a repeater to connect to the operator on the ground and to control of the drones in group flight. To select the effective Drone – Repeater and optimal configuration network of UAV's group, the authors used the method of server selection in computer networks. In chapter criteria of efficiency for the performance of tasks of UAV group flights was analyzed. Graph theory was used for quantitative estimation of effectiveness of UAV group flights. Also, the flight group performance criteria for all types of UAVs connections (e.g., fully connected, a star, ring, tree, with a common tire, mixed, and cellular) were considered and compared. The algorithm of finding central drone in a group of UAVs in flight for relaying control signals from other

UAVs was obtained. Thus, the creation of topology groups of UAVs is advisable to focus on fully connected topology, which was found to be the most effective. Further research should be directed to the solution of practical problems of implementation of group management in driving the UAVs, which leads to the more efficient use of UAVs, namely the possibility of an adjustment plan and the optimization of the flight route, based on data already obtained from other UAVs. The possibility of setting different tasks for members of multi-UAV considering efficiency topology groups was noted.

REFERENCES

Amelin, K. S., Antal, E., & Vasiliev, V. (2013). *Adaptive management of an autonomous group of unmanned aerial vehicles*. St. Petersburg State University.

Austin, R. (2010). *Unmanned Aircraft Systems: UAVs Design, Development and Deployment*. John Wiley & Sons. doi:10.1002/9780470664797

Bondarev, D., Jafarzadeh, R., & Kozub, A. (2014). Effectiveness of Group Flight of Unmanned Aerial Vehicles. Systems of Information Processing, 6, 9-15.

Bondarev, D., Shmelova, T., & Kucherov, D. (2015). Models of group flights of unmanned aerial vehicles using Science and technology of the Air Forces of the Armed Forces of Ukraine. *Scientific Periodicals, 3*, 68-75.

Chekunov, E. (2010). Application of the UAV of US Armed Forces in Military Conflicts. *Foreign Military*. *RE:view*, *7*, 41–50.

Fradkov, A., & Grigoriev, G. (2013). Decentral adaptive control of synchronization of networks of dynamical systems of dynamic systems with limited perturbations. *Automation and Telemechanics, 5*.

Ganin, S., & Karpenko, A. (1999). Unmanned aerial vehicles. St. Petersburg: Academic Press.

Guinness World Record. (2016). *Intel stuns during CES keynote with record for most drones airborne simultaneously*. Retrieved from http://www.guinnessworldrecords.com/news/2016/1/intel-stuns-during-ces-keynote-with-record-for-most-drones-airborne-simultaneousl-411677

Gulevich, S. P., Veselov, Y. G., Pryadkin, S. P., & Tirnov, S. D. (2012). Analysis of factors affecting the safety of the flight of unmanned aerial vehicles. Ru. Causes of accidents drones and methods of preventing them. *Science and Education*, 12.

Ignatiev, O. (2010). UAV: The state and prospects of their use in the interests of NATO. *Foreign Military Review, 12*, 45–52.

International Civil Aviation Organization. (2005). *Global Air Traffic Management Operational Concept. Doc. ICAO Doc 9854-AN/458*. Montreal, Canada: Author.

International Civil Aviation Organization. (2009). *Manual on Global Performance of the Air Navigation System. Doc. ICAO 9883*. Montreal, Canada: Author.

International Civil Aviation Organization. (2015). *Manual of Remoted Piloted Aircraft Systems (RPAS). Doc. ICAO 10019/AN 507*. Montreal, Canada: Author.

Kharchenko, V., Shmelova, T., Bondarev, D., & Stratiy, A. (2017). Modeling of Decision Making of Unmanned Aerial Vehicle's Operator in Emergency Situations. *Proceedings of the National Aviation University, 1*, 20–28.

Kharchenko, V., Shmelova, T., & Sikirda, Y. (2012). *Decision-Making of Operator in Air Navigation System: a monograph*. Kirovograd: KFA of NAU.

Kharchenko, V., Shmelova, T., & Sikirda, Y. (2016). *Decision-Making in Socio-Technical Systems: a monograph*. Kyiv: NAU.

Kreps, S., & Zenko, M. (2014). The Next Drone Wars. *Foreign Affairs*. Retrieved from http://bmpd.livejournal.com/861105.html

Kucherov, D. P., & Kozub, A. N. (2015). Control System Objects with Multiple Stream of Information. *Proceedings IEEE 3rd International Conference "Actual Problems of Unmanned Aerial Vehicles Developments (APUAVD)"*, 290-293. 10.1109/APUAVD.2015.7346623

Kucherov, D. P., & Kozub, A. N. (2016). Group Behavior of UAVs in obstacle presence. *Proceeding IEEE 4th International Conference "Methods and Systems of Navigation and Motion Control (MSNMC)"*, 51-54. 10.1109/MSNMC.2016.7783104

Kucherov, D. P., Kozub, A. N., & Kostyna, A. N. (2017). Group of UAVs Moving on Smooth Control Law with Fixed Obstacles. *Advances in Science, Technology and Engineering Systems, 2*(3), 1034–1041. doi:10.25046/aj0203131

Kuzin, A. V. (2011). Computer networks / A.B. Cousin. Moscow: INFRA-M.

Marsh, L., Calbert, G., Tu, J., Gossink, D., & Kwok, H. (2005) Multi-Agent UAV Path Planning. In *MSSANZ Conference on Modelling and Simulation (MODSIM 05)* (pp. 2188 – 2194). Melbourne, Australia: Academic Press.

Montgomery, J. F., & Bekey, G. A. (1998). Learning helicopter control through "teaching by showing. *IEEE Conference on Decision and Control*. 10.1109/CDC.1998.761747

Nikiforov, A., & Munimaev, V. (2010). *Analysis of foreign unmanned aerial vehicles*. St. Petersburg Forestry Academy.

Olifer, V.G., & Olifer N.A. (2006). *Computer networks: principles, technologies, protocols*. Academic Press.

Shmelova, T., & Bondarev, D. (2016). Graph Theory Applying for Quantitative Estimation of UAV's Group Flight into Aerial Photography. *Electronics and Control Systems*, *4*(46), 128–133.

Shmelova, T., Bondarev, D., & Znakovska, Y. (2016). Modeling of the Decision Making by UAV's Operator in Emergency Situations. In *4th International Conference on Methods and Systems of Navigation and Motion Control (MSNMC-2016)*. Kyiv: NAU.

Shmelova, T., & Stratij, A. (2015). *Computer Program "Aggregation of heterogeneous information flows": Certificate of registration of copyright in the product Nº 59466*. Academic Press.

Sizova, O. A. (2012). The application of graph theory in various types of scientific activity. *Baku*.

Tarasov, V., & Gerasimov, B. (2007). *Intelligent Decision Support Systems: Theory, Synthesis, Efficiency: a monograph*. Kyiv: International Academy of Computer Science and Systems.

KEY TERMS AND DEFINITIONS

Air Traffic Management (ATM): An aviation term encompassing all systems that assist aircraft to depart from an aerodrome, transit airspace, and land at a destination aerodrome, including air traffic services (ATS), airspace management (ASM), and air traffic flow and capacity management (ATFCM).

Command and Control (C2): Denotes the set of organizational and technical attributes and processes by which enterprise marshals and employs human, physical, and information resources to solve problems and accomplish missions.

Communication, Navigation, and Surveillance (CNS): The main functions that form the infrastructure for air traffic management, and ensure that air traffic is safe and efficient.

Decision Support System (DSS): The interactive computer system intended to support different types of activity during the decision making including poorly-structured and unstructured problems.

Distributed Decision Support System (DDSS): A decision support system which supports distributed organizational decision making.

NoSQL (Originally Referring to "Non SQL" or "Non-Relational") Database: Provides a mechanism for storage and retrieval of data that is modeled in means other than the tabular relations used in relational databases.

Remotely Piloted Aircraft (RPA)/Remotely Piloted Aircraft System (RPAS)/ Unmanned Aerial Vehicle (UAV)/Drone: A system that based on cutting-edge developments in aerospace technologies, offering advancements which are opening new and enhanced civil-commercial applications as well as improvements to the safety and efficiency of the entire civil aviation. The terms RPA or UAV are used to describe the aircraft itself, whereas the term RPAS is generally used to describe the entire operating equipment including the aircraft, the control station from where the aircraft is operated and the wireless data link.

Chapter 9
The Changing Face of Airmanship and Safety Culture Operating Unmanned Aircraft Systems

Tracy Lamb
Association for Unmanned Vehicle Systems International, USA

ABSTRACT

The notion of using drones for commercial purposes has evolved in the past 5 years from the initial "boom" of excitement around this, somewhat of a novelty and curiosity, to more calculated and sophisticated use of unmanned aircraft systems (UAS), or drones. In the hands of true professionals, drones can offer highly efficient and profitable solutions for industrial, and commercial inspections and other data capturing tasks. The appetite for safe and efficient collection of data is a changing face of safety cultures and how teams and individuals apply airmanship principles, and how inspection crew and UAS crew interact. UAS are no longer viewed as novelty or useful addition to the inspectors' "toolbox," but as an integrated part of safety critical system. While there is much to be learned from tradition manned aviation, UAS pilots are confronted with different task priorities in order to effectively "aviate," and therefore, like the changing face of airmanship and safety culture, to "aviate" emerges has having different attributes when compared to manned aviation.

DOI: 10.4018/978-1-5225-7900-7.ch009

THE COMMERCIAL DRONE BOOM OF 2013-2018: ITS VICTORIES AND CHALLENGES

Much of the small commercial Unmanned Aerial System (UAS) industry emerged out of the industrial inspection, information technology, photogrammetric survey, 'big data', and technical industry. However, some had their origins in the hobby and 'modellers' area. The emerging applications of small commercial UAS started to increase in prevalence in the public media following Federal Aviation Administration (FAA)'s approval of first commercial UAS flights over land in 2014 (FAA, 2014). The novel use of UAS has rapidly advanced to a variety of industry sectors around the world, and today, in 2018, the FAA has just recorded over 600,000 commercial UAS operators licensed under Code of Federal Regulations (CFR) Part 107, in the United States alone (Elwell, 2017).

Back in 2013, as awareness increased about the innovative uses, and benefits of using small UAS in practical applications, the commercial organizations became highly motivated to focus their attention on learning and capturing this novel and cost-effective way to bring value to their customers, to capture images of their assets, and more importantly, not to miss out on what appeared to be a competitive edge. The Association for Unmanned Vehicle Systems International (AUVSI)'s economic report of 2013 predicted that the small commercial UAS industry would be cumulatively worth USD 1.8 billion, between 2015-2015 (Jenkins & Vasigh, 2013). Today, in 2018, we recognize that those predictions were not conservative, considering the latest studies (e.g., Grand View Research, 2016) indicate that the small commercial UAS industry is expected to reach USD 2.07 billion by 2022.

In many cases prior to 2017, the procurement of small commercial UAS or 'drone' was not formal management or organisational decision, the *C-suite* was generally not informed of the acquisition of the 'drone', as it was viewed as another relatively low cost '*tool*' or solution for day-to-day operations.

Specialised inspection personnel who recognized the value in the data collection ability viewed the UAS as something that would add value a capability to their tool box, usually after learning and reading about the latest technological advancements through blogs, social media, technical papers, and other industry peers to achieve efficient and effective completion of allocated tasks (Lamb, 2017).

The rapid advancement of UAS technology, including supportive platforms, software, First Person View (FPV), and other gadgets to improve performance, quickly became the focus of technical discussions and field stories[1], among the professional community of inspectors. For these professionals, working in industries that rely on data monitoring and imagery to determine the condition of their high-value infrastructure and physical assets such as oil, gas, mining, utilities, rail and road. For these organizations and their stakeholders, the return on investment in utilizing the UAS was obvious. However, the benefits of using this new technology came with follow-on implications and considerations that were yet to be discovered. Some of the questions that needed to be addressed relating to topics include: Standards, certification, training, liability, safety and risk management integration. In the absence of formal processes, protocols, regulations and standards for this new industry, organizations did their best by improvising what was required (Lamb, 2017).

Even today, in 2018, the small commercial drone industry presents a relatively low barrier to entry as UAS platforms are considerably less expensive than a conventional commercial aircraft in terms of initial capital investment and the operating costs. Also, the qualifications and approvals required to pilot a small commercial UAS are easy to obtain. In the United States, there are currently no pilot competency assessments required to gain a small commercial UAS license under Federal Aviation Regulation (FAR) Part 107 (FAA, 2018b).

This low barrier to entry makes it an attractive opportunity for organizations to gain the benefits of UAS. In terms of the diversity in UAS's design, capabilities and their applications, this has led to the dynamic growth in the ubiquitous utilization of UAS (e.g., Palmer & Clothier, 2013; Weibel & Hansman, 2005)

It is not only the United States that is experiencing this tremendous adoption of UAS into the commercial industry. According to Statista (2018), the estimated spending on UAS in US Dollars for financial years between 2017 and 2021 in the top five countries are:

- 17.5 Billion in the United States;
- 4.5 Billion in China;
- 3.9 Billion in Russia;
- 3.5 Billion in the United Kingdom; and,
- 3.1 Billion in Australia

While both the evolution and proliferation of commercial Unmanned Aerial Vehicles (UAVs) is evident there are growing concerns with their use. These concerns include, but not limited to, the combination of lack of training in the operations and technical expertise such as; understanding the control interface complexities, data processing, programming, security issues with transferring and storing and interpretation, and being able to meet the client expectations or requirements from that data. More importantly, having in-depth understandings of the safety and operational requirements of complex environments owned by the client's industry (e.g., mine site, off-shore rig, chemical plant, cell tower, wind turbine, complex power network) are all key considerations for the success of a UAS service provider.

The majority of the small commercial UAS platforms used today have some basic safety mechanisms embedded within their systems to facilitate the operators to use the UAS safely, such a return to land function, or an altitude limiter. However, these safety mechanisms are not required by legislation or standardized by any manufacturer, and frequently not fully understood by the operators (Plioutsias, Karanikas, & Chatzimihildidou, 2018). With the release of the FAA Reauthorization, Bill 302, in October 2018, there are provisions within this that direct small UAS to be equipped with some of the components that support the safety of flight capabilities as part of the certification process (USA Congress, 2018).

With no provision for remote pilot flight competency, training or assessment provided within legislation, a well-balanced and harmonised merger of safety cultures and safety behaviours remain challenges for both the operator and the organisation. Despite this, most small commercial UAS platforms promote the benefits of using '*off the shelf*' equipment, such as an iPad or computer as they are "easy to use". However, the human-induced errors or '*human factors*' remain a major contributor to small UAS mishaps and safety incidents (Lamb, 2018).

The use of an iPad, iPhone or other familiar devices such as a laptop computer should support UAS operations by providing a familiar platform, however; safety incidences involving small drones is increasing (Flatley, 2017). The propensity for human errors increases in complex activities, especially when interacting with new and complex technology such as the interaction required to operate a drone remotely from a computer, iPad or control station (Mouloua, Gilson, Kring, & Hancock, 2001).

This industry challenge is further exacerbated by the fact that there are currently no requirements for flight competency assessments within the regulatory framework. Neil and Griffin (2002) argued that leadership is a critical driver to have a positive impact on safety behaviors in terms of safety compliance and safety participation, at this juncture, there is no leadership in the industry on competency-based training standards or regulatory flight assessment standards. This has a twofold effect where the former refers to individuals performing core activities while maintaining workplace safety while the latter refers to individuals developing an environment that supports safety (Neil & Griffin, 2002). Therefore; safety leadership must be driven and supported by the leaders in the UAS industry and organizations who support the use of UAS in their day-to-day operations, must support, champion and drive a proactive safety culture that integrates drones (Helmerich & Merritt, 2001).

THE CHALLENGES OF INTEGRATING SAFETY CULTURES AND MERGING NEW TECHNOLOGIES

Despite the considerable benefits in utilising UAS, there are significant integration factors that must be considered when using UAS in commercial applications. Research (e.g., Gawron, 1998, as cited in McCarley & Wickens, 2006), has found that having the human separated from machine produces different human performance and risk factors that attract a multitude of potential hazards when compared to conventional aviation. However, there are considerable challenges from the human integration perspective as well as the introduction of small commercial UAS as a cultural shift, both at the organizational level and the personal safety level was postulated (Helmerich & Merritt, 2001; Lamb, 2017).

Another part of the challenge is recognising the falsehood of the inaccurate aggregate risk value, promulgated by the 'probability' factor, that potential hazards associated with small commercial UAS are viewed as '*low risk*', because chances of the UAS causing major damage or loss is low. This perception is largely due to the weakness in the use of the standard Risk Assessment Matrix although the standard risk matrix is widely used and highly promoted and recommended by aviation regulators worldwide (ICAO, 2012).

Two main limitations of the standard risk matrix; the first lies within the human factor realm, this limitation is subjectivity, especially when assessing the 'Low Probability' (LP), and 'High Consequence' (HC), risks (Robinsons & Francis, 2014). The second; the specialized training, or lack thereof, provided to the front-line staff who use this commonly applied tool (Peace, 2017). Most users of the risk matrix estimate the probability factor of a catastrophic event caused by the UAS being extremely rare (*total risk value = probability x consequence).* This is an estimation based on the subjective perspective or frame of reference to the safety representative of the operation.

While the aviation regulators direct operators to consider residual risks and support safety assurance (closed loop systems), the risk matrix as a standalone tool does not direct, or remind the operator to report back into their system or consider additional or residual risks, or confirm the accuracy of their initial aggregate risk values. Therefore; the risk matrix itself as a tool "*...when used with front line staff, who are not properly trained, performs like an open loop system*" (Stolzer, Halford, & Goglia, 2013, pp. 156-168). Therefore, small commercial UAS operators often miss this critical function that is implied but not indicated on a standard risk matrix.

Research has shown that mishap rates involving UAS are up to 300 times greater than that of conventionally piloted aircraft fleets in general aviation (Carrigan, Long, Cummings, & Duffner, 2008) and UAS operations in military is no exception (Williams, 2004), which is surprising if you consider the amount of training that military crew receive compared to that of small commercial UAS pilots. Therefore, this trend of ubiquitous mishap rates should be further investigated given the unprecedented growth in the utilisation of UAS across many industries.

It is not only private commercial organizations that have become captivated by this emerging UAS technology, government and public safety agencies began embracing the potential benefits of using drones as early as 2005 when drones were used to search for survivors of Hurricane Katrina (NSF, n.d.). Implementation and coordination of UAS in public safety are now viewed as a matter of paramount urgency given the increases in natural event occurrences such as flooding, fire and severe storms. The national emergencies in the USA associated with hurricanes and floods in 2017 and 2018 increased the use of designated public safety UAS to an all-time high with the FAA issuing over 300 special flight permits called 'waivers', in response to the demand for UAS search and rescue capabilities, and has since issued over 2500 waivers (Figure 1) to commercial UAS operators (FAA, 2018b).

Humans play a vital and centric role in UAS operations. Research has indicated that this role is more complex than conventional aviation and safety critical activities such as commercial infrastructure inspections, testing and data collection. For instance, McCarley and Wickens (2004) identified that an Unmanned Aerial Vehicle (UAV) operator performed duties "in relative sensory isolation" due to the absence of various sensory cues such as vestibular input and ambient noise (p. 1). As the UAS industry is growing exponentially and UAS operators have become more prevalent in commercial aerospace, safety behaviours, safety culture and the concept of being able to 'Aviate' are evolving.

Figure 1. Waiver types granted to operators
Adapted from AUVSI, 2018. (https://www.auvsi.org/our-impact/waivers-under-part-107-interactive-report)

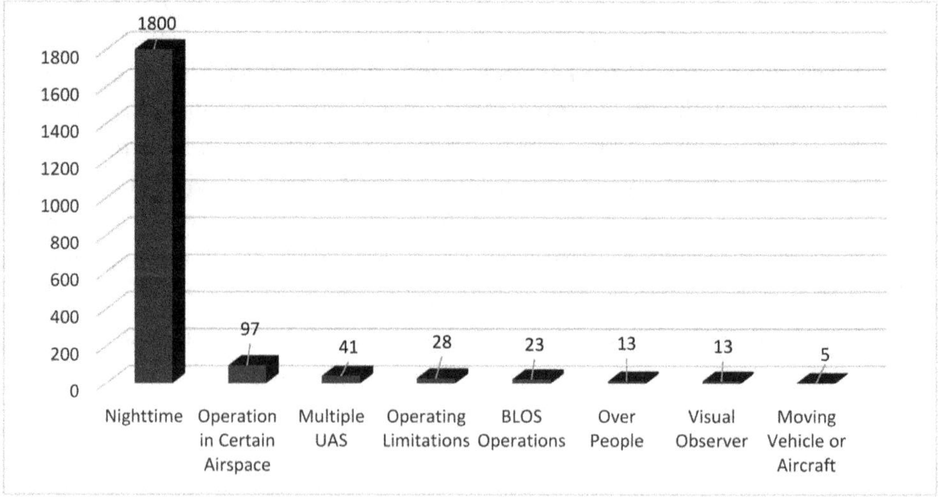

To '*Aviate*': We Used to Know What That Meant

According to Oxford dictionary (2018), 'Aviate', is defined as *"to pilot or fly an aircraft"*. Similarly, the FAA defines 'Aviate' as 'maintaining control of the aircraft' (FAA, 2018a; Hobbs & Lyall, 2016). Based on currently accepted aviation and general definitions to a pilot, 'Aviate' refers to pilots' priority to 'fly the aircraft' and this specifically addresses the pilots' performance tasks of maintaining control and keeping the aircraft in a safe state prior to addressing other tasks of navigating or communicating. Many pilots recite this commonly used phrase; *"Aviate, Navigate, Communicate"* (A-N-C), as a reminder of priorities the pilot-in-command follows particularly in an emergency or non-normal situation.

1. **Aviate**: Maintain control of the aircraft
2. **Navigate**: Know where you are and where you intend to go
3. **Communicate**: Let someone know your plans and needs

It may be assumed that the common definition of 'maintaining control of the aircraft' could easily be transferred to the unmanned aircraft system, except that the techniques and modes of how that is achieved are radically different to a conventionally manned aircraft. In a UAS operation, operators are required to consider numerous factors including; command control (C^2 Link)[2] latency, or interruptions, limited ability to "see and avoid", absence of sensory feedback of aircraft state, and control station configuration that may not relate to conventional pilot task actions. All of these factors directly or indirectly affect the task of controlling the aircraft and maintaining it in a stable powered state.

According to a NASA report (Hobbs & Lyall, 2016), most current designs of advanced Remotely Piloted Aircraft System (RPAS) rely entirely on automated systems for basic flight control and they do not provide options for pilot manual control. Instead, the remote pilot is responsible for supervisory control of the automation. Consequently, manual flight control becomes less of an issue for the remote pilot making automation management issues of critical importance. However, unmanned aircraft systems present unique challenges to effective Human System Integration (HIS), and the various levels of automation may affect the pilot of a remotely piloted aircraft's ability to control the aircraft on certain tasks. Some aircraft systems are on the higher level of automation, to the extent that they approach 'autonomous', the level at which the pilot is excluded from determining and actioning the task, effectively removing and/or limiting the pilot from some of the flight process (Parasuraman & Wickens, 2008).

A pilot of a conventional aircraft can lose almost all on board functions (navigation, communication, even flight control and propulsion) and the pilot may still be able to 'Aviate' or control the aircraft to a safe or safer outcome than a crash (examples include 'Captain Sully Sullenberger landing his crippled Airbus on the Hudson River, the Glimi Glider). Not having those human minds on board the aircraft presents challenges. Clearly, there is a barrier, an interface that sometimes does not support the indirect control 'connections' of the system, as witnessed in examples of this in a high percentage of human factors in remotely piloted aircraft incidents (Tvaryanas, Thompson, & Constable, 2006). Remote pilots' ability to continue to perform the traditional tasks designated as 'Aviate' may be completely removed due to the failure of the C^2 link. Depending on the level of automation, this may occur at a much lower threshold in some UAS platforms, compared to others.

For instance, a complete C2 link failure may remove the operators' ability to Aviate, navigate and communicate all at once, even though the aircraft may continue to maintain stable powered flight. This is referred to as the 'lost link flight profile' where the aircraft flies for a pre-determined time on predetermined programmed flight planning inputs. In this situation, operators are expected to organise backup or alternatives systems or 'capture' methods. Alternatively, the operators may orchestrate the emergency procedures. These potential actions do not fit to a conventional definition of 'Aviate'. Therefore, if the term 'Aviate' is to be used in the context of he remote pilot in control, being responsible for controlling the unmanned aircraft, it must be clear on both its definition and contextual elements to ensure that the correct meaning and intent is incorporated appropriately and accurately in ICAO documents, with a particular focus on new proposed Standards And Recommended Practices (SARPs) for RPAS operations. This is a topic that is currently being discussed among industry experts.

From Figure 2, you may notice that when controlling the aircraft sensory and perception differences have an influence, and there are also environmental and network quality and latency that may also affect the pilot's ability to Aviate effectively in the traditional sense. The Pilot's commands and the aircraft messages relayed back to the control station may suffer degraded C^2 influences; variable lengths of time delay (latency); distortion, interference, switching faults or failure. Any one of these faults with the C2 link will affect a pilot's response time and effective control.

There has been research to explore the issues of the pilot's ability to Aviate at differing levels of automation. An illustration of how the different automated tasks affect a pilot's ability to Aviate is presented in Figure 3 (Hobbs & Lyall, 2016). This figure provides a model of the unique pilot performance tasks that are required to maintain or regain control of the UAS. This high-level model of the responsibilities of the remote pilot, consistent with FAA assumptions, adapted from Mutuel, Wargo and DiFelici (2015).

In addition to varying levels of automation, the layout and displays of the control stations can differ greatly that results in the presentation of ergonomic and cognitive

Figure 2. ICAO RPAS Panel Concept of Operations. The blue indicates the remote pilot activity of Aviate
(Diagram reproduced courtesy of ICAO)

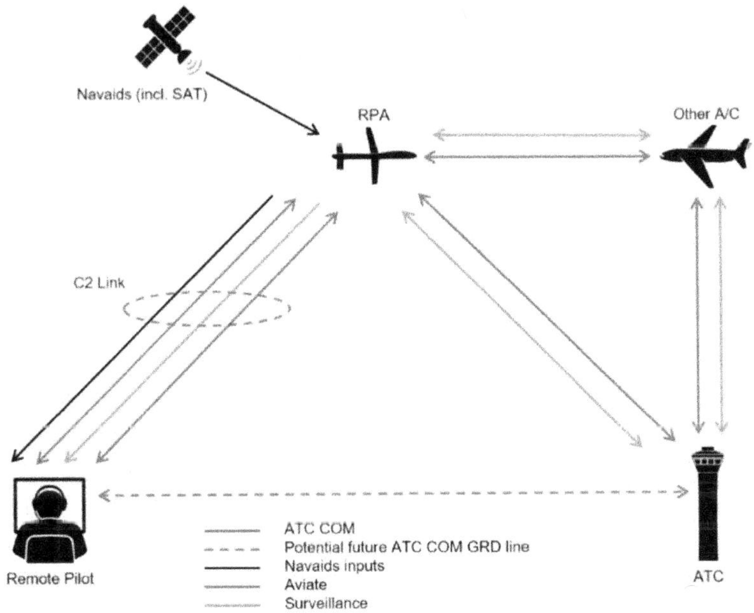

Figure 3. Responsibilities of remote pilots when operating RPAS
Source: adapted from Hobbs and Lyall (2016, p. 18).

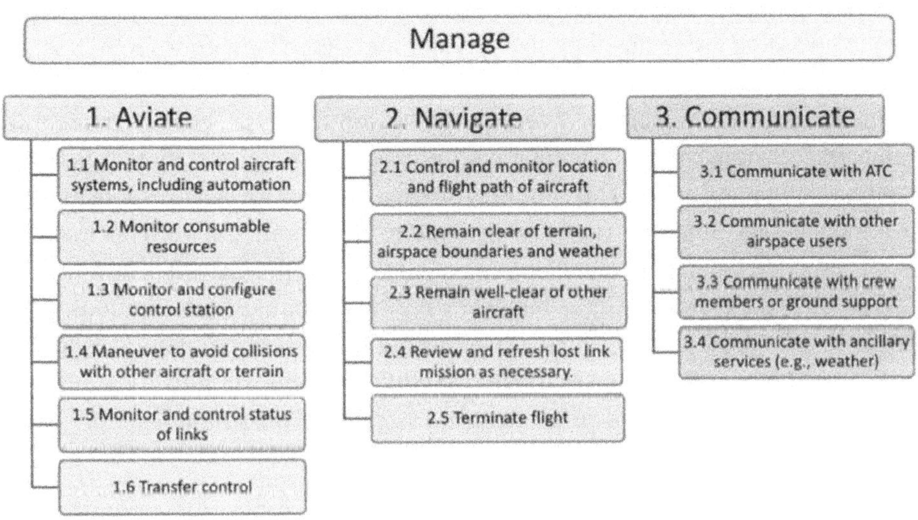

challenges for the remote pilot. A pilot of a conventional aircraft can lose almost all on board functions; navigation, communication, even flight control and propulsion, and may still be able to control the aircraft to a safe or safer outcome than a crash. Examples of the human's ability to Aviate include; 'Capt Sully Sullenberger who controlled his Airbus when both engines failed, on the Hudson River New York City.

With UAS becoming increasingly automated and as we move towards some components of the UAS utilizing machine learning algorithms, deterministic and non-deterministic [autonomous] neural networks, one must question, the boundaries that define human's ability to Aviate and the capability for human resilience to be replaced. It is a complex question with deep implications, *who is responsible for what?*

The ability to safely pilot an aircraft remotely is the foundational premise of the ICAO's, Remotely Piloted Aircraft, Concept of Operations document. This document is the operational reference that the RPAS panel follows to develop the SARPs that will evenly be codified into legislation by the 192 contracting states [countries] who are under the 'Chicago Convention'. At present, with regards to piloting the aircraft safely, it is mentioned throughout the document and continually referred to in the current 'draft SARPs' due to be incorporated into state legislation by 2023. The SARPs have been written with the directive assumption (assumption 8), that '*the remote pilot must be competent, licensed and capable to discharge the responsibility for safe flight'*.[3]

There are also implications on the safety assurance processes connected to safety flying the RPAS to mitigate undesirable events such as mid-air collisions or injury to people and damage to property on the ground[4]. Section; 2.6.2 of the ICAO RPAS CONOPS states that all system designs must ensure that the responsibility and liability for safe operation are retained by the operator and their flight crew (ICAO, 2017). Remote pilots must be able to override or modify automated functions, except where such actions cannot be executed safely due to the immediacy of the situation (e.g., an imminent collision avoidance manoeuvre) or where task complexity makes human intervention unreasonable. This guiding principal indicates a clear responsibility of the remote pilot to be able to firstly safety operate the aircraft, the word operates, and fly is used interchangeably in the context of controlling the aircraft. The CONOPS section 4.5.3 mentions in-flight handover between Remote Pilot Stations (RPS) that there are considerations during handovers that need clarifications as to what constitutes the remote pilot in commands ability and obligation to Aviate (ICAO, 2017). The section states; '*In either case, the safe and effective handover of piloting control from one station to another must be assured'*.[5]

Safety Culture and the Challenges for UAS Operators

Turner, Pidgeon, Blockley and Toft (1989) defined safety culture as "the set of beliefs, norms, attitudes, roles, and social and technical practices within an organisation which is concerned with minimising exposure of employees, managers, customers, suppliers and members of the general public to conditions considered to be dangerous or injurious" (p. 4). Simply put, it is a type of relationship that an organisation has with safety (Helmreich, 1998; Helmerich & Merritt, 2001).

The concepts of safety culture highlight a dynamic and multifaceted nature of humans and their behaviour, especially within the workplace or organization framework. Safety-related behaviour includes, but is not limited to, dynamic interfaces between psychological and organisational factors within the organisation, and occurs at different levels, from the individual to the collective, and pervading from top management. It has been described that safety culture as a composite and dynamic term, wherein the two components, safety and culture, can be defined independently depending on individual perspective. In reality, safety culture is a dynamic configuration of factors interacting on multiple levels within an organisation that influence and measurably determine total safety performance (Helmreich, 1998; Reason, 1990) The safety culture of an organization will directly impact safety performance of the organization.

The term safety culture first appeared in a report in the aftermath of the 1986 Chernobyl disaster (Cooper, 2000), where the errors and violations of operating procedures that contributed to the accident were identified by some experts as being evidence of a 'poor safety culture', (Salas & Maurino, 2010, p. 97). Historically, safety culture has been developed as a result of incidents and accidents. This is what is described as a reactive safety culture' (ICAO, 2012) whereby past accidents and mishaps escalated the evolution of safety policy and practices. Research findings (e.g., Stolzer, Halford, & Goglia, 2013; Helmreich & Merritt, 2001; Helmreich, 1988) suggested the importance of having, at the very least, a 'proactive' approach of safety management. From there, drive towards a predictive approach to safety, involving not just studying past statistics but shaping safety behaviours of the people within the system through developing, adopting and improving the safety policies, procedures and values over time.

Bethune and Huler (1998) added to this concept of safety culture stating that it could take organisations up to 10 years to achieve notable positive changes in behaviours. Hudson (2001) expanded that it was necessary to further develop organisational cultures that supported higher processes such as 'thinking the unthinkable' and being intrinsically motivated to be safe even when there seemed no obvious reason to do this. If having a strong positive, predictive safety culture takes a long time to evolve, where does that leave the exponentially growing vulnerable commercial UAS industry?

Safety culture is not an abstract concept. Components of safety culture can be developed, implemented and improved upon and thus, it is often termed a part of a 'living system'. Cooper (2000) added that the establishment and enhancement of a safety culture relied on the deliberate and targeted manipulation of various organisational characteristics potentially impacting safety management practices. Alternatively, safety culture can start, for instance, with one committed individual and filter through the whole organisation as it can and has been used as a motivating tool for the organisation, ultimately leading to achieving its safety goals. It is recognized that the most influential individual to drive an organization's safety culture is the top-level management, an accountable executive, who is usually the Chief Operating Officer (CEO) of the organization (ICAO, 2012).

Organizations that engage in safety critical activities are often called 'High-Reliability Organisations' (HROs) (Helmerich & Merritt, 2001; Reason, 1990). Examples of HROs' are: nuclear power plants, industrial installations, offshore installations, mine sites, rail organisations, wind farms, aviation and aerospace organisations, and medical facilities such as the emergency departments in hospitals.

HROs are acutely aware of and pay particular attention to their safety cultures as safety culture is a lead indication of safety performance (Helmreich & Merritt, 2001). As the benefits of using UAS are becoming widely accepted, commercial UAS operations are integrated with these HRO industries, and therefore; bring 'outside' influences into the established HRO safety culture (Lamb, 2017). Companies that achieve safety and efficiency in HROs share a common practice. Their safety systems and protocols are the critical focus, forming an integral part of '*how they do business',* in order to ensure a safe, efficient, healthy and productive workplace. Generally, HROs are highly dependent and technology centric, including software, data analytics, navigation technology and new composite hardware such as robotics and now, with the increase in small UAS platform, readily available 'Off The Shelf Solutions' (OTSS).

Once an organisation has decided to replace some of their traditional way of performing certain operations such as site inspection with small commercial UAS capability, there are usually enthusiastic discussions around the benefits of capturing data and imagery with UAS capability, and the tremendous financial benefits that are attached to that for stakeholders. Occupational Health and Safety (OH&S) managers are also enthusiastic about utilizing small commercial UAS, realising considerable 'safety savings', by not exposing their personnel to; working on ropes, scaffolding, at heights or in dangerous, potentially toxic or combustible environments.

However, there is a significant truism that must be raised—*people love gadgets*— especially gadgets that are innovative, fun and novel, and especially in industries that are technical in nature such as the HROs. There is a novelty factor with the introduction of a commercial UAS into an operation, and this factor has been an influence in more than one commercial drone mishap[6].

Introducing a small UAS into an inspection operation can entice curiosity, unintentional (and intentional) non-compliance, excitement and distraction. There is also a misconception that safety risks are extremely low or eliminated by using a UAS. However, this potentially introduces new latent risk producing conditions within the operation, particularly in unfamiliar areas. The danger with these latent conditions is twofold. The potential hazards are hidden within the systems, and the UAS crew and other company safety personnel are also in direct contact with the operation. Integration of the roles and responsibilities of UAS crew with the host team and establishment of how they will work together is a key enabler of UAS safety culture. This highlights the importance of practising non-technical skills in UAS operation such as effective communication.

Communication is one of the complex factors that affect safety at the crew level and the organisational level for both the UAS crew and the host organisation in HROs. Communication modes in UAS also utilise many and varied mediums including verbal, digital, and visual modes. Often verbal communication is augmented by digital channels such as Data-link (Ashdown & Cummings, 2007), VHF communication radio, walkie talkies, mobile phones, and Wi-Fi communications via laptop computers.

Effective communication has a significant impact on how safety culture, policy, procedures and checklists are executed. Effective communication is the essential social element on which a positive safety culture is established and sustained. Effective communication not only facilitates an effective team culture but enables the tone of the safety culture within complex systems. Research (e.g., Calhoun, 2006; Foltz, Martin, Cooke, Kiekel & Gorman, 2003) identified that it was fundamental to have a clear understanding of and to subsequently facilitate crew communication across all phases of a UAS flight for a generic safety culture to be established and maintained.

Communication challenges that are unique to UAS include crews that are geographically dispersed rather than co-located. According to Mouloua (2003), these challenges were primarily the results of time delay issues with satellite and data-link relays, lack of real-time feedback of control responses and interference or distortion of images and data. An awareness of the potential for errors in communications is essential when evaluating the potential risks in the UAS operation. A study conducted by Barshi and Farris (2013) based on analysis of more than 12,000 aviation incident reports from NASA's Aviation Safety Reporting System (ASRS) revealed that over 73% contained evidence of a problem in the transfer of information.

The commercial UAS crew and the host organisation will each have their unique method and style of communication, and there may also be differences in acronyms, terminologies and, often, very different phraseologies. What one phrase or acronym means to the host organisation may have a very different meaning to the UAS crew; CRM can mean 'Customer Relationship Management, or 'Crew Resource Management, and these two are very different concepts. In addition, communication methods, contents and styles often vary according to the operational type and environmental considerations. Smaller teams tend to communicate directly with each other which is commonly described as 'horizontal communication', whereas organisations with a number of teams controlled by a hierarchical structure, such as in many HROs tend to follow more 'vertical communication' style. An example includes the hierarchical vertical communication structure of the front lines workers, up to the foreman, the site manager, director of Health, Safety and Environmental (HSE), then Vice President (VP) of HSE, with communication being formal, structured and not in 'real time', whereas UAS crew tend to communicate horizontally and directly, usually in 'real time' with headphones via direct LTG network (phone) connections. Often UAS crews are smaller in numbers and include the higher level operational crew, who operate within a utilitarian corporate structure.

Research has identified that the most effective teams in complex control environments will engage in 'horizontal' communication of their shared mental models of the situation to ensure task affectivity (e.g., Cooke et al., 2007; Waller, Gupta & Giambastista, 2004). Waller et al. (2004) found that the most effective communication was usually between teams operating on the same level, rather than from a higher or lower hierarchical level and this suggested that the commercial UAS crew and the host organisation must work together to achieve the shared mental model and safety values prior to working together in the HRO to achieve effective safety policy, culture and behaviours. Cooke et al. (2007) concurred that although each member might understandably be attuned to different aspects of the same event, particularly an event in a complex team environment, team members were encouraged to share this information in order for the system to be coordinated.

Commercial UAS teams must be supported and integrated within the host organisation's culture, at both the top level (policy and procedures) and the same level of functional teams with which they will directly interact. This integration will facilitate effective communication that may support generative safety behaviours, especially when considering emergency response plans and roles in an emergency and unexpected situation.

The success of the emerging UAS industry will be determined by the willingness and ability of UAS crew and the end users to employ aviation safety philosophies, disciplines and the proven aviation safety culture model. The UAS industry needs more than a compliance philosophy to be successful (Lamb, 2017). A positive commitment to developing a UAS safety culture is paramount, and the core foundation of this is a widely accepted philosophy and framework of behaviours called Airmanship. The application of this type of safety culture to be adopted and integrated into the commercial UAV industry will lead to creating a new generation of Airmanship for a new generation of aviation.

Airmanship: Where Did It Come From and Where Is It Going?

The aviation safety culture is founded on collective principles called *'Airmanship'*. Kern (1997) described airmanship as an uncompromising discipline developed through systematic skill acquisition and continuing proficiency. The International Civil Aviation Organisation defines 'Airmanship' as… "*the consistent use of good judgement and well-developed knowledge, skills and attitudes to accomplish flight objectives*" (ICAO, 2011, P. 1-1). The principles of airmanship are indoctrinated to the potential aviator from the moment they step through the door of their flight training organisation and continues through every step of every day of their career. Implementation of the principles of Airmanship has provided consistent positive safety outcomes for these industries and forms an integral part of their brand of safety culture.

The airmanship model consists of bedrock principles, five pillars of knowledge and capstone outcomes. The five pillars are (Kern, 1997):

- Know yourself;
- Know your aircraft (UAS in this chapter);
- Know your team;
- Know your environment; and,
- Know your risk

The above pillars of knowledge require good aviators to draw on multiple knowledge bases (Kern, 1997). For instance, every commercial aviation organisation has its own '*brand*' of safety culture which defines their unique relationship with safety and what may differentiate them from other organisations. When experienced flight instructors and captains transfer to a new aviation organisation, indoctrinating them into that particular organisation's *'brand'* of safety culture and systems is of paramount importance and one of the essential 'onboarding' elements. This 'onboarding' or induction into an organisation's safety culture is an essential component in the airmanship knowledge pillar of 'knowing your team'.

Airmanship focuses on the conduct and attributes of the individual, their professionalism, skill, discipline, knowledge base and decision-making ability. This is an important concept to consider within the commercial UAS operation. However, while presented in the context of the individual, the principals of Airmanship can and should be applied to the greater team (i.e., all the crew involved in the UAS operation). This is of importance when it is considered that one of the key ingredients to the success of a UAS operation is having clearly defined roles and responsibilities as part of an effective safety management system (ICAO, 2012).

Airmanship is applicable to every team member and the role played by everyone within the UAS team. The performance of each team member contributes to the total system performance at any given part of the operation. For instance, when the UAS team leaders (e.g., remote pilot in command and managers) apply the principles of airmanship at the micro level (e.g., pre-flight task analysis, flight and task planning) right through to product delivery, the collective result achieves higher safety standards, greater efficiency and value in the end products (i.e., deliverables).

The level of professionalism of the UAS operator is often reflected in the deliverables. Deliverables are usually in a form of data such as images, measurements, items, frequency counts, or other detectable and quantifiable objectives. High quality and efficient service delivery include; safe execution of the operation, accurate data acquisition, processing, and storage. In many cases, interpretation and presentation of those data into relevance and 'value' to the end user is all a by-product of the level of professionalism of the UAS operator. The practice of airmanship principals, by each member of a UAS, will yield tangible benefits in customer satisfaction, safety and the long-term sustainability and reliability of UAS enabled services.

Dedicated Roles and the UAS Operation

Wickens (2007) described commercial UAS operations in HROs as a complex system requiring operators to be responsible for multiple tasks. Understandably, a UAS operator involved in a large commercial UAS operation has their role divided between navigation, flight control, communication, system monitoring, target or defect inspection, and mission management. On the other hand, in the case of the smaller commercial UAS operation conducting, for instance, inspection tasks in a complex environment, a UAS team may consist of only one or two crew members, giving a false impression that they are 'self-contained' or 'self-sufficient'. The host organisation may have the impression that providing the UAS team with the 'standard contractor' briefing before leaving them to work is an adequate safety procedure. However, research (e.g., van Breda, 1995) has shown that assigning all tasks to a single operator has been found to substantially degrade the operator's performance. Therefore, it is important to integrate and involve the UAS crew into the organisation's safety and operational team so that all crew have designated roles and a shared mental model of the operation, including what to do in the event of an emergency or 'non- normal' situation. To highlight the dangers of not providing the environment for UAS crew to interact with other onsite crew, consider the following real examples;

Example 1: During an operational safety audit of a commercial UAS HRO, it was found that the flight task had to be abruptly aborted due to an immediate safety issue. The flight activity of the contracted commercial UAS crew had not been communicated, nor had any action been taken to integrate the UAS operation or its crew into the host organisation's site. The incident caused a major distraction to the host organisation's work area and created the potential for several serious accidents. While no injuries or damage occurred, the cost was incurred due to supply interruption, lost productivity, potential OH&S injuries, and delay in data collection and processing all of which represented two weeks of lost revenue for the host organisation.

Example 2: During an operational safety audit of a commercial UAS operation, on behalf of the host organization, it was found that during one of the flight tasks, one of the site workers accidentally 'ran over' part of the UAS equipment. The driver of the vehicle was unaware of the parameters of the operational (launch and recovery) area of the UAS crew and was distracted by the UAS activity. The damage to the equipment was substantial and had to be replaced. This also meant that the UAS surveillance activity had to cease until new equipment was acquired.

There are a number of procedures involved in commercial UAS operations. One of them is a '*hand over*', also known as a '*transfer of control*' which is described as the role and responsibilities of one crew member (the remote pilot(s)or members of the UAS team) are transferred to another remote pilot(s) or UAS crew members. Such a procedure is a deceptively complex task that may occur at the change of shift during the flight or ground operations. These handovers can be a time of increased exposure to risks, especially those risks associated with system mode errors and coordination breakdowns. Hobbs and Lyall (2016) found that there had been cases of inadvertent transfer of control between remote pilot control stations, due to controls set in error. One of the most cited cases that highlight crew co-ordination factors that lead to a mishap is the crash of the United States Department of Defence, Predator B aircraft in Nogales. The causal factors cited in the accident were; human factors, organizational failures, system design and integration (Carrigan, 2008).

Safety in UAS can only be attained if defined and specialised roles are developed with specific training procedures and philosophies, including precision system-based processes for crew changes and shift handovers (AAIF, 2011). Successful UAS crews appear to be evolving to embrace traditional safety procedures and practices that have proven efficient and increased safety in aviation, medicine and other safety critical operations handovers. The use of precise and appropriate checklists, methodical preparation of both human and material assets is now identified as critical to reduce errors. The power of scenario-based training (simulation) has also proven to reduce handover error in many industries (Flynn, 2008) and UAS operations will yield the same safety benefits if quality scenario-based training is employed.

CONCLUSION

As we drive towards harmonious integration of UAS into both our lower and upper airspace, we face a considerable number of new challenges. Of these challenges, human performance and human systems integration pose the greatest threats to seamless integration. Not surprisingly, the closer to earth (especially below 400ft above ground level), the more challenges arise for UAS integration, the complexities of package delivery, urban mobility and a new Unmanned Traffic Management System (UTM) and these provide not only logistical challenges but safety, certification and training challenges. To achieve this harmonious integration of man, machine, air traffic systems and commercial industry, we must begin at the foundational level, the human operator.

The interaction between human roles and responsibilities, between the technology components, the software, the platforms, and the UTM must be interoperable and harmonised if the system is to be safe and effective. Consider that this is not only at the local area but on a global scale. Achieving this harmonisation will be reliant on the strength of a robust, appropriate safety culture, applied airmanship principles, training, education and certification. Humans will remain the key component to the success of unmanned aviation even as the automation levels continue to increase towards the highest levels of automation, and arguably higher into the levels of non-deterministic system intelligence (autonomy].

Integrating a commercial drone operation into an existing organisation's safety management system and safety culture can provide many challenges on various operational levels. There are human factors challenges, policy and procedural challenges, logistical challenges, possibly financial challenges and the challenge of managing change and expectations. To successfully cope with these challenges and integrate drones into your organisation, the solution may be found by starting at the individual level, educating and cultivating the *Airmanship* principles. This may also have the added benefits of impowering individuals to assist in building the new safety culture.

It is important to understand that safety culture exists in some form or another in any given organisation. For instance, it could be reactive, proactive, punitive, generative, bureaucratic, and predictive. Even the lack of an organised safety culture indicates the organisations' relationship with safety, and usually, it is the key personnel who are responsible for implementing and cultivating the safety behaviours and attitudes of the organization.

Safety cultures can and do evolve unintentionally, usually as a result of drifting away from a disciplined approach to safety, towards the easier, cheaper or faster way of operating. Unintentional safety cultures are usually the by-product of some of the following factors; misalignment of senior management with safety protocols, lack of appropriately qualified safety leader, missing safety oversight and or accountability, deficient regulatory framework, a 'blame culture', lack of education, lack of resources, lack of safety information sharing (and much more).

Many of the aforementioned factors are contributing to the industry culture that is being experienced now, especially in the small commercial UAS industry, operators are protective about disclosing the nature of their mishaps, safety data, contributing factors, and any corrective measures implemented in a highly competitive industry are often viewed as a competitive edge. This is a very different culture to that of manned aviation. Therefore, we must accept this difference and work with these concerns and challenges to unite the industry under common safety interests. With many countries still grappling with UAS legislation, some small commercial UAS operators have little professional industry support or guidance, unlike the traditional aviation community or model aircraft culture. This is a time where regulators and industries in all countries can and must work together to find the solutions to safety challenges.

The success of the emerging commercial UAS industry will be determined by the willingness and ability of organisations to adopt not only conventional aviation safety philosophies and disciplines but work together to develop a specific UAS safety philosophy, supported by Safety Management System (SMS) and Quality Management System (QMS) principals, cultivate UAS safety Culture at the individual level. This UAS safety philosophy will be successful if it embraces and supports the unique challenges of integrating professional UAS crews and their equipment into our national airspace, society and operational environments.

REFERENCES

AAIF. (2011). *AAIF Certification and regulation working group Unmanned aircraft systems sub-committee.* Australian Aerospace Industry Forum, Adelaide, Australia.

Ashdown, M., & Cummings, M. L. (2007). *Asymmetric synchronous collaboration within distributed teams.* Paper presented at the 7th International Conference Engineering Psychology and Cognitive Ergonomics. 10.1007/978-3-540-73331-7_26

Barshi, I., & Farris, C. (2013). *Misunderstandings in ATC Communication: Language, Cognition, and Experimental Methodology*. Surrey, UK: Ashgate.

Bethune, G., & Huler, S. (1998). *From Worst to First*. New York: John Wiley & Sons.

Carrigan, G., Long, D., Cummings, M. L., & Duffner, J. (2008). *Human Factors Analysis of Predator B Crash.* Paper presented at the AUVSI 2008: Unmanned Systems North America, San Diego, CA.

Civil Aviation Authority. (2016). *ATM Automation: Guidance on human technology integration. (CAP 1377)*. West Sussex, UK: Author.

Cooke, N. J., Gorman, J., Pedersen, H., Winner, J., Duran, J., Taylor, A., ... Rowe, L. (2007). *Aquisition and Retention of Team Co-ordination in Command and Control.* Mesa, AZ: Cognitive Engineering Research Institute.

Elwell, D. (2017). *Testimony - Statement of Dan Elwell.* Retrieved from https://www.faa.gov/news/testimony/news_story.cfm?newsId=22354&omniRss=testimonyAoc&cid=105_Testimony

Federal Aviation Administration. (2015). *Safety Management Systems for Aviation Service Providers (AC 120-92B)*. Washington, DC: Author.

Federal Aviation Administration. (2016). *AC107-2 Small Unmanned Aircraft Systems (sUAS)*. Retrieved from https://www.faa.gov/regulations_policies/advisory_circulars/index.cfm/go/document.information/documentID/1019962

Federal Aviation Administration. (2018a). *Pilot's Handbook of Aeronautical Knowledge (FAA-H-8083-25A)*. Washington, DC: Author.

Federal Aviation Administration. (2018b). *Part 107 waivers issued.* Retrieved from www.faa.gov: https://www.faa.gov/uas/request_waiver/waivers_granted/

Flatley, D. (2017). *Safety Incidents Involving Civilian Drones are Surging*. Retrieved from https://www.bloomberg.com/news/articles/2017-12-05/safety-incidents-rise-along-with-popularity-of-civilian-drones

Flin, R., O'Connor, P., & Crichton, M. (2008). *Safety at the Sharp End: A Guide to Non-Technical Skills*. Surrey, UK: Ashgate.

Foltz, P. W., Martin, M. J., Cooke, N. J., Kiekel, P. A., & Gorman, J. (2003). Automated Team Communication Analysis with Latent Semantic Analysis. *Proceedings of the Collaborative Technologies Alliance: Advanced Decision Architectures,* 33-40.

Gawron, V. (2008). *Human Performance, Workload and Situational Awareness Measures Handbook*. Boca Raton, FL: Faylor & Francis. doi:10.1201/9781420064506

Gloria, L., & Calhoun, M. H. (2006). Advances in Human Performance and Cognitive Engineering Research. In H. L. Nancy & J. Cooke (Eds.), *Multi-Sensory Interfaces for Remotely Operated Vehicles* (pp. 149–163). Emerald Group.

Goldman Sachs. (2017). *Estimated Drone Spending in USD FY2017-2021*. Retrieved from https://www.goldmansachs.com/insights/technology-driving-innovation/drones/

Grand View Reserach. (2016). *Comercial UAS market size to reach $2.07 billion by 2022*. Retrieved from https://www.grandviewresearch.com/press-release/commercial-uav-market-analysis

Helmerich, R., & Merritt, A. (2001). *Culture at Work in Aviation and Medicine, Organisational and Professional influences*. Surrey, UK: Ashgate.

Helmreich, R. (1998). Building safety on the three cultures of aviation. *IATA: Human Factors Seminar*, 39-43.

Hobbs, A., & Lyall, B. (2016). *Human Factors Guidelines for Remotely Piloted Aircraft System Remote Pilot Stations (TN-34128)*. San Jose, CA: NASA Ames Research Center.

Hudson, P. (2001). Safety Culture: The Ultimate Goal. *Flight Safety Australia*, 9-31.

ICAO. (2012). *Safety Management Manual*. Montreal: International Civil Aviation Organization.

International Civil Aviation Organisation. (2009). Safety Management Manual (2nd ed.; Doc 9859). Montreal, Canada: Author.

International Civil Aviation Organisation. (2011a). *Unmanned Aircraft Systems (UAS) (ICAO Cir 328)*. Montreal, Canada: Author.

International Civil Aviation Organisation. (2011b). *Personnel Licensing*. Montreal, Canada: Author.

International Civil Aviation Organisation. (2015). *Manual on Remotely Piloted Aircraft Systems (RPAS) (Doc 10019)*. Montreal, Canada: Author.

Jenkins, D. J., & Vasigh, B. (2013). *The economic impact of Umanned Systems integration in the United States*. Arlington, VA: AUVSI.

Kern, T. (1997). *Redefining Airmanship*. New York, NY: McGraw-Hill.

Lamb, T. (2018, October 28). *Factors Influencing Small Commercial UAS Mishaps.* Unpublished work, Embry-Riddle Aeronautical University, PhD program. Daytona Beach, FL: Embry-Riddle Aeronautical University.

Lamb, T. (2017). *Developing a Safety Culutre for Remotely Piloted Aircraft Systems Operations: To boldly go where no drone has gone before.* Society of Petroleum Engineers, SPE-84476-MS-MS.

Mouloua, M., Gilson, R., Kring, J., & Hancock, P. (2001). Workload, situation awareness and teaming issues for UAV/UCAV operations. In *Proceedings of the Human Factors and Ergonomics Society 45th Annual meeting* (p. 162). Sage. 10.1177/154193120104500235

Mouloua, M. G. (2003). Designing controls for future unmanned aerial vehicles. *Ergonomics in Design, 11*(4), 6–11. doi:10.1177/106480460301100103

Orlady, H. W., & Orlady, L. M. (2002). *Human Factors in Multi-Crew Flight Operations.* Aldershot, UK: Ashgate Publishing Company.

Parasuraman, R., & Riley, V. (1997). Humans and Automation: Use, Misuse, Disuse, Abuse. *Human Factors, 39*(2), 230–253. doi:10.1518/001872097778543886

Parasuraman, R., & Wickens, C. (2008). Humans: Vital after all these years of automation. *Human Factors, 50*(3), 511–520. doi:10.1518/001872008X312198 PMID:18689061

Peace, C. (2017). The Risk Matrix; Uncertain results? *Policy and Practice in Health and Safety, 15*(2), 131–144.

Plioutsias, A., Karanikas, N., & Chatzimihildidou, M. M. (2018). Hazard Analysis and Safety Requirements for Small Drone Operations: To What Extent Do Popular Drones Embed safety? *Risk Analysis, 38*(3), 562–584. doi:10.1111/risa.12867 PMID:28768049

Radi, A. (2013). *Human injury model for small unmanned aircraft impacts.* Canberra, Australia: Civil Aviation Safety Authority.

Rash, C. E., LeDuc, P. A., & Manning, S. D. (2006). Human Factors in U.S. Military Unmanned Aerial Vehicle Accidents. In H. L. Nancy & J. Cooke (Eds.), Advances in Human Performance and Cognitive Engineering Research (pp. 117 - 131). New York: Emerald Group.

Rasmussen, J., & Svendung, I. (2000). *Proactive Risk Management in a Dynamic Society.* Karlstad: Swedish Rescue Services Agency.

Reason, J. (1990). *Human Error*. Cambridge, UK: Cambridge University Press. doi:10.1017/CBO9781139062367

Robinsons, R., & Francis, G. (2014). *SFAIRP vs ALARP. In Conference On Railway Excellence* (p. 7). Adelaide: RTSA.

Salas, E., & Maurino, D. (2010). Human Factors in Aviation. Oxford, UK: Elsevier. doi:10.1016/B978-0-12-374518-7.00001-8

Stolzer, A. J., Halford, C. D., & Goglia, J. J. (2013). *Safety Managment Systems in Aviation*. Burlington: Ashgate.

Tvaryanas, A. P., Thompson, W. T., & Constable, S. H. (2006). Human Factors in Remotely Piloted Aircraft Operations: HFACS Analysis of 221 Mishaps over 10 Years. *Aviation, Space, and Environmental Medicine*, *77*(2), 724–732. PMID:16856358

van Breda, L. (1995). *Operator Performance in Multi Maratime Unmanned Air Vehicle Control.* Soesterberg, The Netherlands: Human Factors Research Institute.

Waller, M. J., Gupta, N., & Giambatista, R. C. (2004). Effects of Adaptive Behaviors and Shared Mental Models on Control Crew Performance. *Management Science*, *50*(11), 1534–1544. doi:10.1287/mnsc.1040.0210

Wickens, C., & Colcombe, A. (2007). Dual-Task Performance Consequences of Imperfect alerting associated with a cockpit display of traffic information. *Human Factors*, *49*(5), 839–850. doi:10.1518/001872007X230217 PMID:17915602

ENDNOTES

1. Pilots refer to this as '*Hangar talk*'.
2. The C^2 link is defined as the communications, *command and control* link to the remotely piloted aircraft. It is usually the only direct link between the remote pilot's ability to control the aircraft. The C^2 link is able to be provided for via several mediums; satellite, WIFI, telecommunications networks.
3. ICAO RPAS Concept of OperationS (CONOPS) (March 2017), page 5.
4. ICAO RPAS CONOPS (March 2017) page 22.
5. ICAO RPAS CONOPS (March 2017) page 19.
6. During 3 years as an ISO accredited QMS and Safety auditor, 3 separate incident reports indicated that distraction of workers due to 'watching' a UAS conduct its mission, occurred. One incident resulted in a vehicle colliding with the UAS crew launch equipment. 2014-2016 Australia.

Chapter 10
The Future for Civilian UAV Operations

Tarryn Kille
University of Southern Queensland, Australia

Paul R. Bates
University of Southern Queensland, Australia

Seung Young Lee
University of Southern Queensland, Australia

David Murray Kille
University of Southern Queensland, Australia

ABSTRACT

The future looks bright for unmanned aerial vehicles (UAVs). Their ability to carry sophisticated imaging equipment attached to lightweight vehicles, to hover in position despite incremental weather conditions, to fly simple missions, and takeoff and land automatically, combined with their comparatively (compared to manned aircraft) lower investment and operational costs has driven a paradigm shift in the history of air transport. This chapter is organized around six themes that underscore the current discourse regarding the future of UAVs in civilian commercial operations, as well as highlighting the discussions of the previous chapters regarding policy and certification, technology, training, social and economic forces, air cargo, and the effect of UAVs on other sectors of the air transport industry.

DOI: 10.4018/978-1-5225-7900-7.ch010

INTRODUCTION

The future looks bright for Unmanned Aerial Vehicles (UAVs) (Albaker, 2013; Peterson, 2006). Their ability to carry sophisticated imaging equipment attached to light weight vehicles, to hover in position despite incremental weather conditions, to fly simple missions, and takeoff and land automatically, combined with their comparatively (compared to manned aircraft) lower investment and operational costs, has driven a paradigm shift in the history of air transport (Perritt & Sprague, 2017).

Their current applications will see them provide aviation support to markets and economies where current manned aircraft are unable to operate (Wargo, Church, Glaneueski, & Strout, 2014). Their novelty is already causing heated debate from a political, public and regulatory perspective (Hail & Coyne, 2014). The discourse is often divided between those that are optimistic about the potential applications and solutions to problems that UAVs can offer, and those that oppose based on the fear of perceived risks to safety, security and privacy that UAVs pose (Elias, 2012).

This chapter is organized around six themes that underlie the current discourse about the future of UAVs in civilian commercial operations, as well as the discussions of the previous chapters regarding: policy and certification; technology; training; social and economic forces; air cargo; and the effect of UAVs on other sectors of the air transport industry.

POLICY AND CERTIFICATION

It is apparent from the preceding chapters that Unmanned Aerial Vehicle (UAV) operations in the civilian environment are on a forward trajectory, increasing in numbers and applications with more complex and demanding missions. Currently, system developments in this field are driven by the preferences and inclinations of manufacturers and users. Such development tends to lead, inevitably, to a vast array of control station configurations, internal vehicle software and a perplexing mix of pilot-vehicle interfaces. Predictably, well considered Human Systems Integration (HSI) supported by holistic systems engineering approach is illusive and rarely applied (Bennet, Bridewell, Rowe, & Craig, 2016; Gawron, 1998). Hence, operator and vehicle certification has become a significant cause for concern (Du & Heldeweg, 2018)

The variations in the dimensions and sizes of UAVs range from very small (under 25 kilograms) to very large (over one tonne) (Perritt & Sprague, 2017) provide an additional complication to the situation. Moreover, UAVs are being used to respond to an extensive variety of existing and rapidly emerging needs in commercial and consumer applications. These emerging needs in the commercial and consumer sector include examples such as agricultural surveying and crop inspection, motorway surveillance, bridge inspection, vaccine delivery and package delivery (De la Torre, Ramallo, & Cervantes, 2016).

The global political and regulatory environment encompassing UAV operations will continue to be problematic (Kreps, 2014). The Federal Aviation Administration (FAA) released a Notice of Proposed Rule Making (NPRM) in February 2015, which commits to the development and establishment of rules governing the operation of small remotely piloted aircraft (under 25 kilograms) (Jiang, Geller, Ni, & Collura, 2016). However the NPRM raised a number of questions and does not address the rapid emergence of the UAVs weighing more than 25 kilograms. Since then, Part 107 has been published. "Part 107" refers to Part 107 of Chapter 14 of the Code of Federal Regulations published by the FAA (Olsen, 2017). This rule provides a regulatory framework that every drone pilot must follow in order to commercially fly an unmanned aerial vehicle (UAV), or drone. This rule includes operational limitations, pilot responsibilities, and aircraft requirements.

To be effective, rules need to be realistically enforceable with the appropriate budget and staff provided to the enforcing authority. Part 107 also introduced a new category of pilot license for the UAV operator. The world of aviation and aerospace is facing a paradigm shift in air transportation (Lacher & Maroney, 2012). This shift is grounded in autonomous flight and UAV technology. In previous chapters we have established that today, UAVs are used for such tasks as inspecting tracks or power lines and assessing bushfires. However, the UAVs of the future will be larger models, capable of transforming industries such as the construction and retail industries by carrying freight to difficult to reach places (Valavanis, 2007). In less than a decade, we may see the easing of traffic congestion and urban pollution from the increased use of electric unmanned aircraft transporting people or products around cities (Goodchild & Toy, 2018).

Yet, reaching that potential will require holistic policy approaches and new regulations. Additionally, reaching the future potential of UAV applications will need to be coupled with significant advances in technology to enhance safe operation. Although we have seen a sense of impatience among the business community who perceive that regulators are undertaking a go-slow approach to UAVs (Luppicini & So, 2016), a lack of system or process to sufficiently test and validate these rapidly emerging technologies could severely hinder the future development of this revolution (Rao, Gopi, & Maione, 2016) with the first human death from a UAV mishap. Prior to this technology being embraced for large-scale commercial use, the risks (especially those posed by larger UAVs) must be identified, and strategies to mitigate those hazards need to be developed (Weibel & Hansman, 2005).

In the United States, much of the federal regulations on UAVs restrict their use: UAVs are not permitted to fly over most federal facilities or over people; UAVs are not permitted to fly at night or within five miles of an airport without appropriate permission; UAVs are required to fly below 400 feet (Vincent, Leland, & Ditmer, 2015) and at less than 100 miles per hour; with some exceptions, they must weigh under 55 pounds (25 kilograms); and they must give right of way to manned aircraft (Federal Aviation Administration, 2016). In addition, some states are debating further restrictions on the use of UAVs in hunting (Schneider, 2014); in California, UAVs can not be used to record another person without first receiving consent. Another restriction, that presents one the most significant limitations on UAV operations when it comes to the widespread commercial use of UAVs is that they must be kept in line of sight of the operator at all times.

Recently, a congressionally mandated report from the National Academies of Sciences, Engineering, and Medicine scolded the FAA, for focusing on the risks presented by UAV operations, rather than their potential benefits. "Fear of making a mistake drives a risk culture at the FAA that is too often overly conservative, particularly [with] UAS technologies," (p.20) the National Academies report concluded (National Academies of Sciences, 2018).

Those supporting UAV development are calling for regulation to be relaxed on flying smaller UAVs further than the limit of sight (Chen, 2017). Perhaps the Academies' criticism may be valid here, as there can be no substantial commercial application if businesses are required to seek permission from the FAA on every occasion that business calls for the UAV to fly beyond an operator's line of sight.

There is also some ambiguity in the national and global political environment. In Australia, the public concerns regarding privacy are yet to be addressed (Molnar & Parsons, 2016). The Privacy Act, for example, only applies to organisations with an annual turnover of $3 million or more. With many recreational UAVs now equipped with camera's, the public concern over privacy has become an increasingly vexing issue for governments and commercial UAV operators. Most recreational drone owners, it's fair to say, wouldn't meet the criteria for the Privacy Act. As such, the legality associated with the use of UAVs to spy on others is still not clear.

In the US, 14 CFR Federal Aviation Regulations provide robust specifications and standards regarding airworthiness and certification for manned aircraft. The demand for larger UAVs for commercial applications means that there will be a need to integrate these larger UAVs within the airspace of manned aircraft (Elias, 2012). This realization highlights the important need to apply equally rigorous airworthiness and operational standards to UAV aircraft.

One alternative is to allow UAV certification to be governed by "industry consensus standards" (Riganati & Harrison, 2016) in a similar fashion to that of Light Sport Aircraft category. In Australia, Recreation Aviation Australia is the peak body in Australia responsible for administering ultralight, recreational and Light Sport Aircraft

(LSA) operations. However, this option is not conducive to the appropriate level of rigour and regulation required for global military and commercial UAV activity.

One of the most concerning issues is the pressing matter of air traffic management and collision avoidance (Dalamagkidis, Valavanis, & Piegl, 2008). With respect to manned aircraft, traffic management and collision avoidance are supported by procedural separation and the application of the principle of 'see-and-avoid'. Procedural separation is the favoured mode in comprehensively controlled airspace (e.g., operating above 18,000 feet). However, this also becomes problematic in airspace where controlled traffic (such as aircraft operating under Instrument Flight Rules) and uncontrolled traffic (such as aircraft operating under Visual Flight Rules) are mixed.

The 'see-and-avoid' strategy may not be applicable to UAV operations, and thus a new term 'sense-and-avoid', for UAV operations has emerged (Yu & Zhang, 2015). As we experience denser air space, resulting from increased air traffic and commercial UAV operations proliferate, a 'sense-and-avoid' program of operation must consider the needs of operators and decision makers. As we have discovered in chapter one, the challenges associated with data link latency, data displays and appropriate levels of automation and autonomy must be resolved to the satisfaction of safety regulators, operators and the public (those on the ground and those traveling by air).

The introduction of UAV operations must not compromise the safety of manned aircraft. However, the growth of UAV and their commercial applications will need to be supported with appropriate regulations (Weibel & Hansman, 2005).

TECHNOLOGY

The civilian UAV industry will benefit from the technologies developed in other industries. Likewise, other industries will benefit from drone technology.

New Manned Aircraft Designs

Engineers of manned aircraft are innovating. Sometimes, such innovation is hindered by the burden of FAA airworthiness certification for new systems (Allen & Berg, 2018). However, the miniaturized, light weight, low-power innovative systems developed in the UAV industry will also be of interest to manned-aircraft system designers and engineers. Such UAV technology may provide suitable data to sway FAA certification for broader aviation use. In any event, the innovation born in the manned aircraft sector will certainly be of influence to UAV development.

Aircraft have generally taken on the same appearance since the 1930s in terms of wings with ailerons controlling pitch, and a rudder, controlling yaw. Engines are located in the aircraft nose, or attached to wings. Similarly, helicopters have also held the same appearance since the 1950s with a main rotor controlling pitch blades and a tail rotor to compensate for the engine torque.

Small UAVs have not taken on the appearance of manned aircraft. The miniaturization of electronics and the enhancements in battery technology and other innovative design features of electric propulsion systems have allowed small UAVs to develop in ways that will influence manned aircraft designs over the coming years (Perritt & Sprague, 2017).

Vertical Takeoff and Landing (VTOL)

VTOL aircraft are observed in military operations, yet this technology is receiveing increased attention from the commercial sector. There is already evidence that this technology is influencing UAV configurations (Prisacariu, Boscoianu, & Luchian, 2014).

Tilt-rotor configurations will receive a boost as the Agusta-Westland AW609 commences services with Bristow (Bristow, 2015), a significant helicopter operator in the oil and gas industry, entices additional customers. The AW609 and similar designers will remain more expensive than the standard commercial aircraft and helicopter designs (Perritt & Sprague, 2017). However the market demand will clarify which missions justify the additional costs.

The AW609 markets the flexibility of the aircraft and productivity in the oil and gas industry, as this aircraft type provides capabilities that enable it to substantially reduce trip time while it's helicopter capabilities allow it to operate from pumping and drilling rigs to existing heliports/helipads on the land. The AW609 is also marketed for emergency support services, as the technology provides the aircraft with the ability to not only rescue accident victims from difficult and unprepared areas, but it also has the range and speed to perform over long distances (Udroiu & Blaj, 2016).

These concepts and innovations are already being seen in UAV technological developments. The Aerovel Flexrotor is a hybrid VTOL/fixed wing configuration that weighs just over 20 kilograms. It has a wing span of 3 feet and a length of 2 feet. The Aerovel Flexrotor has the ability to fly for up to 40 hours on ocean and land surveillance missions. It has the capability of carrying an electro-optical camera. Enhancements to the engine may see this aircraft with the ability to carry heavier payloads and other sensors. The single propellor allows the UAV to take off and land vertically and is powered by a small gasoline engine. The manufacturer of the Aerovel Flexrotor is targeting the civilian sector, with a marketing strategy that highlights the flexibility of this UAV in the face of significant absence of ground infrastructure, but also affordability.

It is interesting to note that Amazon's preliminary design for the package delivery trial also featured a hybrid VTOL/helicopter design (Antcliff, Moore, & Goodrich, 2016). This technology and developments is likely to impact both manned and UAV aircraft design. The flexibility of operations, combined with payload and endurance capabilities is promising for commercial UAV applications.

Hybrids

The progressive enhancements in battery technology, combined with the captivating promises of electric propulsion will invigorate interest in hybrid aircraft power. Because battery technology remains a current limitation on realizing the complete advantages that electric propulsion systems have to offer, we are experiencing greater experimentation with regard to hybrid propulsion systems (Capata, Marino, & Sciubba, 2014). Hybrid propulsion technology, satisfying interim needs, is not new to the transport industry. The petrol engine combined with an electric motor has been seen in the automotive industry (Westbrook, 2001). In fact, it appears that a number of major automotive manufacturers are concluding that hybrid approaches have superior performance capabilities over pure electric vehicles in the car and truck markets (Romm, 2006).

Airbus, has a long standing business strategy focusing on innovation in technology (Bowonder, Dambal, Shambhu, & Shirodkar, 2015). This aircraft manufacturer is also emphasizing hybrid approaches through its electric aircraft development (Airbus, 2017). Airbus is making significant investments into a family of hybrid aircraft, in the rotary and fixed wing markets. This strategy has led Airbus to recently release a smaller version of this technology in the trainer and UAV markets. This has sparked the interest of customers and provides a suitable platform from which to attain much needed data about operational reliability, flexibility and costs. Depending on the UAV actual fuel consumption, endurance, and range, these configurations may offer suitable alternatives to conventional designs.

As chapter one and chapter three highlighted, autonomy and autopilot capability will continue to be at the forefront of research and design. In manned aircraft fleets, category 3 systems of autopilots are able to fly entire flight routes from takeoff to landing and ground taxi to the gate. As the price of such technology reduces, it is likely that such technology will be installed on the existing fleets of aircraft and it is fast becoming a regular option for new aircraft.

It is apparent that three-axis autopilots in helicopters (initially excluding the takeoff and landing phases) are penetrating the commercial helicopter fleet (Bower, 2004). While the early adopters were in the larger helicopters supporting the oil and gas industry, there has been a widening expansion to single-engine helicopters. This has had the effect of reducing rates of accidents and inadvertent flights into Instrument Meteorological Conditions (Matthes, Alexander, & Stone, 2018).

The technological development in more autonomy for manned aircraft will provide a useful platform for discoveries and the eventual wider acceptance of larger drone operations in the civilian sector. In August 2015, an incident involving the crew of a Delta Airlines flight used auto-pilot to avert a likely disaster (Hradecky, 2015). After experiencing severe hail damage, the flight crew were unable to see through the windshield. By relying on the aircraft's autoland capability, the crew were able to land the aircraft safely (Perritt & Sprague, 2017). Such case studies will support the argument that autonomous UAV aircraft can operate safely.

NextGen Navigation Systems

A current mandate exists that by 2020, all manned aircraft (with some exceptions) must be equipped with ADS-B out (Zimmerman, 2013). ADS-B in is somewhat cheaper and as such, is likely to motivate the widespread utilization of this technology. ADS-B is a system that allows aircraft to transmit position, speed, and directional data to each other while in flight, which offers the opportunity to avoid other traffic automatically (Zimmerman, 2013).

For smaller general aviation aircraft, the cost of such systems is between US$2000 and US$3000 (Moorman, 2017). It is likely that this cost will fall by 2020 (Perritt & Sprague, 2017). The rapid emergence of UAV operations and UAV fleets combined with safety regulators pursuit for a suitable system to allow safe flight beyond line of sight for UAV operations will provide a strong impetus for developers to reduce weights, and power consumption so they are able to take advantage of the UAV market also (Pahsa, Kaya, Alat, & Baykal, 2011).

UAVs equipped with ADSB-out will be electronically visible to manned aircraft that are equipped with ADS-B in. UAVs that are fitted with ADS-B in will be electronically aware of the position of manned aircraft within its proximity.

The current prices, weights and power requirements may restrict the integration of this technology in small UAVs. However, those operators that wish to operate larger UAVs within manned aircraft airspace will likely recognize the advantages of such equipment. Installing the equipment is relatively straight forward, and resolves any issues regarding collision avoidance that may prevent the useful applications of UAVs now and into the future.

The end game of integrating UAVs into NextGen will effectively respond to the risk-based objections to opening up the airspace more broadly to UAV operations alongside manned aircraft. NextGen data communications will gradually replace

AM voice radio communication for delivering and confirming ATC clearances. Such developments will provide the appropriate environment for the integration of UAVs.

TRAINING

UAV operator selection and training in the commercial sector, as opposed to the military sector, are emerging issues of concern requiring immediate attention and clarification. Our focus, in this book, has been in the use of UAV technology in logistics and supply chain management. Many of the underlying applications require UAVs to occupy and operate in airspace. With this in mind, national government bodies will have regulatory authority over the training outcomes of commercial enterprise operators.

The US military services have highly developed training programs and systems for developing well equipped UAV pilots and systems operators. Even though these operators are qualified to operate in military-controlled airspace, they are not necessarily accepted by the FAA to fly commercial (or civil) UAVs in civil airspace. This certification requirement is similar to the manned aircraft circumstance, (i.e., a US military pilot must receive addition training and certification from the FAA before being permitted to operate civil aircraft). Such rules apply to training as well as medical certification. By extension, it is unlikely that military training standards and procedures will be applied, without some modification and adjustment to the needs of the civil sector.

In Australia, the Civil Aviation Safety Authority has published rigorous rules on commercial UAV operations (Civil Aviation Safety Authority, 2019) . This includes the need for operators to hold a Remote Pilot Licence, which can only be attained after successfully completing training and certification through an approved training organization that holds a Remote Pilot Operating Certificate. Additionally, commercial UAV operations can only be undertaken by operators operating under an approved Remote Pilot Operating Certificate issued by CASA.

In the United States, for manned aircraft, the FAA specifies in 14 CFR 61 and 14 CFR 145, the training required for pilots and the experience levels required for the various levels of qualifications. As demand for UAVs expands in the commercial sector, and the applications become more complex, it is important that safety regulators ensure that training and certification standards are rigorous, and transparent.

While regulation of training standards is an important challenge to UAV operations, of equal concern is the need for further research into non technical skills (NTS) training. Within any flight, there are many complex layers of the operation such as technological and behavioural layers and protocol. NTS are an essential layer as they reduce the likelihood of an error or failure leading to an accident. NTS are important social, mental and personal management skills for safer and more efficient operations. NTS also complement the technical skills of flight crew and contribute to dependable and effective performance in the complex work systems that characterise the aviation industry. NTS in the aviation industry typically include: situational awareness; workload/task management; decision-making; communication; team work; leadership; managing stress; and coping with fatigue.

More research is required which specifically focuses on the human – machine interactions and the human factors associated with such interactions. With a greater awareness of how these challenges impact on UAV operators, the industry will be better prepared to provide the training programs that focus on developing both the required technical and non technical skills for safe and efficient UAV operations.

SOCIAL AND ECONOMIC FORCES

Markets are often influenced by the law, and the law is often propelled by politics, which in turn, is motivated by public opinion. How market-oriented economies operate will generally be dependent on not only technology and economics, but also on the social and environmental forces that shape the decisions made by key stakeholders, and on regulators' attempts to improve the functioning of markets and remove the harmful externalities.

The following subsections highlight the important facets of environmental forces, concerning energy, noise, and privacy, that can influence regulation, aviation markets and UAV operations within those markets.

Energy

Concerns regarding energy usage has been thoroughly embedded in the public mind as well as in the strategy planning of aircraft manufacturers and air transport operators. So embedded is this concern, that it continues to influence aircraft design (Baharozu, Soykan, & Ozerdem, 2017). While the public continues to display concern for hostile regime and military activities across the globe, there continues to also be a growing acceptance of the crisis imposed by climate change. These concerns will further increase pressure to lower our dependence on carbon fuels.

This pressure will likely spur incentives to invest both public and private funding into battery technology (Catenacci, Verdolini, Bsetti, & Fiorese, 2013). It is the battery technology that supports the further development and efficiency gains of alternative energies extending from solar and wind power to electric motor cars. This investment in battery technology will be a great encouragement for aircraft manufacturers and engine manufacturers to enhance the performance capabilities of aircraft engines, to extend the use of electric propulsion, but also to lower aircraft weights with the increased use of composites. Such technology will extend to UAV aircraft design, further enhancing their capabilities, and potential applications (Floreano & Wood, 2015).

Noise

The issue of noise has been the cause of great debate between the public and commercial air transport operators, particularly around airports (Broer & Duyvendak, 2009). Public interest groups have rallied to either restrict operations at airports or challenge the introduction of new operations. Now, noise limitations are well established in statutes and regulations, and this will continue to be of great importance to aircraft designers in their research and development efforts associated with aircraft noise reduction.

Opposition to the broader utilization of UAVs will likely be based on noise (Elias, 2012). However, the public concerns associated with noise will motivate the development of technology which intends to reduce noise in UAV operations, particularly with regards to aircraft engine, or rotor noise.

Privacy

While privacy continues to be a contentious issue requiring further regulatory clarity, it is evident that barriers based on privacy issues alone, are not a significant barrier to future commercial UAV expansion. However, this will only be the case where the public believe that there are substantial benefits to the development of UAV operations. The media often highlights many of the negative aspects of UAV operations, and the significant social, innovation and commercial benefits are often neglected. As such, where the public becomes more informed about the ways in which UAVs can improve economic activity, consumer welfare and public safety, the privacy concerns are likely to remain in the background.

Air Cargo

Considering the challenges discussed above, it is likely that unmanned air cargo flights will occur earlier than the widespread implementation of package delivery by microdrones (Perritt & Sprague, 2017). When autopilot technology is adjusted, unmanned air cargo aircraft can operate within the existing system for controlling flights under Instrument Flight Rules.

ICAO concurs, stating that "Larger and more complex RPA – able to undertake more challenging tasks – will most likely begin to operate in controlled airspace where all traffic is known and where ATC is able to provide separation from other traffic. This could conceivably lead to routine unmanned commercial cargo flights" (International Civil Aviation Organisation, 2011)(pp. 8).

As we have discussed in previous sections, microdrone package delivery requires the resolution of complex challenges such as the development and implementation of a new system of lower-level air space management and 'sense-and-avoid' process and technology.

THE EFFECT OF UAVS ON OTHER SECTORS OF THE AVIATION INDUSTRY

As the previous chapters of this book have indicated, small UAVs will offer aviation support to various parts of the economy that are currently unable to be served by helicopters and/or fixed wing aircraft. This, in part, may be due to cost or the presence of risks not acceptable to flight crew. As technology develops, larger drones will be a direct source of competition for manned aircraft. Comparisons will be drawn based on price and performance.

Although UAVs certainly have a role in complementing manned aircraft in the air transport industry as highlighted in previous chapters, greater potential lies in the future. Logistics is one example. As airspace management challenges are overcome, small-package delivery in specific centers will become a regular feature of the logistics sector. Small-package delivery is not commonly a function performed by helicopters. As such, the threat to helicopter operations will be minimal. Unmanned cargo aircraft are likely to see first introduction in the military sector. Debate continues on the national security issues presented by such transport (Betson, 2012). However, while the debate continues, the call for such technology remains open (Warwick, 2014).

However, discoveries in the military sector and the usefulness of such unmanned cargo aircraft applications will extend to the civilian market. There are two categories of unmanned cargo aircraft. The first category is for short distance small to medium sized unmanned cargo aircraft able to transport small specialized items such as medicines and packages. The payloads for this type of unmanned cargo aircraft would be between 1 to 50 kilograms. They would operate in an urban environment for the delivery of packages 5-10 miles from a centralized distribution point. In some rural areas, such unmanned cargo aircraft may be travelling 20 to 25 miles to deliver the cargo. Examples of this are the Amazon Prime Air multirotor package delivery system.

The second category is for long distance package delivery systems. The long distance unmanned cargo aircraft are able to carry 100 to 25,000 pounds and have a range of 200 to 10,000 miles. Over the past two decades, only the military UAVs have been capable of heavy payloads and long distance. This will soon change as the air transport industry enters a commercial era of unmanned aircraft operating with other manned aircraft in the National Airspace System (NAS). This will have a significant impact on the aircraft cargo industry and the way high value cargo is shipped (Collins, 2017)

Amazon's recent proposal for the segregation of drone airspace (Todorova, 2015) is already creating contention, as it suggests that helicopters and fixed wing aircraft would be prohibited in such low-level airspace. Compromises resulting for the trial are likely. However, airspace management solutions, and/or some form of airspace segregation with sense-and-avoid collision avoidance is likely to be developed in the next 5-10 years (Perritt & Sprague, 2017). This is driven by the underlying philosophy of national regulators such as the FAA who have long term plans to move away from infrastructure-based systems towards a more autonomous, vehicle-based system for collision avoidance (Department of Defense, 2004). The integration of unmanned aircraft into airspace in which manned aircraft currently occupy will inevitably require manned aircraft to carry such collision avoidance equipment that cooperates with that approved for unmanned aircraft.

Some UAV commentators believe that UAVs will not carry passengers any time in the near future (Perritt & Sprague, 2017). However, there appears to be a gradual change in the public acceptance of passenger carrying vehicle transport where the driver is in a remote location. For example, the use of driverless buses, trains and trams at airports. The introduction of driverless taxis and buses in urban areas also appears to be an accepted method of transport. These vehicles, however, do not carry passengers in three dimensions. MacSween-George (2003) conducted a public opinion survey of unmanned aircraft use for applications such as cargo transport, commercial and civil applications (e.g., fire-fighting, crop dusting, etc.), and passenger transportation. The study revealed that the public perceived commercial applications and cargo transportation as acceptable forms of UAV applications. Regarding public perceptions on UAV passenger carrying operations, risk perceptions often involve technology reliability and higher perceived safety with a human pilot onboard (Tam, 2011).

While some may argue that the only sector relatively immune to UAV operations is those carrying passengers (Perritt & Sprague, 2017), current public opinion surveys seem to indicate that the public may be open to such transport opportunities, where appropriate safety statistics and information is provided (Tam, 2011). As such, even current manned passenger carrying operations may be threatened by competition from future UAV operations.

REFERENCES

Airbus. (2017). *Airbus, Rolls-Royce, and Siemans team up for electric future Partnership launches E-Fan X hybrid-electric flight demonstrator*. Retrieved from https://www.airbus.com/newsroom/press-releases/en/2017/11/airbus--rolls-royce--and-siemens-team-up-for-electric-future-par.html

Albakcr, B. M. (2013). *Preliminary architectonic design of a smart solar-powered UAV*. Paper presented at the 2013 IEEE Conference on Clean Energy and Technology. 10.1109/CEAT.2013.6775633

Allen, D., & Berg, G. (2018). Regulation and Technological Change. In D. Allen & C. Berg (Eds.), *Australia's Red Tape Crisis*. Connor Court Publishing.

Antcliff, K. R., Moore, M. D., & Goodrich, K. H. (2016). *Silcon Valley as an Early Adopter for On-Demand Civil VTOL Operations*. Paper presented at the 16th AIAA Aviation Technology, Integration, and Operations Conference.

Baharozu, E., Soykan, G., & Ozerdem, M. B. (2017). Future aircraft concept in terms of energy efficiency and environmental factors. *Energy*, *140*, 1368–1377. doi:10.1016/j.energy.2017.09.007

Bennet, W., Bridewell, J. B., Rowe, L., & Craig, S. D. (2016). Training issues for remotely piloted aircraft systems from a human systems integration perspective. In N. J. Cook, L. J. Rower, W. Bennett, & D. Q. Joralmon (Eds.), *Remotely Piloted Aircraft Systems: A Human Systems Integration Perspective*. West Sussex, UK: John Wiley & Sons Ltd. doi:10.1002/9781118965900.ch7

Betson, A. P. (2012). The Case Against a Cargo Unmanned Aircraft System. *Army Sustainment*, *44*(5), 28–29.

Bower, R. (2004). *No Hands*. Rotor and Wing International.

Bowonder, B., Dambal, A., Shambhu, K., & Shirodkar, A. (2015). Innovation strategies for creating competitive advantage. *Research Technology Management*, *53*(3), 19–32. doi:10.1080/08956308.2010.11657628

Bristow. (2015). *AgustaWestland and Bristow sign exclusive agreement for AW609*. Retrieved from http://bristowgroup.com/bristow-news/latest-news/2015/agustawestland-and-bristow-sign-exclusive-platform/

Broer, C., & Duyvendak, J. (2009). Discursive opportunities, feeling rules, and teh rise of protests against aircraft noise. *Mobilization: An International Quarterly (San Diego, Calif.)*, *14*(3), 337–356.

Capata, R., Marino, L., & Sciubba, E. (2014). A hybrid propulsion system for a high-endurance UAV: Configuration selection, aerodynamic study, and gas turbine bench tests. *Journal of Unmanned Vehicle Systems*, *2*(1), 16–35. doi:10.1139/juvs-2013-0005

Catenacci, M., Verdolini, E., Bsetti, V., & Fiorese, G. (2013). Going electric: Expert survey on the future of battery technologies for electric vehicles. *Energy Policy*, *61*, 403–413. doi:10.1016/j.enpol.2013.06.078

Chen, G. Y. (2017). Reforming the current regulatory framework for commercial drones: Retaining American businesses' competitive advantage in the global economy. *Northerwestern Journal of International Law & Business*, *37*(3), 513–537.

Civil Aviation Safety Authority. (2019). *Flying drones/remotely piloted aircraft in Australia.* Retrieved from https://www.casa.gov.au/aircraft/landing-page/flying-drones-australia

Collins, M. P. (2017). *The future market for large unmanned cargo aircraft in the national airspace system (Master of Science).* Lewis University.

Dalamagkidis, K., Valavanis, K., & Piegl, L. A. (2008). On unmanned aircraft systems issies, challenges and operational restrictions preventing integration into the National Airspace System. *Progress in Aerospace Sciences*, *44*(7), 503–519.

De la Torre, G., Ramallo, M., & Cervantes, E. (2016). Workload perception in drone flight training simulators. *Computers in Human Behavior*, *64*, 449–454. doi:10.1016/j.chb.2016.07.040

Department of Defense. (2004). *Airspace Integration Plan for Unmanned Aircraft.* Author.

Du, H., & Heldeweg, M. A. (2018). An experimental approach to regulating non-military unmanned aircraft systems. *International Review of Law Computers & Technology*, 1–24. doi:10.1080/13600869.2018.1429721

Elias, B. (2012). *Pilotless drones: Background and considerations for congress regarding unmanned aircraft operations in the national airspace system.* Retrieved from https://biotech.law.lsu.edu/crs/r42718.pdf

Federal Aviation Administration. (2016). *Operation and Certification of Small Unmanned Aircraft Systems 14 CFR Part 107.* Retrieved from https://www.faa.gov/uas/media/RIN_2120-AJ60_Clean_Signed.pdf

Floreano, D., & Wood, R. (2015). Science, technology and the future of small autonomous drones. *Nature*, *521*(7553), 460–466. doi:10.1038/nature14542 PMID:26017445

Gawron, V. J. (1998). *Human factors issues in the development, evaluation, and operation of uninhabited aerial vehicles.* Paper presented at the Associations of Unmanned Vehicles Systems International.

Goodchild, A., & Toy, J. (2018). Delivery by drone: An evaluation of unmanned aerial vehicle technology in reducing CO2 emissions in the delivery service industry. *Transportation Research Part D, Transport and Environment*, *61*, 58–67. doi:10.1016/j.trd.2017.02.017

Hail, A. R., & Coyne, C. J. (2014). The political economy of drones. *Defence and Peace Economics*, *25*(5), 445–460. doi:10.1080/10242694.2013.833369

Hradecky, S. (2015). *Incident: Delta A320 near Denver on August 7th 2015, hail strike*. Retrieved from http://avherald.com/h?article=48a8329f

International Civil Aviation Organisation. (2011). *Unmanned Aircraft Systems (UAS): Cir 328 AN/190.* Retrieved from https://docplayer.net/7556384-Unmanned-aircraft-systems-uas.html

Jiang, T., Geller, J., Ni, D., & Collura, J. (2016). Unmanned Aircraft System Traffic management: Concept of operation and system architechture. *International Journal of Transportation Science and Technology, 5*(3), 123-135.

Kreps, S. (2014). Flying under the radar: A study of public attitudes towards unmanned aerial vehicles. *Research and Politics*, *1*(1), 1–7. doi:10.1177/2053168014536533

Lacher, A., & Maroney, D. (2012). *A new paradigm for small UAS*. Retrieved from https://www.mitre.org/sites/default/files/pdf/12_2840.pdf

Luppicini, R., & So, A. (2016). A technoethical review of commercial drone use in the context of governance, ethics and privacy. *Technology in Society*, *46*, 109–119. doi:10.1016/j.techsoc.2016.03.003

MacSween-George, S. L. (2003). *A Public Opinion Survey - Unmanned Aerial Vehicles for Cargo, Commercial, and Passenger Transportation.* Paper presented at the 2nd AIAA "Unmanned Unlimited" Systems, Technologies, and Operations, San Diego, CA.

Matthes, R., Alexander, R., & Stone, R. (2018). Helicopter accident trends in eight ISASI countries. *Journal of the International Society of Air Safety Investigators*, *51*(4), 4–12.

Molnar, A., & Parsons, C. (2016). Unmanned Aerial Vehicles (UAVs) and Law Enforcement in Australia and Canada: Governance through 'Privacy' in an Era of Counter-Law? *National Security, Surveillance and Terror*, 225-247.

Moorman, R. (2017, December). Pay to Play: The cost of ADS-B and where to find financial assistance. *Avionics Digital.*

National Academies of Sciences and Medicine. (2018). Assessing the Risks of Integrating Unmanned Aircraft Systems (UAS) into the National Airspace System. Washington, DC: The National Academies Press.

Olsen, R. G. (2017). Paperweights: FAA Regulation and the banishment of commercial drones. *Berkeley Technology Law Journal*, *32*(4), 621–652.

Pahsa, A., Kaya, P., Alat, G., & Baykal, B. (2011). *Integrating navigation and surveillance of unmanned air vehicles into civilian national airpaces by using ADS-B applications.* Paper presented at the 2011 Integrated Communications, Navigation, and Surveillance Conference.

Perritt, H., & Sprague, E. O. (2017). *Domesticating Drones: The technology, Law, and Economics of Unmanned Aircraft.* New York: Routledge.

Peterson, M. E. (2006). The UAV and the current and future regulatory construct for integration into the national airspace system. *Journal of Air Law and Commerce*, *71*(521).

Prisacariu, V., Boscoianu, M., & Luchian, A. (2014). Innovative solutions and UAS limits. *Review of the Air Foce Academy*, *2*, 26.

Rao, B., Gopi, A. G., & Maione, R. (2016). The societal impact of commercial drones. *Technology in Society*, *45*, 83–90. doi:10.1016/j.techsoc.2016.02.009

Riganati, J. P., & Harrison, G. (2016). The path forward for remotely piloted aircraft systems. In *Remotely Piloted Aircraft Systems: A Human Systems Integration Perspective*. West Sussex, UK: John Wile & Sons Ltd. doi:10.1002/9781118965900.ch10

Romm, J. (2006). The car and fuel of the future. *Energy Policy*, *34*(2609-2614).

Schneider, D. (2014). Open season on drones? *IEEE Spectrum*, *51*(1), 32–33. doi:10.1109/MSPEC.2014.6701428

Tam, A. (2011). Public Perception of Unmanned Aerial Vehicles. Aviation Technology Graduate Student Publications, (3), 1-23.

Todorova, I. (2015). The sky is the limit: UAVs by private actors and the implication to common-law privacy. *FIU Law Review*, *10*(2), 803–838.

Udroiu, R., & Blaj, M. I. (2016). Conceptual design of a vtol remotely piloted aircraft for emergency missions. *Scientific Research and Education in the Air Force - AFASES*. doi:10.19062/2247-3173.2016.18.1.27

Valavanis, K. (2007). *Advances in Unmanned Aerial Vehicles: State of the Art and the Road to Autonomy*. Springer. doi:10.1007/978-1-4020-6114-1

Vincent, J. B., Leland, K. W., & Ditmer, M. A. (2015). Barriers to adding UAVs to the ecologist's toolbox: Peer-reviewed letter. *Frontiers in Ecology and the Environment*, *13*(2), 74–75. doi:10.1890/15.WB.002

Wargo, C. A., Church, G. C., Glaneueski, J., & Strout, M. (2014). *Unmanned Aircraft Systems (UAS) research and future analysis.* Paper presented at the 2014 IEEE Aerospace Conference. 10.1109/AERO.2014.6836448

Warwick, G. (2014). Cargo Shift: Department of Defense planning to develop unmanned cargo resupply aircraft. *Aviation Week & Space Technology, 176*(16), 53-55.

Weibel, R. E., & Hansman, R. J. (2005). *Safety considerations for operation of unmanned aerial vehicles in the national airspace system*. Technical Report ICAT-2005-1.

Westbrook, M. H. (2001). *The electric car: Development and Future of Battery, Hybrid and Fuel-Cell Cars*. London: Institution of Electrical Engineers.

Yu, X., & Zhang, Y. (2015). Sense and avoid technologies with applications to unmanned aircraft systems: Review and prospects. *Progress in Aerospace Sciences*, *74*, 152–166. doi:10.1016/j.paerosci.2015.01.001

Zimmerman, J. (2013). *ADS-B 101: what is it and why you should care*. Air Facts Journal.

Compilation of References

AAIF. (2011). *AAIF Certification and regulation working group Unmanned aircraft systems sub-committee.* Australian Aerospace Industry Forum, Adelaide, Australia.

Abbott, K. A., Slotte, S. M., & Stimson, D. K. (1996). *Federal Aviation Administration Human Factors Team Report on: The Interfaces Between Flightcrews and Modern Flight Deck Systems.* Technical Report. Retrieved August, 2018, from http://www.tc.faa.gov/its/worldpac/techrpt/hffaces.pdf

ABC (Australian Broadcasting Corporation). (2016). *Cyclone Pam: Vanuatu one year on.* Retrieved from: www.abc.net.au/news/2016-03-13/cyclone-pam-vanuatu-one-year-on/7242620

Ackerman, E., & Strickland, E. (2018). Medical delivery drones take flight in east africa. *IEEE Spectrum*, *55*(1), 34–35. doi:10.1109/MSPEC.2018.8241731

Adams, M. J., Tenney, Y. J., & Pew, R. W. (1995). Situation awareness and the cognitive management of complex systems. *Human Factors*, *37*(1), 85–104. doi:10.1518/001872095779049462

Agatz, N., Bouman, P., & Schmidt, M. (2015). *Optimzation Approaches for the Traveling Salesman Problem with Drone, ERIM Report Series Research in Management.* Rotterdam: Erasmus Research Institute of Management.

Agrawal, A. J. (2017). *5 Ways Marketers Can Take Advantage of Drone Technology.* Retrieved from: https://www.forbes.com/sites/ajagrawal/2017/06/10/5-ways-marketers-can-take-advantage-of-drone-technology/#2c3eb5e658cc

Aimi, G. (2016). Last-mile logistics. *MHD Supply Chain Solutions*, *46*(4), 60.

Airbus. (2017). *Airbus, Rolls-Royce, and Siemans team up for electric future Partnership launches E-Fan X hybrid-electric flight demonstrator*. Retrieved from https://www.airbus.com/newsroom/press-releases/en/2017/11/airbus--rolls-royce--and-siemens-team-up-for-electric-future-par.html

Airservices Australia. (2018). *Designated airspace handbook*. Retrieved 22 May 2018 from http://www.airservicesaustralia.com/aip/pending/dah/dah.pdf

Albaker, B. M. (2013). *Preliminary architectonic design of a smart solar-powered UAV.* Paper presented at the 2013 IEEE Conference on Clean Energy and Technology. 10.1109/CEAT.2013.6775633

Alex, C., & Vijaychandra, A. (2016). Autonomous cloud based drone system for disaster response and mitigation. In *2016 International Conference on Robotics and Automation for Humanitarian Applications (RAHA)* (pp. 1–4). Academic Press. 10.1109/RAHA.2016.7931889

Allen, D., & Berg, G. (2018). Regulation and Technological Change. In D. Allen & C. Berg (Eds.), *Australia's Red Tape Crisis*. Connor Court Publishing.

Amazon Prime Air. (2016). *First Prime Air Delivery*. Retrieved from https://www.amazon.com/Amazon-Prime-Air/b?ie=UTF8&node=8037720011

Ambrosia, V. G., Wegener, S., & Zajkowliski, T. (2011). The Ikhana unmanned airborne system (UAS) western states fire imaging missions: From concept to reality (2006-2010). *Geocarto International*, *26*(2), 85–101. doi:10.1080/10106049.2010.539302

Amelin, K. S., Antal, E., & Vasiliev, V. (2013). *Adaptive management of an autonomous group of unmanned aerial vehicles*. St. Petersburg State University.

American Red Cross. (2015). *Drones for Disaster Response and Relief Operations*. Retrieved from: https://www.zurichna.com/en/search?q=Drones+for+Disaster+Response+and+Relief+Operations&page=2#

Amukele, T. K., Hernandez, J., Snozck, C. L., Wyatt, R. G., Douglas, M., Amini, R., & Street, J. (2017). Drone Transport of Chemistry and Hematology Samples Over Long Distances. *American Journal of Clinical Pathology*, *148*(5), 427–435. doi:10.1093/ajcp/aqx090 PMID:29016811

Amukele, T. K., Street, J., Carroll, K., Miller, H., & Zhang, S. X. (2016). Drone Transport of Microbes in Blood and Sputum Laboratory Specimens. *Journal of Clinical Microbiology*, *54*(10), 2622–2625. doi:10.1128/JCM.01204-16 PMID:27535683

Anand, S. (2007). Domestic use of unmanned aircraft systems: An evaluation of policy constraints and the role of industry consensus standards. *ASTM Standardization News*, *35*(9), 30.

Anbaroğlu, B. (2017). Parcel Delivery in an Urban Environment Using Unmanned Aerial Systems: A Vision Paper. In ISPRS Annals of Photogrammetry, Remote Sensing and Spatial Information Sciences (Vol. IV-4-W4, pp. 73–79). Copernicus GmbH. doi:10.5194/isprs-annals-IV-4-W4-73-2017

Antcliff, K. R., Moore, M. D., & Goodrich, K. H. (2016). *Silcon Valley as an Early Adopter for On-Demand Civil VTOL Operations.* Paper presented at the 16th AIAA Aviation Technology, Integration, and Operations Conference.

Artemenko, O., Rubina, A., Golokolenko, O., Simon, T., Romisch, J., & Mitschele-Thiel, A. (2014). Validation and evaluation of the chosen path planning algorithm for localization of nodes using an unmanned aerial vehicle in disaster scenarios. *Lecture Notes of the Institute for Computer Sciences Social-Informatics and Telecommunications Engineering*, *140*, 192–203.

Ashdown, M., & Cummings, M. L. (2007). *Asymmetric synchronous collaboration within distributed teams.* Paper presented at the 7th International Conference Engineering Psychology and Cognitive Ergonomics. 10.1007/978-3-540-73331-7_26

ASSURE. (2018). *Annual Report.* Retrieved from http://www.assureuas.org/annual-reports/ASSURE_2018AR.pdf

Astuti, G., Longo, D., Melita, D., Muscato, G., & Orlando, A. (2008). HIL tuning of UAV for exploration of risky environments. *International Journal of Advanced Robotic Systems*, *5*(4), 419–424. doi:10.5772/6227

Austin, R. (2010). *Unmanned Aircraft Systems: UAVs Design, Development and Deployment.* John Wiley & Sons. doi:10.1002/9780470664797

Australian Government. (n.d.). Retrieved from: https://www.casa.gov.au/aircraft/standard-page/commercial-unmanned-flight-gaining-your-remotely-piloted-aircraft-pilot

Bačelić, M. (2016, August 1). Kako se u biznisu možete koristiti bespilotnim letjelicama, Lider, 24-28.

Baharozu, E., Soykan, G., & Ozerdem, M. B. (2017). Future aircraft concept in terms of energy efficiency and environmental factors. *Energy*, *140*, 1368–1377. doi:10.1016/j.energy.2017.09.007

Bainbridge, L. (1987). Ironies of automation. In J. Rasmussen, K. Duncan, & J. Leplat (Eds.), *New technology and human error* (pp. 271–283). New York: Wiley.

Balasingam, M. (2017). Drones in medicine—The rise of the machines. *International Journal of Clinical Practice*, *71*(9), e12989. doi:10.1111/ijcp.12989 PMID:28851081

Barmpounakis, E. N., Vlahogianni, E. I., & Golias, J. C. (2016). Unmanned Aerial Aircraft Systems for transportation engineering: Current practice and future challenges. *International Journal of Transportation Science and Technology*, *5*(3), 111–122. doi:10.1016/j.ijtst.2017.02.001

Barnard Microsystems. (1999). *First Atlantic crossing by an unmanned aircraft.* Retrieved from: http://www.barnardmicrosystems.com/UAV/milestones/atlantic_crossing_1.html

Barnes, M. J., & Matz, M. F. (1998). *Crew simulations for Unmanned Aerial Vehicles (UAV) applications: Sustained effects, shift factors, interface issues, and crew size.* Paper presented at the Human Factors and Ergonomics Society Annual Meeting, Santa Monica, CA. 10.1177/154193129804200132

Barshi, I., & Farris, C. (2013). *Misunderstandings in ATC Communication: Language, Cognition, and Experimental Methodology*. Surrey, UK: Ashgate.

Bartzen Culver, K. (2014). From Battlefield to Newsroom: Implications of Drone Technology in Journalism. *Journal of Mass Media Ethics*, *29*(1), 52–64. doi:10.1080/08900523.2013.829679

Bashshur, R. L. (1995). On the definition and evaluation of telemedicine. *Telemedicine Journal: The Official Journal of the American Telemedicine Association*, *1*(1), 19–30. doi:10.1089/tmj.1.1995.1.19 PMID:10165319

Belan, V., Yun, G., & Gnashuk, A. (2018). Problem of manned aircraft conversion into unmanned aerial vehicles. In *Proceedings of the National Aviation University* (N 2, pp. 57 61). Kyiv, Ukraine: National Aviation University.

Benjamin, E. J., Virani Salim, S., Callaway Clifton, W., Chamberlain Alanna, M., Chang Alexander, R., & Cheng, S. (2018). Heart Disease and Stroke Statistics—2018 Update: A Report From the American Heart Association. *Circulation*, *137*(12), e67–e492. doi:10.1161/CIR.0000000000000558 PMID:29386200

Benjamin, M., & Ehrenreich, B. (2013). *Drone Warfare: Killing by Remote Control (Fully revised and updated edition)*. London: Verso.

Bennet, W., Bridewell, J. B., Rowe, L., & Craig, S. D. (2016). Training issues for remotely piloted aircraft systems from a human systems integration perspective. In N. J. Cook, L. J. Rower, W. Bennett, & D. Q. Joralmon (Eds.), *Remotely Piloted Aircraft Systems: A Human Systems Integration Perspective*. West Sussex, UK: John Wiley & Sons Ltd. doi:10.1002/9781118965900.ch7

Bernama. (2017). S. Korean gov't unveils new safety rules on drone flight. *New Straits Times*. Retrieved from: https://www.nst.com.my/world/2017/11/303032/skorean-govt-unveils-new-safety-rules-drone-flight

Bethune, G., & Huler, S. (1998). *From Worst to First*. New York: John Wiley & Sons.

Betson, A. P. (2012). The Case Against a Cargo Unmanned Aircraft System. *Army Sustainment*, *44*(5), 28–29.

Bhatt, K., Pourmand, A., & Sikka, N. (2018). Targeted Applications of Unmanned Aerial Vehicles (Drones) in Telemedicine. *Telemedicine Journal and e-Health*, *24*(11), 833–838. doi:10.1089/tmj.2017.0289 PMID:29489441

Blain, L. (2016). *NOVAerial's helicopter drones: Better than multicopters?* Retrieved 28 May 2018 from https://newatlas.com/novaerial-procyon-800e-helicopter-drone/47013/

Blain, L. (2017). *AT Transporter multirotor "air car" carries cargo or people in detachable pods*. Retrieved from https://newatlas.com/advanced-tactics-transporter-at-manned-multirotor/52042/

Blain, L. (2017). *Russian unmanned, long-range, hybrid multirotor wants to do the heavy lifting*. Retrieved 28 May 2018 from https://newatlas.com/ardn-russia-skyf-heavy-lift-hybrid-multirotor/52300/

Blau, M. (2017). *Condom drops and airborne meds: 6 ways drones could change health care*. Retrieved from: https://www.statnews.com/2017/06/13/drones-health-care/

Böckstiegel. (2001). *Zeitschrift Für Luftund Weltraumrecht*. Köln: Institut für Luft- und Weltraumrecht der Köln Universität.

Bolstad, C. A., Endsley, M. R., Costello, A. M., & Howell, C. D. (2010). Evaluation of computer-based situation awareness training for general aviation pilots. *The International Journal of Aviation Psychology*, *20*(3), 269–294. doi:10.1080/10508414.2010.487013

Bondarev, D., Jafarzadeh, R., & Kozub, A. (2014). Effectiveness of Group Flight of Unmanned Aerial Vehicles. Systems of Information Processing, 6, 9-15.

Bondarev, D., Shmelova, T., & Kucherov, D. (2015). Models of group flights of unmanned aerial vehicles using Science and technology of the Air Forces of the Armed Forces of Ukraine. *Scientific Periodicals, 3*, 68-75.

Boucher, P. (2015). Domesticating the Drone: The Demilitarisation of Unmanned Aircraft for Civil Markets. *Science and Engineering Ethics*, *21*(6), 1393–1412. doi:10.100711948-014-9603-3 PMID:25371277

Bower, R. (2004). *No Hands*. Rotor and Wing International.

Bowonder, B., Dambal, A., Shambhu, K., & Shirodkar, A. (2015). Innovation strategies for creating competitive advantage. *Research Technology Management*, *53*(3), 19–32. doi:10.1080/08956308.2010.11657628

Boyle, A. (2012). The US and its UAVs: A Cost Benefit Analysis. *American Security Project*. Retrieved from http://www.americansecurityproject.org/the-us-and-its-uavs-a-cost-benefit-analysis/

Brady, J. M., Stokes, M. D., Bonnardel, J., & Bertram, T. H. (2016). Characterization of a Quadrotor Unmanned Aircraft System for Aerosol-Particle-Concentration Measurements. *Environmental Science & Technology*, *50*(3), 1376–1383. doi:10.1021/acs.est.5b05320 PMID:26730457

Brar, S., Rabbat, R., Raithatha, V., Runcie, G., & Yu, A. (2015). *Drones for Deliveries. Sutardja Center for Entrepreneurship & Technology Technical Report*. Retrieved from http://scet.berkeley.edu/wp-content/uploads/ConnCarProjectReport-1.pdf

Brar, S., Rabbat, R., Raithatha, V., Runcie, G., & Yu, A. (2015). *Drones for Deliveries*. Berkeley, CA: Sutardja Center Entrepreneurship Technol.

Bristow. (2015). *AgustaWestland and Bristow sign exclusive agreement for AW609*. Retrieved from http://bristowgroup.com/bristow-news/latest-news/2015/agustawestland-and-bristow-sign-exclusive-platform/

Broadbent, D. (1958). *Perception and Communication*. London: Pergamon Press. doi:10.1037/10037-000

Broer, C., & Duyvendak, J. (2009). Discursive opportunities, feeling rules, and teh rise of protests against aircraft noise. *Mobilization: An International Quarterly (San Diego, Calif.)*, *14*(3), 337–356.

Brown, J. (2017). *How Drones Will Revolutionize Marketing Practices*. Retrieved from: https://www.digitaldoughnut.com/articles/2017/march/how-drones-will-revolutionize-marketing-practices

Bryan, V., & Char, P. (2014). *Drone delivery: DHL 'parcelcopter' flies to German isle*. Retrieved from https://www.reuters.com/article/us-deutsche-post-drones-idUSKCN0HJ1ED20140924?feedType=RSS

Burton, E., Meier, C., Olarte, R., Skeini, H., & Zahan, F. (2017). Airshare: A Food Sharing Concept. In *Proceedings of the 29th Australian Conference on Computer-Human Interaction* (pp. 634–639). New York, NY: ACM. 10.1145/3152771.3154859

Burzichelli, C. D. (2016). Delivery Drones: Will amazon air see the national airspace. *Rutgers Computer & Technology Law Journal*, *42*, 162.

Buttigieg, M., & Sanderson, P. (1991). Emergent features in visual display design for two types of failure detection tasks. *Human Factors*, *33*(6), 631–351. doi:10.1177/001872089103300603 PMID:1800293

CAA. (2014). *Criteria for an application for and grant of, or a variation to, an ATOL: Fitness, competence and accountable person*. Retrieved 17 May 2018 from https://publicapps.caa.co.uk/docs/33/Fitness%20note%2022814.pdf

CAA. (2015). *Unmanned aircraft system operations in UK airspace: Guidance*. Retrieved 18 May 2018 from http://publicapps.caa.co.uk/docs/33/CAP%20722%20Sixth%20Edition%20March%202015.pdf

CAANZ. (2011). *Alcohol consumption and medical aviation safety*. Retrieved 28 May 2018 from http://www.caa.govt.nz/assets/legacy/medical/Med_Info_Sheets/MIS014.pdf

CAANZ. (2015a). *Fit and proper person assessment handbook: Procedures and guidance*. Retrieved 17 May 2018 from https://www.caa.govt.nz/Forms/FPP_Handbook.pdf

CAANZ. (2015b). *Sample exposition: Part 102 unmanned aircraft operator's certificate*. Retrieved 15 May 2018 from https://www.caa.govt.nz/rpas/Part_102_sample_exposition.doc

CAANZ. (2015c). *Unmanned aircraft: Operator certification* [AC102-1]. Retrieved from https://www.caa.govt.nz/assets/legacy/Advisory_Circulars/AC102-1.pdf

CAANZ. (2016a). *Advisory circular: Safety management* [AC100-1]. Retrieved 17 May 2018 from https://www.caa.govt.nz/assets/legacy/Advisory_Circulars/AC100-1.pdf

CAANZ. (2016b). *Civil Aviation Rules*. Retrieved from https://www.caa.govt.nz/rules/rules.htm

CAANZ. (2018). *Civil aviation rules part 1: Definitions and abbreviations*. Retrieved 22 May 2018 from https://www.caa.govt.nz/rules/Rule_Consolidations/Part_001_Consolidation.pdf

CAANZ. (n.d.). *Part 39: Airworthiness directives*. Retrieved 29 May 2018 from https://www.caa.govt.nz/rules/part-039-brief/

Calhoun, G. L., Draper, M. H., Abernathy, M. F., Patzek, M., & Delgado, F. (2005). Synthetic vision systems for improving unmanned aerial vehicle operator situation awareness. *Proceedings of SPIE: Enhanced and Synthetic Vision*. 10.1117/12.603421

Calhoun, G. L., Draper, M. H., & Ruff, H. A. (2002). Utility of a tactile display for cue faults. *Proceedings of the Human Factors and Ergonomics Society 46th Annual Meeting*.

Calishan, N. (2018). *İzinsiz uçuşa 11 bin 447 TL'lik ceza kesilecek*. Retrieved from: https://www.haberturk.com/izinsiz-drone-ucurana-11-bin-447-tl-lik-para-cezasi-2131702-ekonomi

Capata, R., Marino, L., & Sciubba, E. (2014). A hybrid propulsion system for a high-endurance UAV: Configuration selection, aerodynamic study, and gas turbine bench tests. *Journal of Unmanned Vehicle Systems*, *2*(1), 16–35. doi:10.1139/juvs-2013-0005

Capolupo, A., Pindozzi, S., Okello, C., Fiorentino, N., & Boccia, L. (2015). Photogrammetry for environmental monitoring: The use of drones and hydrological models for detection of soil contaminated by copper. *The Science of the Total Environment, 514*, 298–306. doi:10.1016/j.scitotenv.2015.01.109 PMID:25668282

Carrigan, G., Long, D., Cummings, M. L., & Duffner, J. (2008). *Human Factors Analysis of Predator B Crash.* Paper presented at the AUVSI 2008: Unmanned Systems North America, San Diego, CA.

CASA. (2014). *Chief pilot guide.* Retrieved 17 May 2018 from https://www.casa.gov.au/sites/g/files/net351/f/_assets/main/pilots/download/chief-pilot-guide.pdf

CASA. (2016). *Enforcement manual.* Retrieved 17 May 2018 from https://www.casa.gov.au/files/009rfullpdf

CASA. (2016). Red tape cut for remotely piloted aircraft. *Australian Government: Civil Aviation Safety Authority.* Retrieved from: https://www.casa.gov.au/publications-and-resources/media-release/red-tape-cut-remotely-piloted-aircraft

CASA. (2017). *Remotely piloted aircraft systems: Licensing and operations* [AC 101-01]. Retrieved 14 May 2018 from https://www.casa.gov.au/files/101c01pdf

CASA. (2018). *Air operator's certificate handbook: Volume 2 - flying operations.* Retrieved 17 May 2018 from https://www.casa.gov.au/file/174881/download?token=Tn6gR-ti

Casbeer, D., Beard, R., McLain, T., Li, S., & Mehra, R. (2005). *Forest fire monitoring with multiple small UAVs.* Paper presented at the American control conference. 10.1109/ACC.2005.1470520

Catenacci, M., Verdolini, E., Bsetti, V., & Fiorese, G. (2013). Going electric: Expert survey on the future of battery technologies for electric vehicles. *Energy Policy, 61*, 403–413. doi:10.1016/j.enpol.2013.06.078

Çeker, D. (2014). Adli Antropolojide Perimortem ve Postmortem Kırıkların Ayırımı ve Travma Analizlerindeki Önemi. *Ankara Üniversitesi Dil ve Tarih Coğrafya Fakültesi Antropoloji Dergisi., 27*, 47–64.

Ceker, D. (2016). Olay yeri inceleme ve Calismalarinda adli arkeolog ve adli antropologlarin rolu: Kuzey kibris ve Turkiyede'ki guncel durum. *AÜDTCF. Antropoloji Dergisi, 32*, 13–21.

Cetin, O. (2016). *Türkiye'de Drone kullanmanın şartları.* Retrieved from: https://mediatrend.mediamarkt.com.tr/turkiyede-drone-kullanmanin-sartlari/

Cevap, H. (2016). *Drone kullanmak yasal mıdır?* Retrieved from: https://www.cnnturk.com/video/hazir-cevap/hukuk/drone-kullanmak-yasal-midir

Chapman, A. (2016). Types of drones: Multi-rotor vs fixed-wing vs single rotor vs hybrid VTOL. *Drone Magazine.* Retrieved 28 May 2018 from https://www.auav.com.au/articles/drone-types/

Chekunov, E. (2010). Application of the UAV of US Armed Forces in Military Conflicts. *Foreign Military. RE:view*, *7*, 41–50.

Cheney, M. (2001). *Tesla: Man Out of Time.* New York: Simon &Schuster.

Cheng & Kim. (2000). *The Utilization of the World's Air Space and Free Outer Space in the 21st Century.* The Hague: Kluwer Law International.

Cheng. (1998). *The Use of Airspace and Outer Space Cooperation and Competition.* The Hague: Kluwer Law International.

Chen, G. Y. (2017). Reforming the current regulatory framework for commercial drones: Retaining American businesses' competitive advantage in the global economy. *Northerwestern Journal of International Law & Business*, *37*(3), 513–537.

Choi, C. H., Jang, H. J., Lim, S. G., Lim, H. C., Cho, S. H., & Gaponov, I. (2016). Automatic wireless drone charging station creating essential environment for continuous drone operation. In *2016 International Conference on Control, Automation and Information Sciences (ICCAIS)* (pp. 132–136). Academic Press. 10.1109/ICCAIS.2016.7822448

Chowdhary, G., Johnson, E. N., Magree, D., Wu, A., & Shein, A. (2013). GPS denied Indoor and Outdoor Monocular Vision Aided Navigation and Control of Unmanned Aircraft. *Journal of Field Robotics*, *30*(3), 415–438. doi:10.1002/rob.21454

Christian, A., & Cabell, R. (2017). *Initial investigation into the psychoacoustic properties of small unmanned aerial system noise.* Presented at the *23rd AIAA/CEAS Aeroacoustics Conference.*

Christopher, M. G. (2016). *Logistics and Supply Chain Management* (5th ed.). Harlow, UK: Pearson Education Ltd.

Činčurak Erceg, B., Vasilj, A., & Erceg, A. (2017). Unmanned Aircrafts ('Drones') – The Need for Legal Regulation and Harmonization of Legislation in the European Union. In M. Froechlich & C. Voda (Eds.), *South Eastern Europe and the European Union - Legal Perspectives - Series of Papers* (Vol. 3, pp. 9–22). Saarbrucken: SEE EU Cluster of Excellence in European and International Law.

Civil Aviation Authority. (2016). *ATM Automation: Guidance on human technology integration. (CAP 1377).* West Sussex, UK: Author.

Civil Aviation Safety Authority. (2019). *Flying drones/remotely piloted aircraft in Australia.* Retrieved from https://www.casa.gov.au/aircraft/landing-page/flying-drones-australia

Claesson, A., Fredman, D., Svensson, L., Ringh, M., Hollenberg, J., Nordberg, P., ... Ban, Y. (2016). Unmanned aerial vehicles (drones) in out-of-hospital-cardiac-arrest. *Scandinavian Journal of Trauma, Resuscitation and Emergency Medicine, 24*. doi:10.2514/6.2017-4051

Claesson, A., Svensson, L., Nordberg, P., Ringh, M., Rosenqvist, M., Djarv, T., ... Hollenberg, J. (2017). Drones may be used to save lives in out of hospital cardiac arrest due to drowning. *Resuscitation, 114*, 152–156. doi:10.1016/j.resuscitation.2017.01.003 PMID:28110000

Clothier, R. A., Fulton, N. L., & Walker, R. A. (2008). Pilotless aircraft: The horseless carriage of the twenty-first century? *Journal of Risk Research, 11*(8), 999–1023. doi:10.1080/13669870802323353

Clothier, R. A., Greer, D. A., Greer, D. G., & Mehta, A. M. (2015). Risk Perception and the Public Acceptance of Drones. *Risk Analysis, 35*(6), 1167–1183. doi:10.1111/risa.12330 PMID:25689883

Collins, M. P. (2017). *The future market for large unmanned cargo aircraft in the national airspace system (Master of Science).* Lewis University.

Connors, M., Harrison, A., & Akins, F. (Producer). (1985). *Living aloft: Human requirements for extended spaceflight.* Retrieved from https://ntrs.nasa.gov/search.jsp?R=19850024459

Consortiq. (2016). *A Short History of Unmanned Aerial Vehicles (UAVs).* Retrieved from: https://consortiq.com/short-history-unmanned-aerial-vehicles-uavs/

Cooke, N. J., Gorman, J., Pedersen, H., Winner, J., Duran, J., Taylor, A., ... Rowe, L. (2007). *Aquisition and Retention of Team Co-ordination in Command and Control*. Mesa, AZ: Cognitive Engineering Research Institute.

Corcoran, M. (2014). *Drone journalism: Newsgathering applications of Unmanned Aerial Vehicles (UAVs) in covering conflict, civil unrest and disaster*. Retrieved from: http://www.flinders.edu.au/ehl/fms/law_files/Drone%20Journalism%20During%20Conflict,%20Civil%20Unrest%20and%20Disasters%20March%201%202014.pdf

Coulton, P., Lindley, J., Sturdee, M., & Stead, M. (2017). Design Fiction as World Building. Presented at the Research Through Design.

Coxworth, B. (2016). *Heavy-lift drone could carry a person*. Retrieved from https://newatlas.com/griff-300-heavy-lift-drone/47049/

Crandall, J. W., & Cummings, M. L. (2007). *Developing performance metrics for the supervisory control of multiple robots*. Paper presented at the ACM/IEEE international conference on human-robot interaction. 10.1145/1228716.1228722

Cuevas, H. M., & Aguiar, M. (2017). Assessing Situation Awareness in Unmanned Aircraft Systems Operations. *International Journal of Aviation, Aeronautics, and Aerospace*, *4*(4).

Cummings, M. L., Bruni, S., Mercier, S., & Mitchell, P. J. (2007). Automation Architecture for Single Operator, Multiple UAV Command and Control. *The International C2 Journal, 1*(2), 1-24.

Cummings, M. L., & Mitchell, P. J. (2007). Operator scheduling strategies in supervisory control of multiple UAVs. *Aerospace Science and Technology*, *4*(11), 339–348. doi:10.1016/j.ast.2006.10.007

Cuthbertson, A. (2016). *Australia post to launch drone delivery service*. Retrieved 15 February, 2019, from https://www.newsweek.com/australia-post-drone-delivery-service-drones-449442

Dalamagkidis, K., Valavanis, K., & Piegl, L. A. (2008). On unmanned aircraft systems issies, challenges and operational restrictions preventing integration into the National Airspace System. *Progress in Aerospace Sciences*, *44*(7), 503–519.

De Florio, F. (2011). *Airworthiness: An introduction to aircraft certification* (2nd ed.). Burlington, MA: Butterworth Heinemann. doi:10.1016/B978-0-08-096802-5.10002-9

De la Torre, G., Ramallo, M., & Cervantes, E. (2016). Workload perception in drone flight training simulators. *Computers in Human Behavior*, *64*, 449–454. doi:10.1016/j.chb.2016.07.040

de Voorde, P. V., Gautama, S., Momont, A., Ionescu, C. M., Paepe, P. D., & Fraeyman, N. (2017). The drone ambulance [A-UAS]: Golden bullet or just a blank? *Resuscitation*, *116*, 46–48. doi:10.1016/j.resuscitation.2017.04.037 PMID:28487253

Dempsey & Milde. (2005). International Air Carriage Liability: The Montreal Convention of 1999. Institute of Air and Space Law.

Dempsey. (2008). Air Law. Institute of Air and Space Law, McGill University.

Department of Defense. (2004). *Airspace Integration Plan for Unmanned Aircraft*. Author.

Diederiks-Vershoor, I. H. P. (2001). An Introduction to Air Law. Kluwer Law and Taxation Publishers.

Dixon, S. R., Wickens, C. D., & Chang, D. (2003). Comparing quantitative model predictions to experimental data in multiple-UAV flight control. *Proceedings of the Human Factors and Ergonomics Society 47th Annual Meeting*. 10.1177/154193120304700122

DJI. (2018). *Agras MG-1*. Retrieved 09 May 2018 from https://www.dji.com/mg-1/info

Domino's. (2016). *Pizza-by-drone a reality with world-first customer deliveries in New Zealand*. Retrieved from https://www.dominos.com.au/inside-dominos/media/november-2016-pizza-by-drone-a-reality-with-world-first-customer-deliveries-in-new-zealand

Dorling, K., Heinrichs, J., Messier, G. G., & Magierowski, S. (2017). Vehicle routing problems for drone delivery. *IEEE Transactions on Systems, Man, and Cybernetics. Systems*, *47*(1), 70–85. doi:10.1109/TSMC.2016.2582745

DR1 Racing. (2017). *DHL Champions Series*. Retrieved from: https://www.dr1racing.tv/champions-series

Draper, M. H., Ruff, H. A., Repperger, D. W., & Lu, L. G. (2000). Multi-sensory interface concepts supporing turbulence detection by UAV controllers. *Proceedings of the First Human Performance, Situational Awareness and Automation Conference*.

Drury, J. L., & Scott, S. D. (Producer). (2008). *Awareness in UAV Operations*. Retrieved from http://www.cs.uml.edu/~holly/91.550/papers/Drury_Scott_UAV-Awareness.pdf

Du, H., & Heldeweg, M. A. (2018). An experimental approach to regulating non-military unmanned aircraft systems. *International Review of Law Computers & Technology*, 1–24. doi:10.1080/13600869.2018.1429721

Dwyer, F. (2018). *Medical delivery drones may soon be helping rural Americans*. Retrieved from https://www.wetalkuav.com/medical-delivery-drones-could-help-rural-americans/

EASA. (2015). *Introduction of a regulatory framework for the operation of unmanned aircraft*. Retrieved from https://www.easa.europa.eu/sites/default/files/dfu/Introduction%20of%20a%20regulatory%20framework%20for%20the%20operation%20of%20unmanned%20aircraft.pdf

EASA. (2018). *Introduction of a regulatory framework for the operation of unmanned aircraft systems in the 'open' and 'specific' categories*. Retrieved 18 May 2018 from https://www.easa.europa.eu/sites/default/files/dfu/Opinion%20No%2001-2018.pdf

ECAA. (2013). *Advisory circular: Maintenance control manual (MCM)*. Retrieved 17 May 2018 from http://www.ecaa.gov.et/upload/ac/MAINTENANCE%20CONTROL%20MANUAL(MCM).pdf

Eckelkamp. (2018). *Drone ROI realized by crop advisers*. Retrieved 29 May 2018 from https://www.agprofessional.com/article/drone-roi-realized-crop-advisers

Economist. (2009). *Attack of the drones*. Retrieved from: http://www.economist.com/node/14299496/print

Economist. (2011). *Flight of the drones*. Retrieved from: http://www.economist.com/node/21531433/print

Eisenbeiss, H. (2004). A mini unmanned aerial vehicle (UAV): System Overview and Image Acquisition. *Presented at International Workshop on Processing and Visualization Using High-Resolution Imagery*, 1-7.

Elias, B. (2012). *Pilotless drones: Background and considerations for congress regarding unmanned aircraft operations in the national airspace system*. Retrieved from https://biotech.law.lsu.edu/crs/r42718.pdf

Elias, B. (2012). *Pilotless drones: Background and considerations for congress regarding unmanned aircraft operations in the national airspace system.* Congressional Research Service.

Elwell, D. (2017). *Testimony - Statement of Dan Elwell.* Retrieved from https://www.faa.gov/news/testimony/news_story.cfm?newsId=22354&omniRss=testimonyAoc&cid=105_Testimony

Emir, A., & Eren, A. (2016). *Gökyüzünün Yeni Sakinleri.* Retrieved from: https//:archive.ismmmo.org.tr

ENAC. (2016). *Regulation: Remotely piloted aerial vehicles.* Retrieved 15 May 2018 from http://www.enac.gov.it/repository/contentmanagement/information/n1220929004/regulation_rpas_issue_2_rev%202_eng.pdf

Endsley, M. R. (2000). Theoretical Underpinnings of Situation Awareness: a Critical Review. In M. R. Endsley & D. J. Garland (Eds.), *Situation Awareness Analysis and Measurement.* Mahwah, NJ: Lawrence Erlbaum Associates.

Engvers, A. (2001). *The Principle of Sovereignty in the Air - To what extent can it be upheld against aerial intruders?* Retrieved from http://lup.lub.lu.se/student-papers/record/1557232

Erceg, A., Činčurak Erceg, B., & Vasilj, A. (2017). Unmanned aircraft systems in logistics – legal regulation and worldwide examples toward use in Croatia. In *Proceedings of the 17th International Scientific Conference Business Logistics in Modern Management.* Osijek: Ekonomski fakultet.

Erceg, A., Erceg, B. C., & Vasilj, A. (2017). Unmanned aircraft systems in logistics legal regulation and worldwide examples toward use in Croatia. *Business Logistics in Modern Management,* 43-62.

Estampe, D. (2015). *Planning the Future Supply Chain Together - New Technologies: Big data, 3d Printers, Drones.* White paper. Bordeaux: Kedge Business School.

Etherington, D. (2017). *Google's Project Wing team takes a key step towards making drone delivery real.* Retrieved from: https://techcrunch.com/2017/06/07/googles-project-wing-team-takes-a-key-step-towards-making-drone-delivery-real/

FAA (Federal Aviation Administration). (2016). FAA Aerospace Forecast: Fiscal Years 2016-2036. *FAA*. Retrieved from: http://www.faa.gov/data_research/aviation/aerospace_forecasts/media/FY2016-36_FAA_Aerospace_Forecast.pdf

FAA. (2016a). *Advisory circular: Small unmanned aircraft systems (sUAS)* [AC No: 107-2]. Retrieved 09 May 2018 from https://www.faa.gov/uas/media/AC_107-2_AFS-1_Signed.pdf

FAA. (2016b). *Pilot's handbook of aeronautical knowledge*. Retrieved 21 May 2018 from https://www.faa.gov/regulations_policies/handbooks_manuals/aviation/phak/

FAA. (2018). *Aeronautical information services: Aeronautical chart user's guide*. Retrieved 22 May 2018 from https://www.faa.gov/air_traffic/flight_info/aeronav/digital_products/aero_guide/media/editions/cug-complete.pdf

FAA. (2018). *UAS Integration Pilot Program* [template]. Retrieved 10 December 2018, from https://www.faa.gov/uas/programs_partnerships/uas_integration_pilot_program/

Fahlstrom, P., & Gleason, T. (2012). *Introduction to UAV Systems* (4th ed.). Wiley.

Federal Aviation Administration. (2015). *Safety Management Systems for Aviation Service Providers (AC 120-92B)*. Washington, DC: Author.

Federal Aviation Administration. (2016). *AC107-2 Small Unmanned Aircraft Systems (sUAS)*. Retrieved from https://www.faa.gov/regulations_policies/advisory_circulars/index.cfm/go/document.information/documentID/1019962

Federal Aviation Administration. (2016). *Operation and Certification of Small Unmanned Aircraft Systems 14 CFR Part 107*. Retrieved 12 December, 2018, from https://www.faa.gov/uas/media/RIN_2120-AJ60_Clean_Signed.pdf

Federal Aviation Administration. (2016). *Operation and Certification of Small Unmanned Aircraft Systems 14 CFR Part 107*. Retrieved from https://www.faa.gov/uas/media/RIN_2120-AJ60_Clean_Signed.pdf

Federal Aviation Administration. (2018a). *Pilot's Handbook of Aeronautical Knowledge (FAA-H-8083-25A)*. Washington, DC: Author.

Federal Aviation Administration. (2018b). *Part 107 waivers issued*. Retrieved from www.faa.gov: https://www.faa.gov/uas/request_waiver/waivers_granted/

Feng, D., & Yuan, X. (2016). *Automatic construction of aerial corridor for navigation of unmanned aircraft systems in class G airspace using LiDAR* (Vol. 9828, pp. 98280I-98280I – 8). doi:10.1117/12.2224359

Fetisov, V. S. (2014). *Unmanned Aviation: Terminology, Classification, Current State: Monograph.* Ufa, Russia: PHOTON.

Finnegan, P. (2015). *Press release: UAV Production Will Total $93 Billion.* Retrieved from: http://tealgroup.com/index.php/teal-group-news-media/item/press-release-uav-production-will-total-93-billion

Finn, R. L., & Wright, D. (2012). Unmanned aircraft systems: Surveillance, ethics and privacy in civil applications. *Computer Law & Security Review*, *28*(2), 184–194. doi:10.1016/j.clsr.2012.01.005

Fisher, J. (2017). *DJI Inspire 2.* Retrieved from http://au.pcmag.com/dji-inspire-2/44831/review/dji-inspire-2

Fisher, R. D., Jr. (2017). *China's AT200 cargo UAV to supply military installations in South China Sea.* Retrieved from https://www.janes.com/article/75309/china-s-at200-cargo-uav-to-supply-military-installations-in-south-china-sea

Flahault, A., Geissbuhler, A., Guessous, I., Guérin, P., Bolon, I., Salathé, M., & Escher, G. (2017). Precision global health in the digital age. *Swiss Medical Weekly*, *147*, w14423. PMID:28421566

Flatley, D. (2017). *Safety Incidents Involving Civilian Drones are Surging.* Retrieved from https://www.bloomberg.com/news/articles/2017-12-05/safety-incidents-rise-along-with-popularity-of-civilian-drones

Flin, R., O'Connor, P., & Crichton, M. (2008). *Safety at the Sharp End: A Guide to Non-Technical Skills.* Surrey, UK: Ashgate.

Floreano, D., & Wood, R. J. (2015). Science, technology and the future of small autonomous drones. *Nature*, *521*(7553), 460–466. doi:10.1038/nature14542 PMID:26017445

Foltz, P. W., Martin, M. J., Cooke, N. J., Kiekel, P. A., & Gorman, J. (2003). Automated Team Communication Analysis with Latent Semantic Analysis. *Proceedings of the Collaborative Technologies Alliance: Advanced Decision Architectures,* 33-40.

Fornace, K. M., Drakeley, C. J., William, T., Espino, F., & Cox, J. (2014). Mapping infectious disease landscapes: Unmanned aerial vehicles and epidemiology. *Trends in Parasitology*, *30*(11), 514–519. doi:10.1016/j.pt.2014.09.001 PMID:25443854

Fradkov, A., & Grigoriev, G. (2013). Decentral adaptive control of synchronization of networks of dynamical systems of dynamic systems with limited perturbations. *Automation and Telemechanics, 5.*

Franke, U. E. (2015). Civilian drones: Fixing the problems. *THE CSS Blog Network. Center for Security Studies.* Retrieved from: https://isnblog.ethz.ch/security/civilian-drones-fixing-an-image-problem

Franke. (2015, March 5). *Civilian Drones: Fixing an Image Problem?* ISN Blog.

Freeman, P. K., & Freeland, R. S. (2014). Politics & Technology: US policies restricting unmanned aerial systems in agriculture. *Food Policy*, *49*(1), 302–311. doi:10.1016/j.foodpol.2014.09.008

Frost & Sullivan. (2007). *Study Analyzing the Current Activities in the Field of UAV.* Retrieved from: https://ec.europa.eu/home-affairs/sites/homeaffairs/files/e-library/documents/policies/security/pdf/uav_study_element_2_en.pdf

Funk, K., Suroteguh, C., Wilson, J., & Lyall, B. (1998). *Flight deck automation and task management.* Paper presented at the IEEE 1998 International Conference on Systems, Man, and Cybernetics, New York, NY. 10.1109/ICSMC.1998.725523

Ganin, S., & Karpenko, A. (1999). Unmanned aerial vehicles. St. Petersburg: Academic Press.

GAO (United States Government Accountability Office). (2013). Defense Acquisitions: Assessments of Selected Weapon Programs. *GAO*. Retrieved from: http://www.gao.gov/assets/660/653379.pdf

Gavi. (2016). *Rwanda launches world's first national drone delivery service powered by Zipline*. Retrieved 20 December 2018, from https://www.gavi.org/library/news/gavi-features/2016/rwanda-launches-world-s-first-national-drone-delivery-service-powered-by-zipline/

Gawron, V. J. (1998). *Human factors issues in the development, evaluation, and operation of uninhabited aerial vehicles.* Paper presented at the Associations of Unmanned Vehicles Systems International.

Gawron, V. (2008). *Human Performance, Workload and Situational Awareness Measures Handbook*. Boca Raton, FL: Faylor & Francis. doi:10.1201/9781420064506

Global Grand Challenges. (2012). *On-Demand Vaccine Delivery via Low-Cost UAVs | Grand Challenges*. Retrieved 1 January 2019, from https://gcgh.grandchallenges.org/grant/demand-vaccine-delivery-low-cost-uavs

Gloria, L., & Calhoun, M. H. (2006). Advances in Human Performance and Cognitive Engineering Research. In H. L. Nancy & J. Cooke (Eds.), *Multi-Sensory Interfaces for Remotely Operated Vehicles* (pp. 149–163). Emerald Group.

Goldman Sachs. (2017). *Estimated Drone Spending in USD FY2017-2021*. Retrieved from https://www.goldmansachs.com/insights/technology-driving-innovation/drones/

Goldman Sachs. (2018). *Drones: Reporting for work*. Retrieved 28 May 2018 from http://www.goldmansachs.com/our-thinking/technology-driving-innovation/drones/

Gonzalez-Aguilera, D., & Rodriguez-Gonzalvez, P. (2017). Drones editorial. *Drones*, *1*(1), 1–5. doi:10.3390/drones1010001

Goodchild, A., & Toy, J. (2018). Delivery by drone: An evaluation of unmanned aerial vehicle technology in reducing CO2 emissions in the delivery service industry. *Transportation Research Part D, Transport and Environment*, *61*, 58–67. doi:10.1016/j.trd.2017.02.017

Grand View Reserach. (2016). *Comercial UAS market size to reach $2.07 billion by 2022*. Retrieved from https://www.grandviewresearch.com/press-release/commercial-uav-market-analysis

Green, S., Billinghurst, M., Chen, X. Q., & Chase, J. G. (2008). Human-Robot Collaboration: A Literature Review and Augmented Reality Approach in Design. *International Journal of Advanced Robotic Systems*, *5*(1), 1–18. doi:10.5772/5664

Guan, X., Bo, C., Li, Z., & Yu, Y. (2017). ST-hash: An efficient spatiotemporal index for massive trajectory data in a NoSQL database. In *2017 25th International Conference on Geoinformatics* (pp. 1–7). Academic Press. 10.1109/GEOINFORMATICS.2017.8090927

Guinness World Record. (2016). *Intel stuns during CES keynote with record for most drones airborne simultaneously*. Retrieved from http://www.guinnessworldrecords.com/news/2016/1/intel-stuns-during-ces-keynote-with-record-for-most-drones-airborne-simultaneousl-411677

Gulevich, S. P., Veselov, Y. G., Pryadkin, S. P., & Tirnov, S. D. (2012). Analysis of factors affecting the safety of the flight of unmanned aerial vehicles. Ru. Causes of accidents drones and methods of preventing them. *Science and Education*, 12.

Gupta, S. G., Ghonge, M. M., & Jawandhiya, P. M. (2013). Review of unmanned aircraft system (UAS). *Technology*, *2*(4).

Gu, Q., Michanowicz, D. R., & Jia, C. (2018). Developing a Modular Unmanned Aerial Vehicle (UAV) Platform for Air Pollution Profiling. *Sensors (Basel)*, *18*(12), 4363. doi:10.339018124363 PMID:30544691

Gye, H. (2013). Now that's a special delivery: Domino's builds DRONE to deliver pizzas by air and beat the traffic. *Daily Mail*. Retrieved from: http://www.dailymail.co.uk/news/article-2336324/Dominos-builds-DRONE-deliver-pizzas-air-beat-traffic.html#ixzz44XJEkfMO

Haanappel, P. P. C. (2003). *The Law and Policy of Air Space and Outer Space: A Comparative Approach*. The Hague: Kluwer Law International.

Haidari, L., Brown, S., Ferguson, M., Bancroft, E., Spiker, M., Wilcox, A., & Lee, B. (2016). The economic and operational value of using drones to transport vaccines. *Vaccine*, *34*(34), 4062–4067. doi:10.1016/j.vaccine.2016.06.022 PMID:27340098

Hail, A. R., & Coyne, C. J. (2014). The political economy of drones. *Defence and Peace Economics*, *25*(5), 445–460. doi:10.1080/10242694.2013.833369

Hainley, J. C. J. Jr, Duda, K. R., Oman, M., & Natapoff, A. (2013). Pilot performance, workload, and situation awareness during lunar landing mode transitions. *Journal of Spacecraft and Rockets*, *50*(4), 793–801. doi:10.2514/1.A32267

Handford, C., Reeves, F., & Parker, P. (2018). Prospective use of unmanned aerial vehicles for military medical evacuation in future conflicts. *Journal of the Royal Army Medical Corps*, *164*(4), 293–296. doi:10.1136/jramc-2017-000890 PMID:29523753

Hanna, D., Ferworn, A., Lukaczyn, M., Abhari, A., & Lum, J. (2018). Using Unmanned Aerial Vehicles (UAVs) in locating wandering patients with dementia. In *2018 IEEE/ION Position, Location and Navigation Symposium (PLANS)* (pp. 809–815). IEEE. 10.1109/PLANS.2018.8373457

Hardy, A., Makame, M., Cross, D., Majambere, S., & Msellem, M. (2017). Using low-cost drones to map malaria vector habitats. *Parasites & Vectors*, *10*(1), 29. doi:10.118613071-017-1973-3 PMID:28088225

Harrington, A. (2015). Who controls the drones? *Engineering Technology*, *10*(2), 80–83. doi:10.1049/et.2015.0211

Hart, S. G., & Hauser, J. R. (1987). Inflight application of three pilot workload measurement techniques. *Aviation, Space, and Environmental Medicine*. PMID:3593142

Heffernan, D., & Urban, J. (2017). US regulation of commercial uses of drones. *Expert Guides*. Retrieved from: https://www.expertguides.com/articles/us-regulation-of-commercial-uses-of-drones/ARVLNBXG

Helmerich, R., & Merritt, A. (2001). *Culture at Work in Aviation and Medicine, Organisational and Professional influences*. Surrey, UK: Ashgate.

Helmreich, R. (1998). Building safety on the three cultures of aviation. *IATA: Human Factors Seminar*, 39-43.

Helmreich, R. L., & Merritt, A. C. (2000). Safety and error management: The role of crew resource management. In B. J. Hayward & A. C. Lowe (Eds.), *Aviation resource management* (pp. 107–119). Aldershot, UK: Ashgate.

Helmreich, R. L., Merritt, A. C., & Wilhelm, J. A. (1999). The evolution of crew resouce management. *The International Journal of Aviation Psychology*, *9*, 19–32. doi:10.120715327108ijap0901_2 PMID:11541445

Helmreich, R. L., & Wilhelm, J. A. (1991). Outcomes of crew resource management training. *The International Journal of Aviation Psychology*, *1*(4), 287–300. doi:10.120715327108ijap0104_3 PMID:11537899

Heutger, M. (2014). *Unmanned Aerial Vehicles in Logistics – A DHL Perspective on Implications and Use Cases for the Logistics Industry*. DHL Customer Solutions & Innovation. Retrieved from: http://www.dhl.com/content/dam/downloads/g0/about_us/logistics_insights/DHL_TrendReport_UAV.pdf

History of Cabinet Bill No. 75 of 189th Diet Session, HR, Japan.

History of House of Representatives (HR) Bill No. 24 of 189th Diet Session, HR, Japan.

Hobbs, A., & Lyall, B. (2016). *Human Factors Guidelines for Remotely Piloted Aircraft System Remote Pilot Stations (TN-34128)*. San Jose, CA: NASA Ames Research Center.

Holt, J., Biaz, S., Yilmaz, L., & Afi, C. A. (2014). Asymbiotic simulation architecture for evaluating UAV collision avoidance techniques. *Journal of Simulation*, *8*(1), 64–75. doi:10.1057/jos.2013.5

Holz, R. L. (2018). In Rwanda drones deliver medical supplies to remote areas. *Wall Street Journal*. Retrieved from: https://www.wsj.com/articles/in-rwanda-drones-deliver-medical-supplies-to-remote-areas-1512124200

Hong, Z., Tong, X., Cao, W., Jiang, S., Chen, P., & Liu, S. (2015). Rapid three-dimensional detection spproach for building damage due to earthquakes by the use of parallel processing of unmanned aerial vehicle imagery. *Journal of Applied Remote Sensing*, *9*(1), 097292. doi:10.1117/1.JRS.9.097292

Hopcroft, R., Burchat, E., & Vince, J. (2006). Unmanned aerial vehicles for maritime patrol: Human factors issues (No. DSTO-GD-0463). Air operations division: Defence sicenice and technology organisation Edinbrgh (Australia).

Horowitz, M. C., & Fuhrmann, M. (2015). *Droning On: Explaining the Proliferation of Unmanned Aerial Vehicles*. doi:10.2139srn.2514339

Howell, C. T. (2016). *The First Government Sanctioned Delivery of Medical Supplies by Remotely Controlled Unmanned Aerial System (UAS)*. Presented at the Xponential 2016, New Orleans, LA. Retrieved from https://ntrs.nasa.gov/search.jsp?R=20160009120

Hradecky, S. (2015). *Incident: Delta A320 near Denver on August 7th 2015, hail strike*. Retrieved from http://avherald.com/h?article=48a8329f

Huang, Y., Yi, S., & Li, Z. (2011). Design of highway landslide warning and emergency response systems based on UAV. *Proceedings of the 17th Chinese conference on remote sensing*.

Hudson, P. (2001). Safety Culture: The Ultimate Goal. *Flight Safety Australia*, 9-31.

IATA. (2018). *IATA Cargo Strategy*. Retrieved from https://www.iata.org/whatwedo/cargo/Documents/cargo-strategy.pdf

ICAO. (2005). *Doc 9854. Global Air Traffic Management Operational Concept* (1st ed.). Montreal, Quebec, Canada: International Civil Aviation Organization.

ICAO. (2011). ICAO's circular 328 AN/190: Unmanned Aircraft Systems. *ICAO*. Retrieved from: https://www.icao.int/Meetings/UAS/Documents/Circular%20 328_en.pdf

ICAO. (2012). *Safety Management Manual*. Montreal: International Civil Aviation Organization.

ICAO. (2013). *Safety Management Manual (SMM)* [Doc 9859, AN/474] (3rd ed.). Montreal, Canada: International Civil Aviation Organisation.

ICAO. (2016). *Long-Term Traffic Forecasts: Passenger and Cargo*. Retrieved from https://www.icao.int/Meetings/aviationdataseminar/Documents/ICAO-Long-Term-Traffic-Forecasts-July-2016.pdf

ICAO. (2017). Drone enable, ICAO's unmanned aircraft systems (UAS) industry symposium. *ICAO*. Retrieved from: https://www.icao.int/Meetings/UAS2017/Pages/default.aspx

ICAO. (2018). Drone enable/2 symposium. *ICAO*. Retrieved from: https://www.icao.int/Meetings/DRONEENABLE2/Pages/default.aspx

ICAO. (2018a). *Remotely Piloted Aircraft Systems Panel (RPASP)*. Retrieved 09 May 2018 from https://www.icao.int/safety/UA/Pages/Remotely-Piloted-Aircraft-Systems-Panel-(RPASP).aspx

ICAO. (2018b). *Unmanned Aircraft Systems Advisory Group (UAS-AG)*. Retrieved 08 May 2018 from https://www.icao.int/safety/UA/Pages/Unmanned-Aircraft-Systems-Advisory-Group-(UAS-AG).aspx

ICAO. (n.d.). *AIRAC*. Retrieved 22 May 2018 from https://www.icao.int/safety/information-management/Pages/AIRACAdherence.aspx

ICAO. (n.d.). Unmanned aircraft systems advisory group (UAS-AG). *ICAO*. Retrieved from: https://www.icao.int/safety/UA/Pages/Unmanned-Aircraft-Systems-Advisory-Group-(UAS-AG).aspx

IDC. (n.d.). Legal issues surrounding the use of drones. *IDC Legal Clinic for Start-Ups*. Retrieved from: http://legalstart.idc.ac.il/legal-issues-surrounding-the-use-of-drones/

Ignatiev, O. (2010). UAV: The state and prospects of their use in the interests of NATO. *Foreign Military Review*, *12*, 45–52.

Improdone. (2018). *The Pros and Cons of Drones*. Retrieved from: https://improdrone.com/the-pros-and-cons-of-drones/

INSARAG. (2015). *International Search and Rescue Guidelines*. Retrieved from http://www.insarag.org/en/methodology/guidelines.html

International Civil Aviation Organisation. (2009). Safety Management Manual (2nd ed.; Doc 9859). Montreal, Canada: Author.

International Civil Aviation Organisation. (2011). *Unmanned Aircraft Systems (UAS): Cir 328 AN/190*. Retrieved from https://docplayer.net/7556384-Unmanned-aircraft-systems-uas.html

International Civil Aviation Organisation. (2011a). *Unmanned Aircraft Systems (UAS) (ICAO Cir 328)*. Montreal, Canada: Author.

International Civil Aviation Organisation. (2011b). *Personnel Licensing*. Montreal, Canada: Author.

International Civil Aviation Organisation. (2015). *Manual on Remotely Piloted Aircraft Systems (RPAS) (Doc 10019)*. Montreal, Canada: Author.

International Civil Aviation Organization. (2005). *Global Air Traffic Management Operational Concept. Doc. ICAO Doc 9854-AN/458*. Montreal, Canada: Author.

International Civil Aviation Organization. (2009). *Manual on Global Performance of the Air Navigation System. Doc. ICAO 9883*. Montreal, Canada: Author.

International Civil Aviation Organization. (2015). *Manual of Remoted Piloted Aircraft Systems (RPAS). Doc. ICAO 10019/AN 507*. Montreal, Canada: Author.

Isaak, J., & Hanna, M. J. (2018). User Data Privacy: Facebook, Cambridge Analytica, and Privacy Protection. *Computer*, *51*(8), 56–59. doi:10.1109/MC.2018.3191268

Jaberidoost M., Nikfar S., Abdollahiasl A., & Dinarvand R. (2013). Pharmaceutical supply chain risks: a systematic review. *DARU Journal of Pharmaceutical Sciences, 1*(21), 1-7.

Jarnot, G. (2012). History. In R. Barnhart, S. Hottman, D. Marshall, & E. Shappee (Eds.), *Introduction to Unmanned Aircraft Systems*. New York: CRC Press.

JCAB. (2015). *Japan's safety rules on unmanned aircraft (UA)/drone*. Retrieved 14 May 2018 from http://www.mlit.go.jp/en/koku/uas.html

Jenkins, B. (2013). Watching the watchmen: Drone Privacy and the need for oversight. *Kentucky Law Journal (Lexington, Ky.)*, *102*, 161.

Jenkins, D. J., & Vasigh, B. (2013). *The economic impact of Umanned Systems integration in the United States*. Arlington, VA: AUVSI.

Jenkins, D., & Vasigh, B. (2013). *The Economic Impact of Unmanned Aircraft Systems Integration in the United States. Association for Unmanned Vehicle Systems International*. AUVSI.

Jiang, T., Geller, J., Ni, D., & Collura, J. (2016). Unmanned Aircraft System Traffic management: Concept of operation and system architechture. *International Journal of Transportation Science and Technology, 5*(3), 123-135.

Johnson, T. (2017). *Amazon Prime Air: The Evolution of Amazon's Drone Delivery*. Retrieved from: http://www.cpcstrategy.com/blog/2017/07/amazon-prime-air/

Joshi, D. (2017). Commercial Unmanned Aerial Vehicle (UAV) Market Analysis – Industry trends, companies and what you should know. *Business Insider*. Retrieved from: http://www.businessinsider.com/commercial-uav-market-analysis-2017-8/?r=AU&IR=T

Karaca, Y., Cicek, M., Tatli, O., Sahin, A., Pasli, S., Beser, M. F., & Turedi, S. (2018). The potential use of unmanned aircraft systems (drones) in mountain search and rescue operations. *The American Journal of Emergency Medicine*, *36*(4), 583–588. doi:10.1016/j.ajem.2017.09.025 PMID:28928001

Kaupp, T., & Makerenko, A. (2008). *Measuing human-robot team effectiveness to determine an appropriate autonomy level.* Paper presented at the Robotics and automation, 2008, ICRA 2008, IEEE international conference.

KBLA. (n.d.). New Korean drone laws. *KBLA*. Retrieved from: http://www.kbla.info/index.php/the-kbla-blog/1079-new-korean-drone-laws

Kelek, B. S. (2015). Blue Sky Birds Come to the World. *Journal of International Trade. Logistics and Law*, *1*(1), 41–49.

Kern, T. (1997). *Redefining Airmanship*. New York, NY: McGraw-Hill.

Kharchenko, V., Shmelova, T., & Sikirda, Y. (2012). *Decision-Making of Operator in Air Navigation System: a monograph*. Kirovograd: KFA of NAU.

Kharchenko, V., Shmclova, T., Bondarev, D., & Stratiy, A. (2017). Modeling of Decision Making of Unmanned Aerial Vehicle's Operator in Emergency Situations. *Proceedings of the National Aviation University*, *1*, 20–28.

Kharchenko, V., Shmelova, T., & Sikirda, Y. (2016). *Decision-Making in Socio-Technical Systems: a monograph*. Kyiv: NAU.

Kille, T., Bates, P., & Murray, P. S. (2013a). *Farm to Table: Air Transport Issues for an Island State*. Paper presented at the 2013 Australasian Transport Research Forum Conference, Queensland, Australia.

Kille, T., Bates, P., & Murray, P. S. (2013b). Measuring Success: The Search for Assessment Criteria in Determining the Impact of Deregulation in Regional Aviation. *International Journal of Aviation Management*, *2*(1/2), 4–34. doi:10.1504/IJAM.2013.053046

Kille, T., Bates, P., & Murray, P. S. (2014). A systematic review of the role of regional aviation in the airfreight industry. *Science-based Technologies*, *24*(4), 471–485. doi:10.18372/2310-5461.24.7520

Kille, T., Bates, P., & Murray, P. S. (2018). Regional Devlopment: Air Freight Service Needs for Regional Communities. In M. Khosrwo-Pour (Ed.), *Encyclopedia of Information Science and Technology* (4th ed.). Hershey, PA: IGI Global. doi:10.4018/978-1-5225-2255-3.ch684

Kim. (2000). *The Innovation of the Warsaw System and the IATA lnter-carrier Agreement. In The Utilization of the World's Air Space and Free Outer in the 21st Century*. Kluwer Law International.

Kim. (2008). Essays for the Study of the International Air and Space Law. Korean Studies Information Co. Ltd.

Kim, K., & Davidson, J. (2015). Unmanned Aircraft Systems Used for Disaster Management. *Transportation Research Record: Journal of the Transportation Research Board*, *2532*(1), 83–90. doi:10.3141/2532-10

Kim, S. J., Lim, G. J., Cho, J., & Côté, M. J. (2017). Drone-Aided Healthcare Services for Patients with Chronic Diseases in Rural Areas. *Journal of Intelligent & Robotic Systems*, *88*(1), 163–180. doi:10.100710846-017-0548-z

Kinnison, H. A., & Siddiqui, T. (2013). *Aviation maintenance management* (2nd ed.). New York, NY: McGraw-Hill.

Kintsch, W., & Ericsson, K. A. (1995). Long term working memory. *Psychological Review*, *102*(2), 211–245. doi:10.1037/0033-295X.102.2.211 PMID:7740089

Kipper, G., & Rampolla, J. (2013). *Augmented Reality: An emerging technologies guide to AR*. Waltham, MA: Elsevier.

Klingberg, T. (2000). Limitations in information processing in the human brain: Neuroimaging of dual task performance and working memory tasks. *Progress in Brain Research*, *126*, 95–102. doi:10.1016/S0079-6123(00)26009-3 PMID:11105642

Knight, Z. (2017). *What Germany's first armed drones could do*. Retrieved from: https://www.dw.com/en/what-germanys-first-armed-drones-could-do/a-39355009

Koeniger, N., Koeniger, G., Gries, M., & Tingek, S. (2005). Drone Competition at Drone Congregation Areas in Four Apis Species. *Apidologie*, *36*(2), 211–221. doi:10.1051/apido:2005011

Koh, C. H., Low, K. H., Li, L., Zhao, Y., Deng, C., Tan, S. K., ... Li, X. (2018). Weight threshold estimation of falling UAVs (Unmanned Aerial Vehicles) based on impact energy. *Transportation Research Part C, Emerging Technologies*, *93*, 228–255. doi:10.1016/j.trc.2018.04.021

Kohlios, C. P., & Hayajneh, T. (2018). A Comprehensive Attack Flow Model and Security Analysis for Wi-Fi and WPA3. *Electronics (Basel)*, *7*(11), 284. doi:10.3390/electronics7110284

Kontitsis, P., Tsourveloudis, G., & Valavanis, P. (2003). *A UAV based automated airborne survellance system*. Paper presented at the 11th Mediterranean conference on control and automation, Rhodes, Greece.

Korda, Z. (2018). *U Osijeku otvoren prvi sajam dronova u Hrvatskoj*. Retrieved from: https://www.tportal.hr/tehno/clanak/u-osijeku-otvoren-prvi-sajam-dronova-u-hrvatskoj-20180512

Kothari, M., Postlethwaite, I., & Gu, D. (2009). Multi-UAV path planning in obstacle rich environments using Rapidly-exploring Random Trees. In *Proceedings of the 48h IEEE Conference on Decision and Control (CDC) held jointly with 2009 28th Chinese Control Conference* (pp. 3069–3074). IEEE. 10.1109/CDC.2009.5400108

Kreps, S., & Zenko, M. (2014). The Next Drone Wars. *Foreign Affairs*. Retrieved from http://bmpd.livejournal.com/861105.html

Kreps, S. (2014). Flying under the radar: A study of public attitudes towards unmanned aerial vehicles. *Research and Politics*, *1*(1), 1–7. doi:10.1177/2053168014536533

Krey, M. (2018). *Drones in healthcare: Application in Swiss hospitals*. Paper presented at the 51st Hawaii International Conference on System Sciences, Waikoloa, HI. Retrieved from https://scholarspace.manoa.hawaii.edu/bitstream/10125/50277/1/paper0390.pdf

Kucherov, D. P., & Kozub, A. N. (2015). Control System Objects with Multiple Stream of Information. *Proceedings IEEE 3rd International Conference "Actual Problems of Unmanned Aerial Vehicles Developments (APUAVD)"*, 290-293. 10.1109/APUAVD.2015.7346623

Kucherov, D. P., & Kozub, A. N. (2016). Group Behavior of UAVs in obstacle presence. *Proceeding IEEE 4th International Conference "Methods and Systems of Navigation and Motion Control (MSNMC)"*, 51-54. 10.1109/MSNMC.2016.7783104

Kucherov, D. P., Kozub, A. N., & Kostyna, A. N. (2017). Group of UAVs Moving on Smooth Control Law with Fixed Obstacles. *Advances in Science, Technology and Engineering Systems*, *2*(3), 1034–1041. doi:10.25046/aj0203131

Kunz, N., & Reiner, G. (2012). A meta-analysis of humanitarian logistics research. *Journal of Humanitarian Logistics and Supply Chain Management*, *2*(2), 116 147. doi:10.1108/20426741211260723

Kuzin, A. V. (2011). Computer networks / A.B. Cousin. Moscow: INFRA-M.

Lacher, A., & Maroney, D. (2012). *A new paradigm for small UAS*. Retrieved from https://www.mitre.org/sites/default/files/pdf/12_2840.pdf

Lamb, T. (2017). *Developing a Safety Culutre for Remotely Piloted Aircraft Systems Operations: To boldly go where no drone has gone before*. Society of Petroleum Engineers, SPE-84476-MS MS.

Lamb, T. (2018, October 28). *Factors Influencing Small Commercial UAS Mishaps.* Unpublished work, Embry-Riddle Aeronautical University, PhD program. Daytona Beach, FL: Embry-Riddle Aeronautical University.

Landhuis, E. (2018). *Medical cargo could be the gateway for routine drone deliveries.* Retrieved from https://www.npr.org/sections/health-shots/2018/03/10/592059175/medical-cargo-could-be-the-gateway-for-routine-drone-deliveries

Larsen, M. P., Eisenberg, M. S., Cummins, R. O., & Hallstrom, A. P. (1993). Predicting survival from out-of-hospital cardiac arrest: A graphic model. *Annals of Emergency Medicine*, *22*(11), 1652–1658. doi:10.1016/S0196-0644(05)81302-2 PMID:8214853

Lauber, J. K. (1987). Cockpit resource management: Background and overview. In H. W. Orlady & H. C. Foushee (Eds.), *Cockpit resource management training: Proceedings of the NASA/MAC workshop* (pp. 5-14). Moffett Field, CA: NASA-Ames Research Center.

Lee, B. Y., Connor, D. L., Wateska, A. R., Norman, B. A., Rajgopal, J., Cakouros, B. E., ... Brown, S. T. (2015). Landscaping the structures of GAVI country vaccine supply chains and testing the effects of radical redesign. *Vaccine*, *33*(36), 4451–4458. doi:10.1016/j.vaccine.2015.07.033 PMID:26209835

Lee, S. (1997). *Aviation Accident Law*. New York: II, Matthew Bender & Co., Inc.

Lichtenstein, S., Slovic, P., Fischhoff, B., Layman, M., & Combs, B. (1978). Judged frequency of lethal events. *Journal of Experimental Psychology. Human Learning and Memory*, *4*(6), 551–578. doi:10.1037/0278-7393.4.6.551 PMID:731196

Lidynia, C., Philipsen, R., & Ziefle, M. (2017). Droning on About Drones—Acceptance of and Perceived Barriers to Drones in Civil Usage Contexts. In *Advances in Human Factors in Robots and Unmanned Systems* (pp. 317–329). Cham: Springer; doi:10.1007/978-3-319-41959-6_26

Lim, H., Park, J., Lee, D., & Kim, H. J. (2012). Build Your Own Quadrotor: Open-Source Projects on Unmanned Aerial Vehicles. *IEEE Robotics & Automation Magazine*, *19*(3), 33–45. doi:10.1109/MRA.2012.2205629

Lippi, G., & Mattiuzzi, C. (2016). Biological samples transportation by drones: Ready for prime time? *Annals of Translational Medicine*, *4*(5), 92. doi:10.21037/atm.2016.02.03 PMID:27047951

Liu, P., Chen, A. Y., Huang, Y.-N., Han, J.-Y., Lai, J.-S., Kang, S.-C., ... Tsai, M.-H. (2014). A review of rotorcraft unmanned aerial vehicle (UAV) developments and applications in civil engineering. *Smart Structures and Systems*, *13*(6), 1065–1094. doi:10.12989ss.2014.13.6.1065

Lorenz, A., von Mittelstaedt, J., & Schmitz, G. P. (2011). *Messengers of Death - Are Drones Creating a New Global Arms Race?* Retrieved from: http://www.spiegel.de/international/world/messengers-of-death-are-drones-creating-a-new-global-arms-race-a-792590-2.html

Low, I. (2004). A Tool for the Assessment of the Impact of change in Automated ATM Systems on Mental Wokload (Report: HRS/HSP-005-REP-03). Brussels: European Air Traffic Management (EATM).

Lu, L., & Wang G. (2008). A study on multi-agent supply chain framework based on network economy. *Computers & Industrial Engineering, 2*(54), 288-300.

Luppicini, R., & So, A. (2016). A technoethical review of commercial drone use in the context of governance, ethics, and privacy. *Technology in Society*, *46*, 109–119. doi:10.1016/j.techsoc.2016.03.003

Lutz, B. (2014). *Multirotor service-drone market-leading developments*. Retrieved from: http://www.marketwired.com/press-release/multirotor-service-drone-market-leading-developments-in-may-2014-1915272.htm

MacSween-George, S. L. (2003). *A Public Opinion Survey - Unmanned Aerial Vehicles for Cargo, Commercial, and Passenger Transportation.* Paper presented at the 2nd AIAA “Unmanned Unlimited” Systems, Technologies, and Operations, San Diego, CA.

Mailey, C. (2013). Are UAS More Cost Effective that Manned Flights? *AUVSI*. Retrieved from www.auvsi.org

Maltz, M., & Meyer, J. (2003). Use of warnings in an attentionally demanding detection task. *Human Factors*, *43*, 217–226.

Mangiameli, M., & Mussumeci, G. (2014). Real Time Integration of Field Data into a GIS Platform for the Management of Hydrological Emergencies. In ISPRS - International Archives of the Photogrammetry, Remote Sensing and Spatial Information Sciences (Vol. XL-5-W3, pp. 153–158). Copernicus GmbH. doi:10.5194/isprsarchives-XL-5-W3-153-2013

Manning, S. D., Rash, C. E., LeDuc, P. A., Noback, R. K., & McKeon, J. (2004). *The Role of human causal factors in US army unmanned aerial vehicle accdients (No. USAARL-2004-11)*. Fort Rucker, AL: Army Aeromedical Research Lab. doi:10.21236/ADA421592

Marsh, L., Calbert, G., Tu, J., Gossink, D., & Kwok, H. (2005) Multi-Agent UAV Path Planning. In *MSSANZ Conference on Modelling and Simulation (MODSIM 05)* (pp. 2188 – 2194). Melbourne, Australia: Academic Press.

Martin, P. G., Payton, O. D., Fardoulis, J. S., Richards, D. A., Yamashiki, Y., & Scott, T. B. (2016). Low altitude unmanned aerial vehicle for characterising remediation effectiveness following the FDNPP accident. *Journal of Environmental Radioactivity*, *151*, 58–63. doi:10.1016/j.jenvrad.2015.09.007 PMID:26410790

Matte. (1981). *Treatise on Air-Aeronautical Law*. Montreal: Institute and Centre of Air and and Space Law, McGill University.

Matthes, R., Alexander, R., & Stone, R. (2018). Helicopter accident trends in eight ISASI countries. *Journal of the International Society of Air Safety Investigators*, *51*(4), 4–12.

Matthews, M. D., Eid, J., Johnsen, B. H., & Boe, O. C. (2011). A comparison of expert ratings and self-assessments of situation awareness during a combat fatigue course. *Military Psychology*, *23*(2), 125–136. doi:10.1080/08995605.2011.550222

Mazur, M. (2016). *Six Ways Drones Are Revolutionizing Agriculture*. Retrieved from: https://www.technologyreview.com/s/601935/six-ways-drones-are-revolutionizing-agriculture/

Mazur, M., & Wisniewski, A. (2016). *Clarity from above - PwC global report on the commercial applications of drone technology*. PricewaterhouseCoopers. Retrieved from: http://www.pwc.pl/en/publikacje/2016/clarity-from-above.htm

McCarley, J. S., & Wickens, C. D. (2005). *Human factors implications of UAVs in the national airspace*. Aviation Human Factors Division: University of Illinois at Urbana-Champaign.

Médecins Sans Frontières. (2014). *Papua New Guinea: Innovating to reach remote TB patients and improve access to treatment*. Retrieved 15 May 2018 from http://www.msf.org/article/papua-new-guinea-innovating-reach-remote-tb-patients-and-improve-access-treatment

Meiches, B. (2018). Non-human humanitarians. *Review of International Studies*, 1–19. doi:10.1017/S0260210518000281

Meier, P. (2015). UAVs and Humanitarian Response. In *UAVs and Humanitarian Response. Drones and Aerial Observation: New Technologies for Property Rights, Human Rights, and Global Development. A Primer*. Academic Press. Retrieved from http://drones.newamerica.org/primer

Meola, A. (2017). Drone market shows positive outlook with strong industry growth and trends. *Business Insider.* Retrieved from: https://www.businessinsider.com/drone-industry-analysis-market-trends-growth-forecasts-2017-7

Meshkati, N., Hancock, P., & Rahimi, M. (1992). Techniques in mental workload assessment. In J. Wilson & E. Corlett (Eds.), *Evaluation of human work. A practical ergonomics methodology* (pp. 605–627). London: Taylor & Francis.

Mileham, R. (2016). Prime movers. *Engineering & Technology*, *11*(4), 70–72. doi:10.1049/et.2016.0408

Miljković, D. (2018). Methods for attenuation of unmanned aerial vehicle noise. In *2018 41st International Convention on Information and Communication Technology, Electronics and Microelectronics (MIPRO)* (pp. 0914 0919). Academic Press. 10.23919/MIPRO.2018.8400169

Milliken, J., & Linton, D. (2016). Prioritisation of citizen-centric information for disaster response. *Disasters*, *40*(3), 476–493. doi:10.1111/disa.12168 PMID:26434664

Moiseev, V.S., Gushchina, D.S., & Moiseev, G.V. (2010). *Основы теории создания и применения информационных беспилотных авиационных комплексов: Монография* [Fundamentals of Theory of Information Unmanned Aerial Systems Design and Application: Monograph]. Kazan, Russia: Publishing house of the Ministry of Education and Science of the Republic of Tatarstan.

Molnar, A., & Parsons, C. (2016). Unmanned Aerial Vehicles (UAVs) and Law Enforcement in Australia and Canada: Governance through 'Privacy' in an Era of Counter-Law? *National Security, Surveillance and Terror*, 225-247.

Montgomery, J. F., & Bekey, G. A. (1998). Learning helicopter control through "teaching by showing. *IEEE Conference on Decision and Control*. 10.1109/CDC.1998.761747

Moorman, R. (2017, December). Pay to Play: The cost of ADS-B and where to find financial assistance. *Avionics Digital*.

Moses, L. B. (2007). *Recurring Dilemmas: The Law's Race to Keep Up With Technological Change (SSRN Scholarly Paper No. ID 979861)*. Rochester, NY: Social Science Research Network. Retrieved from https://papers.ssrn.com/abstract=979861

Mouloua, M., Gilson, R., Daskarolis-Kring, E., Kring, J., & Hancock, P. (2001). *Ergonomics of UAV/UCAV mission success: Considerations for data link, control and display issues*. Paper presented at the Human Factors and Ergonomics Society 45th Annual Meeting, Santa Monica, CA. 10.1177/154193120104500231

Mouloua, M., Gilson, R., Kring, J., & Hancock, P. (2001). *Workload, Situation Awareness, and Teaming Issues for UAV/UCAV Operations*. Paper presented at the Human Factors and Ergonomics Society 45th Annual Meeting. 10.1177/154193120104500235

Mouloua, M. G. (2003). Designing controls for future unmanned aerial vehicles. *Ergonomics in Design*, *11*(4), 6–11. doi:10.1177/106480460301100103

Mudrić, M., & Katulić, T. (2016). Regulacija sustava bespilotnih zrakoplova u hrvatskom, europskom i međunarodnom pravnom okviru. *Pravo u gospodarstvu*, *55*(2), 123-136.

Murai, S. (2016). Man who landed drone on roof of Japanese prime minister's office gets suspended sentence. *Japan Times*. Retrieved from: http://www.japantimes.co.jp/news/2016/02/16/national/crime-legal/man-landed-drone-roof-japanese-prime-ministers-office-gets-suspended-sentence

Murray, C. C., & Chu, A. G. (2015). The flying sidekick traveling salesman problem: Optimization of drone-assisted parcel delivery. *Transportation Research Part C, Emerging Technologies*, *54*, 86–109. doi:10.1016/j.trc.2015.03.005

NASA. (2007). *NASA and NOAA Fly Unmanned Aircraft into Hurricane Noel.* Retrieved from: http://www.nasa.gov/centers/wallops/news/story105.html

National Academies of Sciences and Medicine. (2018). Assessing the Risks of Integrating Unmanned Aircraft Systems (UAS) into the National Airspace System. Washington, DC: The National Academies Press.

NCSL. (n.d.). Transportation. *National Conference of State Legislatures (NCSL)* Retrieved from: http://www.ncsl.org/research/transportation/2017-unmanned-aircraft-systems-uas-state-legislation- update.aspx

Nedjati, A., Vizvari, B., & Izbirak, G. (2016). Post-earthquake response by small UAV helicopters. *Natural Hazards*, *80*(3), 1669–1688. doi:10.100711069-015-2046-6

Nex, F., & Remondino, F. (2014). UAV for 3D mapping applications: A review. *Applied Geomatics*, *6*(1), 1–15. doi:10.100712518-013-0120-x

Nguyen, T., Lim, C. P., Duy Nguyen, N., Gordon-Brown, L., & Nahavandi, S. (2018). A Review of Situation Awareness Assessment Approaches in Aviation Environments. *Eprint ArXiv 1803 (08067)* Retrieved 29 December, 2018, from https://arxiv.org/ftp/arxiv/papers/1803/1803.08067.pdf

Nikiforov, A., & Munimaev, V. (2010). *Analysis of foreign unmanned aerial vehicles*. St. Petersburg Forestry Academy.

Nisser, T., & Westin, C. (2006). *Human Factors Challenges in Unmanned Aerial Vehicles (UAVs): A Literature Review*. Lund University: School of Aviation.

Official Gazette (2014). *Air Traffic Act*. Official Gazette. 92/2014.

Official Gazette (2015). *Ordinance on unmanned aircraft systems*. Official Gazette. 77/2015.

Olifer, V.G., & Olifer N.A. (2006). *Computer networks: principles, technologies, protocols*. Academic Press.

Oliver, J. M. (2017). Current Status of the Use of Small Unmanned Aerial Systems for Environmental Monitoring. *Environmental Claims Journal*, *29*(2), 159–170. doi:10.1080/10406026.2017.1307009

Olsen, R. G. (2017). Paperweights: FAA Regulation and the banishment of commercial drones. *Berkeley Technology Law Journal*, *32*(4), 621–652.

Olson, K. (2018). *The 3 main categories of drones and their advantages and disadvantages*. Retrieved 28 May 2018 from https://www.botlink.com/blog/the-3-main-categories-of-drones-and-what-they-do

Operation and Certification of Small Unmanned Aircraft Systems, Final Rule, 81 Fed. Reg. 42,063 (June 28, 2016), codified at 14 C.F.R. Part 107.

Orlady, H. W., & Orlady, L. M. (2002). *Human Factors in Multi-Crew Flight Operations*. Aldershot, UK: Ashgate Publishing Company.

Pahsa, A., Kaya, P., Alat, G., & Baykal, B. (2011). *Integrating navigation and surveillance of unmanned air vehicles into civilian national airpaces by using ADS-B applications*. Paper presented at the 2011 Integrated Communications, Navigation, and Surveillance Conference.

Parasuraman, R., & Riley, V. (1997). Humans and Automation: Use, Misuse, Disuse, Abuse. *Human Factors*, *39*(2), 230–253. doi:10.1518/001872097778543886

Parasuraman, R., & Wickens, C. (2008). Humans: Vital after all these years of automation. *Human Factors*, *50*(3), 511–520. doi:10.1518/001872008X312198 PMID:18689061

ParaZero. (2018). *ParaZero SafeAir™ Drone Safety System*. Retrieved from https://parazero.com/

Park, J., Kim, S., & Suh, K. (2018). A Comparative Analysis of the Environmental Benefits of Drone-Based Delivery Services in Urban and Rural Areas. *Sustainability*, *10*(3), 888. doi:10.3390u10030888

Passchier-Vermeer, W., & Passchier, W. F. (2000). Noise exposure and public health. *Environmental Health Perspectives*, *108*(Suppl 1), 123–131. PMID:10698728

Pasztor, A. (2018). FAA projects fourfold increase in commercial drones by 2022. *The Wall Street Journal.* Retrieved 28 May 2018 from https://www.wsj.com/articles/faa-projects-fourfold-increase-in-commercial-drones-by-2022-1521407110

Patrick, W. G. (2016). *Request apparatus for delivery of medical support implement by UAV.* Retrieved from https://patents.google.com/patent/US9307383B1/en

Peace, C. (2017). The Risk Matrix; Uncertain results? *Policy and Practice in Health and Safety, 15*(2), 131–144.

Perritt, H., & Sprague, E. O. (2017). *Domesticating Drones: The technology, Law, and Economics of Unmanned Aircraft.* New York: Routledge.

Peterson, M. E. (2006). The UAV and the current and future regulatory construct for integration into the national airspace system. *Journal of Air Law and Commerce, 71*(521).

Plimp. (2018). *Plimp airship: Revolutionizing flight.* Retrieved 28 May 2018 from http://plimp.com/

Plioutsias, A., Karanikas, N., & Chatzimihildidou, M. M. (2018). Hazard Analysis and Safety Requirements for Small Drone Operations: To What Extent Do Popular Drones Embed safety? *Risk Analysis, 38*(3), 562–584. doi:10.1111/risa.12867 PMID:28768049

Postman, N. (1993). *Technopoly: The Surrender of Culture to Technology.* New York: Vintage.

Prexl, M., Struebig, K., Harder, J., & Hoehn, A. (2017). *User studies of a head-mounted display for search and rescue teleoperation of UAVs via satellite link.* IEEE Aerospace Conference; doi:10.1109/AERO.2017.7943596

PricewaterhouseCoopers (PwC). (2016). *Global Market for Commercial Applications of Drone Technology Valued at over $127 bn.* Retrieved from https://press.pwc.com/News-releases/global-market-for-commercial-applications-of-drone-technology-valued-at-over--127-bn/s/ac04349e-c40d-4767-9f92-a4d219860cd2

Prisacariu, V., Boscoianu, M., & Luchian, A. (2014). Innovative solutions and UAS limits. *Review of the Air Foce Academy, 2*, 26.

Prisacariu, V., Pop, S., & Circiu, I. (2016). Recovery system of the multi-helicopter UAV. *Review of the Air Force Academy, 1*(31), 91–98. doi:10.19062/1842-9238.2016.14.1.13

Pulver, A., & Wei, R. (2018). Optimizing the spatial location of medical drones. *Applied Geography (Sevenoaks, England)*, *90*, 9–16. doi:10.1016/j.apgeog.2017.11.009

Radi, A. (2013). *Human injury model for small unmanned aircraft impacts*. Canberra, Australia: Civil Aviation Safety Authority.

Rao, B., Gopi, A. G., & Maione, R. (2016). The societal impact of commercial drones. *Technology in Society*, *45*, 83–90. doi:10.1016/j.techsoc.2016.02.009

Rash, C. E., LeDuc, P. A., & Manning, S. D. (2006). Human Factors in U.S. Military Unmanned Aerial Vehicle Accidents. In H. L. Nancy & J. Cooke (Eds.), Advances in Human Performance and Cognitive Engineering Research (pp. 117 - 131). New York: Emerald Group.

Rasmussen, J., & Svendung, I. (2000). *Proactive Risk Management in a Dynamic Society*. Karlstad: Swedish Rescue Services Agency.

Rathinam, S., Almeida, P., Kim, Z., Jackson, S., Tinka, A., & Grossman, W. (2007). *Automous searching and tracking of a river using an UAV*. Paper presented at the American control conference.

Reason, J. (1990). *Human Error*. Cambridge, UK: Cambridge University Press. doi:10.1017/CBO9781139062367

Reason, J. (1997). *Managing the risks of organizational accidents*. Aldershot, UK: Ashgate.

Rees, M. (2016). *New 24-hour endurance hybrid drone developed for monitoring & inspection*. Retrieved 28 May 2018 from http://www.unmannedsystemstechnology.com/2016/12/defiant-labs-announces-new-dx-3-drone-for-monitoring-inspection/

Riganati, J. P., & Harrison, G. (2016). The path forward for remotely piloted aircraft systems. In *Remotely Piloted Aircraft Systems: A Human Systems Integration Perspective*. West Sussex, UK: John Wile & Sons Ltd. doi:10.1002/9781118965900.ch10

Robinson. (2016). *R44 Raven II 2016 Estimated Operating Costs*. Retrieved from: https://www.helispot.be/hs/documents/heli/r44_eoc.pdf

Robinsons, R., & Francis, G. (2014). *SFAIRP vs ALARP. In Conference On Railway Excellence* (p. 7). Adelaide: RTSA.

Robson, D. (2017). *Human being pilot: Human factors for aviation professionals* (2nd ed.). Hamilton, Australia: Aviation Theory Centre.

Rod, D. (1989). *Aviation Insurance*. London: Butterworths.

Romm, J. (2006). The car and fuel of the future. *Energy Policy*, *34*(2609-2614).

Rosen, J. W. (2017). *Blood from the sky: an ambitious medical drone delivery system hits Rwanda*. Retrieved 20 December 2018, from https://www.technologyreview.com/s/608034/blood-from-the-sky-ziplines-ambitious-medical-drone-delivery-in-africa/

Rosen, J. W. (2017). *Zipline's Ambitious Medical Drone Delivery in Africa*. Retrieved from https://www.technologyreview.com/s/608034/blood-from-the-sky-ziplines-ambitious-medical-drone-delivery-in-africa/

Rosenberg, A. S. (2009). *An evaluation of a UAV guidance system with consumer grade GPS receivers* (Ph.D. dissertation). University of Arizona. Retrieved from: http://www.casa.arizona.edu/data/abigail/Abigail_Rosenberg_Final_Dissertation.pdf

Rosser, J. C., Vignesh, V., Terwilliger, B. A., & Parker, B. C. (2018). Surgical and Medical Applications of Drones: A Comprehensive Review. *JSLS : Journal of the Society of Laparoendoscopic Surgeons*, *22*(3), e2018.00018. doi:10.4293/JSLS.2018.00018 PMID:30356360

Ruano, S., Cuevas, C., Gallego, G., & Garcia, N. (2017). Augmented Reality Tool for the Situational Awareness Improvement of UAV Operators. *Sensors (Basel)*, *17*(2), 297. doi:10.339017020297 PMID:28178189

Rubio, S., Diaz, E., Martin, J., & Puente, J. (2004). Evaluation of subjective mental workload: A comparison of SWAT, NASA-TLX, and workload profile methods. *Applied Psychology*, *53*(1), 61–86. doi:10.1111/j.1464-0597.2004.00161.x

Ruff, H. A., Narayanan, S., & Draper, M. H. (2002). Human interaction with levels of automation and decision-aid fidelity in the supervisory control of mulitple simulated unmanned air vehicles. *Presence (Cambridge, Mass.)*, *11*(4), 335–351. doi:10.1162/105474602760204264

SACAA. (n.d.). *RPAS regulations (Part 101)*. Retrieved 29 May 2018 from http://www.caa.co.za/Documents/RPAS/Part%20101%20-%20RPAS%20Workshops.pdf

Sachan, D. (2016). The age of drones: What might it mean for health? *Lancet*, *387*(10030), 1803–1804. doi:10.1016/S0140-6736(16)30361-0 PMID:27203491

Sahin, Y. E. (2016). *Drone nedir ve hangi alanlarda kullanılır?* Retrieved from: https://www.gzt.com/teknoloji/drone-nedir-ve-hangi-alanlarda-kullanilir-2576541

Salas, E., & Maurino, D. (2010). Human Factors in Aviation. Oxford, UK: Elsevier. doi:10.1016/B978-0-12-374518-7.00001-8

Sandbrook, C. (2015). The Social Implications of Using Drones for Biodiversity Conservation. *Ambio*, *44*(4), 636–647. doi:10.100713280-015-0714-0 PMID:26508350

Sandvik, K. (2015). African drone stories. *BEHEMOTH: A Journal on Civilization, 8*(2), 73-96.

Sarter, N. B., Woods, D. D., & Billings, C. E. (1999). *Automations Surprises*. Columbus, OH: The Ohio State University, Cognitive Systems Engineering Laboratory.

Sauer, J., Wastell, D., & Hockey, G. (1996). Skill maintenance in extended spaceflight: A human factors analysis of space and analogue work environments. *Acta Astronautica*, *39*(8), 579–587. doi:10.1016/S0094-5765(97)00006-4 PMID:11540781

Sayfa, A. (2016). *Orman yangını alanlarında 'drone' dönemi*. Retrieved from: http://www.trthaber.com/haber/turkiye/orman-yangini-alanlarinda-drone-donemi-286228.html

Sayfa, A. (2017). *Türkiye Drone Şampiyonası İstanbul'da nefes kesti*. Retrieved from: https://www.emlakpencerem.com/turkiye-drone-sampiyonasi-istanbul-da-nefes-kesti/97807/

Schneider, D. (2014). Open season on drones? *IEEE Spectrum*, *51*(1), 32–33. doi:10.1109/MSPEC.2014.6701428

Schneider, D. (2017). Air traffic control for delivery drones. *IEEE Spectrum*, *54*(1), 32–33. doi:10.1109/MSPEC.2017.7802742 PMID:30636790

Schutte, P. C., & Trujillo, A. C. (1996). *Flight crew task management in non-normal situations.* Paper presented at the 40th Annual Meeting of the Human Factors and Ergonomics Society, Santa Monica, CA. 10.1177/154193129604000422

Scott, J. E., & Scott, C. H. (2017). *Drone delivery models for healthcare*. Paper presented at the 50th Hawaii International Conference on System Sciences, Waikoloa, HI. Retrieved from http://aisel.aisnet.org/cgi/viewcontent.cgi?article=1442&context=hicss-50

Scott, J. E., & Scott, C. H. (2017). Drone Delivery Models for Healthcare. In *Proceedings of the 50th Hawaii International Conference on System Sciences* (pp. 3297-3304). Honolulu, HI: University of Hawai'i at Mānoa.

Shawcross & Beaumont. (1986). *Air Law*. Montreal: De Daro Publishing.

Sheets, P., Rowling, C. M., & Jones, T. M. (2015). The view from above (and below): A comparison of American, British and Arab news coverage of US Drones. *Media. War and Conflict*, *8*(3), 289–311. doi:10.1177/1750635215593973

Shin & Kim. (2017). New Korean drone laws. *Shin & Kim*. Retrieved from: https://www.shinkim.com/eng/media/newsletter/927

Shively, R. J. (2014). Breaking Down Barriers to UAS Integration: What are government agencies doing? *Proceedings of the Human Factors and Ergonomics Society Annual Meeting*, *58*(1), 41–43. doi:10.1177/1541931214581009

Shmelova, T., & Stratij, A. (2015). *Computer Program "Aggregation of heterogeneous information flows": Certificate of registration of copyright in the product N° 59466*. Academic Press.

Shmelova, T., Bondarev, D., & Znakovska, Y. (2016). Modeling of the Decision Making by UAV's Operator in Emergency Situations. In *4th International Conference on Methods and Systems of Navigation and Motion Control (MSNMC-2016)*. Kyiv: NAU.

Shmelova, T., & Bondarev, D. (2016). Graph Theory Applying for Quantitative Estimation of UAV's Group Flight into Aerial Photography. *Electronics and Control Systems*, *4*(46), 128–133.

Silvagni, M., Tonoli, A., Zenerino, E., & Chiaberge, M. (2017). Multipurpose UAV for search and rescue operations in mountain avalanche events. *Geomatics, Natural Hazards & Risk*, *8*(1), 18–33. doi:10.1080/19475705.2016.1238852

Sizova, O. A. (2012). The application of graph theory in various types of scientific activity. *Baku.*

Skinner, S. W., Urdahl, S., Harrington, T., Balchanos, M. G., Garcia, E., & Mavris, D. N. (2018). UAV Swarms for Migration Flow Monitoring and Search and Rescue Mission Support. In 2018 AIAA Information Systems-AIAA Infotech @ Aerospace (Vols 1–0). American Institute of Aeronautics and Astronautics. doi:10.2514/6.2018-1489

Slabinac, M. (2015). Innovative Solutions for a "Last Mile" Delivery – European Experience. In *Proceedings of 15th international scientific conference Business Logistics in Modern Management.* Osijek: Ekonomski fakultet.

Smith, J., Mazur, M., & Wisniewski, A. (2017). *Clarity from above: transport infrastructure.* PricewaterhouseCoopers. Retrieved from: http://www.pwc.pl/en/publikacje/2016/clarity-from-above-transport-infrastructure.html

Smyth, D. L., Glavin, F. G., & Madden, M. G. (2018). Using a Game Engine to Simulate Critical Incidents and Data Collection by Autonomous Drones. In *2018 IEEE Games, Entertainment, Media Conference (GEM)* (pp. 1–9). IEEE. 10.1109/GEM.2018.8516527

Soeliso, D., Meier, P., Lessard-Fontaine, A., Du Plessis, J., & Stuhlberger, C. (2016). Drones in Humanitarian Action. *Swiss Foundation for Mine Action (FSD).* Retrieved from: http://drones.fsd.ch/wp-content/uploads/2016/11/Drones-in-Humanitarian-Action.pdf

Solodov, A., Williams, A., Al Hanaei, S., & Goddard, B. (2018). Analyzing the threat of unmanned aerial vehicles (UAV) to nuclear facilities. *Security Journal, 31*(1), 305–324. doi:10.105741284-017-0102-5

Sorenson, L. J., Stanton, N. A., & Banks, A. P. (2011). Back to SA school: Contrasing three approaches to situation awareness in the cockpit. *Theoretical Issues in Ergonomics Science, 12*(6), 451–471. doi:10.1080/1463922X.2010.491874

Spinka, O., Kroupa, S., & Hanzálek, Z. (2007). *Control system for unmanned aerial vehicles.* Paper presented at the 5th IEEE International Conference on Industrial Informatics, Vienna, Austria. Retrieved from https://ieeexplore.ieee.org/abstract/document/4384800/

Stacy, N. J. S., Craig, D. W., Staromlynska, J., & Smith, R. B. (2002). *The Global Hawk UAV Australian deployment: imaging radar sensor modifications and employment for maritime surveillance.* Paper presented at the Geoscience and Remote Sensing Symposium 2002. 10.1109/IGARSS.2002.1025623

Staffan, M. (2002). Similarities and differences in psychophysiological reactions between simulated and real Air-to-ground missions. *The International Journal of Aviation Psychology, 12*(1), 47–61.

Start Up World. (n.d.). Flying a drone in India: rules & regulations – things to know. *Start Up World.* Retrieved from: https://www.startupworld.com/news/technology-news/flying-a-drone-in-india-rules-and-regulations-things-to-know

Stewart, J. (2017). *A drone-slinging UPS van delivers the future.* Retrieved from https://www.wired.com/2017/02/drone-slinging-ups-van-delivers-future/

Stipanović, D. M., Inalhan, G., Teo, R., & Tomlin, C. J. (2004). Decentralized overlapping control of a formation of unmanned aerial vehicles. *Automatica, 40*(8), 1285–1296. doi:10.1016/j.automatica.2004.02.017

Stolaroff, J. K., Samaras, C., O'Neill, E. R., Lubers, A., Mitchell, A. S., & Ceperley, D. (2018). Energy use and life cycle greenhouse gas emissions of drones for commercial package delivery. *Nature Communications, 9*(1), 409. doi:10.103841467-017-02411-5 PMID:29440638

Stolzer, A. J., Halford, C. D., & Goglia, J. J. (2013). *Safety Managment Systems in Aviation.* Burlington: Ashgate.

Stroh, M. B. (2006). *A practical guide to transportation and logistics.* Dumant: Logistics Network.

Sudip, S. (2017). *For Press Release.* Retrieved from: https://forpressrelease.com/forpressrelease-195898-production-of-multirotor-drones-for-non-military-applications-has-positively-impacted-its-market.html

Swiss Foundation for Mine Action. (2016). *Drones in humanitarian action: A guide to the use of airborne systems in humanitarian crises.* Geneva, Switzerland: Swiss Foundation for Mine Action.

Swiss Post. (2018). *Drone to transport laboratory samples across Lake Zurich.* Retrieved 20 December 2018, from https://www.post.ch/en/about-us/company/media/press-releases/2018/drone-to-transport-laboratory-samples-across-lake-zurich

Tam, A. (2011). Public Perception of Unmanned Aerial Vehicles. Aviation Technology Graduate Student Publications, (3), 1-23.

Tanzi, T. J., Chandra, M., Isnard, J., Camara, D., Sebastien, O., & Harivelo, F. (2016). Towards 'drone-borne' disaster management: future application scenarios. In ISPRS Annals of Photogrammetry, Remote Sensing and Spatial Information Sciences (Vol. III–8, pp. 181–189). Copernicus GmbH. doi:10.5194/isprs-annals-III-8-181-2016

Tarasov, V., & Gerasimov, B. (2007). *Intelligent Decision Support Systems: Theory, Synthesis, Efficiency: a monograph*. Kyiv: International Academy of Computer Science and Systems.

Tatham, P. H. (2009). An Initial Investigation into the Suitability of the use of Unmanned Aerial Vehicle Systems (UAVS) to Support the Emergency Assessment Process in Rapid Onset Humanitarian Disasters. *International Journal of Risk Assessment and Management, 13*(1), 60–78. doi:10.1504/IJRAM.2009.026391

Tatham, P. H., Ball, C., Wu, Y., & Diplas, P. (2017). Remotely Piloted Aircraft Systems (RPAS) Support for Humanitarian Logistic Operations: The Current Position and A Proposed Way Ahead. *Journal of Humanitarian Logistics and Supply Chain Management, 7*(1), 2–25. doi:10.1108/JHLSCM-05-2016-0018

Tatham, P. H., & Christopher, M. G. (2017). An Introduction to Humanitarian Logistics. In P. H. Tatham & M. G. Christopher (Eds.), *Humanitarian Logistics: Meeting the Challenge of Preparing for and Responding to Disasters* (3rd ed.). London: Kogan Page.

Tatham, P. H., & Kovács, G. (2010). The impact of gender on humanitarian logistics. *International Journal of Mass Emergencies and Disasters, 28*(2), 148–169.

Tatham, P. H., Kovács, G., & Spens, K. M. (2016a). The humanitarian common logistic operating picture: A solution to the inter-agency coordination challenge. *Disasters: The Journal of Disaster Studies, Policy and Management.*

Tatham, P. H., & Pettit, S. J. (2010). Humanitarian Logistics: The Journey to Supply Network Management. *International Journal of Physical Distribution & Logistics Management, 40*(8/9), 609–622. doi:10.1108/09600031011079283

Tatham, P. H., & Spens, K. M. (2016). Cracking the humanitarian logistics coordination challenge: Lessons from the urban search and rescue community. *Disasters: The Journal of Disaster Studies. Policy and Management*, *40*(2), 246–261. PMID:26283578

Tatham, P., Stadler, F., Murray, A., & Shaban, R. Z. (2017). Flying maggots: A smart logistic solution to an enduring medical challenge. *Journal of Humanitarian Logistics and Supply Chain Management*, *7*(2), 172–193. doi:10.1108/JHLSCM-02-2017-0003

Teknolo. (2015). *Dronelarin Kullanim Alanlari*. Retrieved from: http://www.teknolo.com/wp-content/uploads/2015/11/DRONELARIN-KULLANIM-ALANLARI.pdf

The Guardian. (2013). *Flying aid drones tested in Haiti and Dominican Republic.* Retrieved from https: //www.theguardian.com/global-development/2013/jan/09/flying-aid-drones-haiti-dominican-republic

Thiels, C. A., Aho, J. M., Zictlow, S. P., & Jenkins, D. H. (2015). Use of Unmanned Aerial Vehicles for Medical Product Transport. *Air Medical Journal*, *34*(2), 104–108. doi:10.1016/j.amj.2014.10.011 PMID:25733117

Thomas, M. J. W. (2004). Predictors of Threat and Error Management: Identification of Core Nontechnical Skills and implications for Training Systems Design. *The International Journal of Aviation Psychology*, *14*(2), 207–231. doi:10.120715327108ijap1402_6

Tice, B. (1991). Unmanned Aerial Vehicles – The Force Multiplier of the 1990s. *Airpower Journal.*

Tice, B. P. (2009, July 24). Unmanned Aerial Vehicles – The Force Multiplier of the 1990s. *Airpower Journal*.

Tilley, A. (2016). *UPS Experiments with Drone Delivery In Partnership With Zipline*. Retrieved 20 February, 2019, from https://www.forbes.com/sites/aarontilley/2016/05/09/ups-experiments-with-drone-delivery-in-partnership-with-zipline/#3a424e7e45a4

TIME. (2018). *Welcome to the Drone Age*. Retrieved 20 December 2018, from http://time.com/collection/drones/

Titcomb, J. (2016). Amazon to step up UK tests of delivery drones. *The Telegraph*. Retrieved from: http://www.telegraph.co.uk/technology/2016/07/26/amazon-to-step-up-uk-tests-of-delivery-drones/

Todorova, I. (2015). The sky is the limit: UAVs by private actors and the implication to common-law privacy. *FIU Law Review*, *10*(2), 803–838.

Torresan, C., Berton, A., Carotenuto, F., Gennaro, S. F. D., Gioli, B., Matese, A., ... Wallace, L. (2017). Forestry applications of UAVs in Europe: A review. *International Journal of Remote Sensing*, *38*(8–10), 2427–2447. doi:10.1080/01431161.2016.1252477

Transport Canada. (2015). *Advisory circular: Quality assurance programs*. Retrieved 17 May 2018 from http://www.tc.gc.ca/media/documents/ca-opssvs/AC_QUA-001-ISSUE_01_-_QUALITY_ASSURANCE_PROGRAMS.pdf

Transport Canada. (2016). *Guidance material for operating unmanned air vehicle systems under an exemption*. Retrieved 14 May 2018 from http://www.tc.gc.ca/media/documents/ca-opssvs/AC-600-004.pdf

Tubitak. (2017). *Welcome to the International UAV TURKEY Competition!* Retrieved from: https://www.tubitak.gov.tr/en/announcements/welcome-to-the-international-uav-turkey-competition

Tuna, G., Nezji, B., & Conte, G. (2014). Unmanned aerial vehicle-aided communications system for disaster recovery. *Journal of Network and Computer Applications*, *41*, 27–36. doi:10.1016/j.jnca.2013.10.002

Türkiye İnsansız Hava Aracı Sistemleri Yol Haritası. (2011). *Türkiye İnsansız Hava Aracı Sistemleri Yol Haritası*. Author.

Tvaryanas, A. P., Thompson, B. T., & Constable, S. H. (2005). *U.S. Military Unmanned Aerial Vehicle Mishaps: Assessment of the Role of Human Factors using HFACS, Technical Report*. Brooks City-Base, TX: USAF Human Performance Directorate.

Tvaryanas, A. P., Thompson, W. T., & Constable, S. H. (2006). Human Factors in Remotely Piloted Aircraft Operations: HFACS Analysis of 221 Mishaps over 10 Years. *Aviation, Space, and Environmental Medicine*, *77*(2), 724–732. PMID:16856358

UAV Systems. (n.d.). South Korea drone laws. *UAV Systems*. Retrieved from: https://www.uavsystemsinternational.com/drone-laws-by-country/south-korea-drone-laws

UAVE. (2018). *Large fixed wing UAVs with long flight endurance*. Retrieved 28 May 2018 from http://www.unmannedsystemstechnology.com/company/uave-ltd/

UAViators. (2016). *Case Studies of drones deployed for humanitarian purposes*. Retrieved from: https://docs.google.com/document/d/1KlJyxGLIhGaCmaA8ehzOaS9Yue8zj2PcR3TlxAQME3o/edit#

UAViators. (2018). *Humanitarian UAV Code of Conduct*. Retrieved from: https://uavcode.org/code-of-conduct/

Udroiu, R., & Blaj, M. I. (2016). Conceptual design of a vtol remotely piloted aircraft for emergency missions. *Scientific Research and Education in the Air Force - AFASES*. doi:10.19062/2247-3173.2016.18.1.27

UNICEF. (2016). *Malawi tests first unmanned aerial vehicle flights for HIV early infant diagnosis*. Retrieved 15 May 2018 from https://www.unicef.org/media/media_90462.html

United Nations Logistics Cluster. (2015). *Vanuatu situation updates*. Retrieved from: www.logcluster.org/document/vanuatu-situation-update-18-march-2015

United Nations Logistics Cluster. (2015a). *Vanuatu – Cyclone Pam: concept of operations*. Retrieved from: www.logcluster.org/sites/default/files/logistics_cluster_vanuatu_conops_150320.pdf

United Nations Logistics Cluster. (2015b). *Vanuatu meeting minutes*. Retrieved from: www.logcluster.org/sites/default/files/logistics_cluster_vanuatu_meeting_minutes_150331_0.pdf

UNOCHA (United Nations Office for the Coordination of Humanitarian Affairs). (2014). Unmanned aerial vehicles in humanitarian response. *UNOCHA Policy and Studies Series, Occasional Paper No 10*. Retrieved from: https://docs.unocha.org/sites/dms/Documents/Unmanned%20Aerial%20Vehicles%20in%20Humanitarian%20Response%20OCHA%20July%202014.pdf

US Army. (2010). U.S. Army Unmanned Aircraft Systems Roadmap 2010-2035. *U.S. Army Center of Excellence*. Retrieved from: http://www.rucker.army.mil/usaace/uas/US%20Army%20UAS%20RoadMap%202010%202035.pdf

Valavanis, K. (2007). *Advances in Unmanned Aerial Vehicles: State of the Art and the Road to Autonomy*. Springer. doi:10.1007/978-1-4020-6114-1 van Breda, L. (1995). *Operator Performance in Multi Maratime Unmanned Air Vehicle Control.* Soesterberg, The Netherlands: Human Factors Research Institute.

van de Merwe, K., van Dijk, H., & Zon, R. (2012). Eye movements as an indicator of situation awareness in a flight simulator experiment. *The International Journal of Aviation Psychology*, *22*(1), 78–95. doi:10.1080/10508414.2012.635129

Van Erp, J. B. F. (2000). *Controlling unmanned vehicles: The human factors solution.* Paper presented at the RTO Meeting Proceedings 44 (RTO-MP-44).

Van Erp, J. B. F., & Van Breda, L. (1999). Human factors issues and advanced interface design in maritime unmanned aerial vehicles: A Project Overview (Report TNO TM-99-A004). Soesterberg, The Netherlands.

VandenBos, G., Knapp, S., & Doe, J. (2001). *Role of reference elements in the selection of resources by psychology undergraduates.* Retrieved from http://jbr.org/articles.html

Vanhoef, M., & Piessens, F. (2018). Release the Kraken: New KRACKs in the 802.11 Standard. In *Proceedings of the 2018 ACM SIGSAC Conference on Computer and Communications Security* (pp. 299–314). New York: ACM. 10.1145/3243734.3243807

Velthuijsen, H., & Frank, J. G. (2014). *Entrepreneurship in the Network Economy.* Retrieved from https://www.researchgate.net/publication/267920775_Entrepreneurship_in_the_Network_Economy

Vertical Technologies. (2018). *DeltaQuad rising: The next generation autonomous VTOL UAV.* Retrieved 28 May 2018 from https://www.verticaltechnologies.com/

Vincent, J. B., Leland, K. W., & Ditmer, M. A. (2015). Barriers to adding UAVs to the ecologist's toolbox: Peer-reviewed letter. *Frontiers in Ecology and the Environment*, *13*(2), 74–75. doi:10.1890/15.WB.002

Viroo. (2017). Buy drones in India! Drone legality in India (2018). DGCA, WPC, customs new rules & regulations. *YouTube.* Retrieved from: https://www.youtube.com/watch?v=UVhRt2m2Soc

Vlahović, N., Knežević, B., & Batalić, B. (2016). Implementing Delivery Drones in Logistics Business Process: Case of Pharmaceutical Industry. *International Journal of Social, Behavioral, Educational, Economic, Business and Industrial Engineering*, *10*(12), 3948–3953.

Waller, M. J., Gupta, N., & Giambatista, R. C. (2004). Effects of Adaptive Behaviors and Shared Mental Models on Control Crew Performance. *Management Science*, *50*(11), 1534–1544. doi:10.1287/mnsc.1040.0210

Wargo, C. A., Church, G. C., Glaneueski, J., & Strout, M. (2014). *Unmanned Aircraft Systems (UAS) research and future analysis.* Paper presented at the 2014 IEEE Aerospace Conference. 10.1109/AERO.2014.6836448

Warwick, G. (2014). Cargo Shift: Department of Defense planning to develop unmanned cargo resupply aircraft. *Aviation Week & Space Technology, 176*(16), 53-55.

Wei, J., Ellet, E. T., Ellman, L., Rizzi, P. R., & Zhou, L. (2016). China launches first operational rules for civil unmanned aircraft. *Hogan Lovells.* Retrieved from: http://www.hlmediacomms.com/2016/01/21/china-launches-first-operational-rules-for-civil-unmanned-aircraft

Weibel, R. E., & Hansman, R. J. (2005). *Safety considerations for operation of unmanned aerial vehicles in the national airspace system.* Technical Report ICAT-2005-1.

Weibel, R. E., & Hansman, R. J. (2005). *Safety considerations for operation of unmanned aerial vehicles in the national airspace system.* Technical Report ICAT-2005-1. MIT International Center for Air Transportation.

Weimer, J. (1995). *Research techniques in human engineering*. Englewood Cliffs, NJ: Prentice Hall.

Weissbach, D., & Tebbe, K. (2016). Drones in sight: Rapid growth through M&A's in a soaring new industry. *Strategic Direction*, *32*(6), 37–39. doi:10.1108/SD-04-2016-0044

Welch, C. (2013). *Zookal will deliver textbooks using drones in Australia next year.* Retrieved from: http://edition.cnn.com/2013/10/18/tech/innovation/zookal-will-deliver-textbooks-using-drones/

Westbrook, M. H. (2001). *The electric car: Development and Future of Battery, Hybrid and Fuel-Cell Cars*. London: Institution of Electrical Engineers.

West, J. P., Klofstad, C. A., Uscinski, J. E., & Connolly, J. M. (2019). Citizen Support for Domestic Drone Use and Regulation. *American Politics Research*, *47*(1), 119–151. doi:10.1177/1532673X18782208

WHO. (2013). *Classification and minimum standards for foreign medical teams in sudden onset disasters*. Retrieved from: http://www.who.int/hac/global_health_cluster/fmt_guidelines_september2013.pdf

Wickens, C. D. (1999). Cognitive factors in aviation. In Handbook of applied cognition (pp. 247-282). John Wiley & Sons.

Wickens, C. D., & Dixon, S. R. (2006). *Human machine interface analysis of unmanned vehicle systems* (Technical report HFD-06-1/MAAD-06-01). Savoy, IL: University of Illinois, Institute of Aviation, Aviation Human Factors Division.

Wickens, C. D. (2002). Multiple resources and performance prediction. *Theoretical Issues in Ergonomics Science*, *3*(2), 159–177. doi:10.1080/14639220210123806

Wickens, C. D. (2003). Pilot actions and tasks: Selection, execution, and control. In P. Tsang & M. Vidulich (Eds.), *Principles and Practice of Aviation Psychology* (pp. 239–263). Mahwah, NJ: Erlbaum.

Wickens, C. D., & Hollands, J. G. (2000). *Engineering psychology and human performance* (3rd ed.). Upper Saddle River, NJ: Prentice Hall.

Wickens, C., & Colcombe, A. (2007). Dual-Task Performance Consequences of Imperfect alerting associated with a cockpit display of traffic information. *Human Factors*, *49*(5), 839–850. doi:10.1518/001872007X230217 PMID:17915602

Wiegmann, D. A., & Shappell, S. A. (2001). Human error perspectives in aviation. *The International Journal of Aviation Psychology*, *11*(4), 341–357. doi:10.1207/S15327108IJAP1104_2

Wiener, E. L., Kanki, B. G., & Helmreich, R. L. (1993). *Cockpit resource management*. San Diego, CA: Academic.

Wikens, C. D., Dixon, S., & Chang, D. (2003). *Using Interference Models to Predict Performance in Multiple-Task UAV Environment - 2 UAVs: AHFD-03-9/MAAD-03-1*. Savoy, IL: Aviation Human Factors Division, Institute of Aviation, Universit of Illinois at Urbana-Champaign. doi:10.21236/ADA496812

Wikipedia. (n.d.). China aerospace science and technology corporation. *Wikipedia.* Retrieved from: https://en.wikipedia.org/wiki/China_Aerospace_Science_and_Tcchnology_Corporation

Wu, J., Wang, H., Li, N., Yao, P., Huang, Y., & Yang, H. (2018). Path planning for solar-powered UAV in urban environment. *Neurocomputing*, *275*, 2055–2065. doi:10.1016/j.neucom.2017.10.037

Wurbel, H. (2017). *Framework for the evaluation of cost-effectiveness of drone use for the last-mile delivery of vaccines* (Master of Global Health). University of Barcelona. Retrieved from https://logcluster.org/sites/default/files/gm_files/master_final_project_heike_wurbel_13_jun2017_003.pdf

Wyllie, T. (2001). Parachute recovery for UAV systems. *Aircraft Engineering and Aerospace Technology*, *73*(6), 542–551. doi:10.1108/00022660110696696

Xinhua. (2018). China to fly new member of Wing Loong UAS family in 2018. *China Daily*. Retrieved from: http://www.chinadaily.com.cn/a/201801/25/WS5a696d6ea3106e7dcc1368ed.html

Yeh, M., & Wickens, C. D. (2001). Display signalling in augmented reality: Effects of cue reliability and image realism on attention allocation and trust calibration. *Human Factors*, *43*(3), 355–365. doi:10.1518/001872001775898269 PMID:11866192

Yoo, W., Yu, E., & Jung, J. (2018). Drone delivery: Factors affecting the public's attitude and intention to adopt. *Telematics and Informatics*, *35*(6), 1687–1700. doi:10.1016/j.tele.2018.04.014

Yu, X., & Zhang, Y. (2015). Sense and avoid technologies with applications to unmanned aircraft systems: Review and prospects. *Progress in Aerospace Sciences*, *74*, 152–166. doi:10.1016/j.paerosci.2015.01.001

Zègre-Hemsey, J. K., Bogle, B., Cunningham, C. J., Snyder, K., & Rosamond, W. (2018). Delivery of Automated External Defibrillators (AED) by Drones: Implications for Emergency Cardiac Care. *Current Cardiovascular Risk Reports*, *12*(11), 25. doi:10.100712170-018-0589-2 PMID:30443281

Zhang, Z., Cao, Y., Ding, M., Zhuang, L., & Yao, W. (2016). An intruder detection algorithm for vision based sense and avoid system. In *2016 International Conference on Unmanned Aircraft Systems (ICUAS)* (pp. 550–556). Academic Press. 10.1109/ICUAS.2016.7502521

Zimmerman, J. (2013). *ADS-B 101: what is it and why you should care*. Air Facts Journal.

Zipline. (2018). *Zipline launches fastest delivery drone in the world*. Retrieved 21 September 2018 from http://www.flyzipline.com/uploads/Zipline%20Fastest%20Drone%20Press%20Release.pdf

Zuboff, S. (2015). Big other: Surveillance Capitalism and the Prospects of an Information Civilization. *Journal of Information Technology*, *30*(1), 75–89. doi:10.1057/jit.2015.5

Related References

To continue our tradition of advancing information science and technology research, we have compiled a list of recommended IGI Global readings. These references will provide additional information and guidance to further enrich your knowledge and assist you with your own research and future publications.

Abtahi, M. S., Behboudi, L., & Hasanabad, H. M. (2017). Factors Affecting Internet Advertising Adoption in Ad Agencies. *International Journal of Innovation in the Digital Economy*, *8*(4), 18–29. doi:10.4018/IJIDE.2017100102

Agrawal, S. (2017). The Impact of Emerging Technologies and Social Media on Different Business(es): Marketing and Management. In O. Rishi & A. Sharma (Eds.), *Maximizing Business Performance and Efficiency Through Intelligent Systems* (pp. 37–49). Hershey, PA: IGI Global. doi:10.4018/978-1-5225-2234-8.ch002

Alnoukari, M., Razouk, R., & Hanano, A. (2016). BSC-SI: A Framework for Integrating Strategic Intelligence in Corporate Strategic Management. *International Journal of Social and Organizational Dynamics in IT*, *5*(2), 1–14. doi:10.4018/IJSODIT.2016070101

Alnoukari, M., Razouk, R., & Hanano, A. (2016). BSC-SI, A Framework for Integrating Strategic Intelligence in Corporate Strategic Management. *International Journal of Strategic Information Technology and Applications*, *7*(1), 32–44. doi:10.4018/IJSITA.2016010103

Altındağ, E. (2016). Current Approaches in Change Management. In A. Goksoy (Ed.), *Organizational Change Management Strategies in Modern Business* (pp. 24–51). Hershey, PA: IGI Global. doi:10.4018/978-1-4666-9533-7.ch002

Alvarez-Dionisi, L. E., Turner, R., & Mittra, M. (2016). Global Project Management Trends. *International Journal of Information Technology Project Management, 7*(3), 54–73. doi:10.4018/IJITPM.2016070104

Anantharaman, R. N., Rajeswari, K. S., Angusamy, A., & Kuppusamy, J. (2017). Role of Self-Efficacy and Collective Efficacy as Moderators of Occupational Stress Among Software Development Professionals. *International Journal of Human Capital and Information Technology Professionals, 8*(2), 45–58. doi:10.4018/IJHCITP.2017040103

Aninze, F., El-Gohary, H., & Hussain, J. (2018). The Role of Microfinance to Empower Women: The Case of Developing Countries. *International Journal of Customer Relationship Marketing and Management, 9*(1), 54–78. doi:10.4018/IJCRMM.2018010104

Arsenijević, O. M., Orčić, D., & Kastratović, E. (2017). Development of an Optimization Tool for Intangibles in SMEs: A Case Study from Serbia with a Pilot Research in the Prestige by Milka Company. In M. Vemić (Ed.), *Optimal Management Strategies in Small and Medium Enterprises* (pp. 320–347). Hershey, PA: IGI Global. doi:10.4018/978-1-5225-1949-2.ch015

Aryanto, V. D., Wismantoro, Y., & Widyatmoko, K. (2018). Implementing Eco-Innovation by Utilizing the Internet to Enhance Firm's Marketing Performance: Study of Green Batik Small and Medium Enterprises in Indonesia. *International Journal of E-Business Research, 14*(1), 21–36. doi:10.4018/IJEBR.2018010102

Atiku, S. O., & Fields, Z. (2017). Multicultural Orientations for 21st Century Global Leadership. In N. Baporikar (Ed.), *Management Education for Global Leadership* (pp. 28–51). Hershey, PA: IGI Global. doi:10.4018/978-1-5225-1013-0.ch002

Atiku, S. O., & Fields, Z. (2018). Organisational Learning Dimensions and Talent Retention Strategies for the Service Industries. In N. Baporikar (Ed.), *Global Practices in Knowledge Management for Societal and Organizational Development* (pp. 358–381). Hershey, PA: IGI Global. doi:10.4018/978-1-5225-3009-1.ch017

Ávila, L., & Teixeira, L. (2018). The Main Concepts Behind the Dematerialization of Business Processes. In M. Khosrow-Pour, D.B.A. (Ed.), Encyclopedia of Information Science and Technology, Fourth Edition (pp. 888-898). Hershey, PA: IGI Global. doi:10.4018/978-1-5225-2255-3.ch076

Bartens, Y., Chunpir, H. I., Schulte, F., & Voß, S. (2017). Business/IT Alignment in Two-Sided Markets: A COBIT 5 Analysis for Media Streaming Business Models. In S. De Haes & W. Van Grembergen (Eds.), *Strategic IT Governance and Alignment in Business Settings* (pp. 82–111). Hershey, PA: IGI Global. doi:10.4018/978-1-5225-0861-8.ch004

Bashayreh, A. M. (2018). Organizational Culture and Organizational Performance. In W. Lee & F. Sabetzadeh (Eds.), *Contemporary Knowledge and Systems Science* (pp. 50–69). Hershey, PA: IGI Global. doi:10.4018/978-1-5225-5655-8.ch003

Bedford, D. A. (2018). Sustainable Knowledge Management Strategies: Aligning Business Capabilities and Knowledge Management Goals. In N. Baporikar (Ed.), *Global Practices in Knowledge Management for Societal and Organizational Development* (pp. 46–73). Hershey, PA: IGI Global. doi:10.4018/978-1-5225-3009-1.ch003

Benmoussa, F., Nakara, W. A., & Jaouen, A. (2016). The Use of Social Media by SMEs in the Tourism Industry. In I. Lee (Ed.), *Encyclopedia of E-Commerce Development, Implementation, and Management* (pp. 2159–2170). Hershey, PA: IGI Global. doi:10.4018/978-1-4666-9787-4.ch155

Berger, R. (2016). Indigenous Management and Bottom of Pyramid Countries: The Role of National Institutions. In U. Aung & P. Ordoñez de Pablos (Eds.), *Managerial Strategies and Practice in the Asian Business Sector* (pp. 107–123). Hershey, PA: IGI Global. doi:10.4018/978-1-4666-9758-4.ch007

Bharwani, S., & Musunuri, D. (2018). Reflection as a Process From Theory to Practice. In M. Khosrow-Pour, D.B.A. (Ed.), Encyclopedia of Information Science and Technology, Fourth Edition (pp. 1529-1539). Hershey, PA: IGI Global. doi:10.4018/978-1-5225-2255-3.ch132

Bhatt, G. D., Wang, Z., & Rodger, J. A. (2017). Information Systems Capabilities and Their Effects on Competitive Advantages: A Study of Chinese Companies. *Information Resources Management Journal, 30*(3), 41–57. doi:10.4018/IRMJ.2017070103

Bhushan, M., & Yadav, A. (2017). Concept of Cloud Computing in ESB. In R. Bhadoria, N. Chaudhari, G. Tomar, & S. Singh (Eds.), *Exploring Enterprise Service Bus in the Service-Oriented Architecture Paradigm* (pp. 116–127). Hershey, PA: IGI Global. doi:10.4018/978-1-5225-2157-0.ch008

Bhushan, S. (2017). System Dynamics Base-Model of Humanitarian Supply Chain (HSCM) in Disaster Prone Eco-Communities of India: A Discussion on Simulation and Scenario Results. *International Journal of System Dynamics Applications, 6*(3), 20–37. doi:10.4018/IJSDA.2017070102

Biswas, A., & De, A. K. (2017). On Development of a Fuzzy Stochastic Programming Model with Its Application to Business Management. In S. Trivedi, S. Dey, A. Kumar, & T. Panda (Eds.), *Handbook of Research on Advanced Data Mining Techniques and Applications for Business Intelligence* (pp. 353–378). Hershey, PA: IGI Global. doi:10.4018/978-1-5225-2031-3.ch021

Bücker, J., & Ernste, K. (2018). Use of Brand Heroes in Strategic Reputation Management: The Case of Bacardi, Adidas, and Daimler. In A. Erdemir (Ed.), *Reputation Management Techniques in Public Relations* (pp. 126–150). Hershey, PA: IGI Global. doi:10.4018/978-1-5225-3619-2.ch007

Bureš, V. (2018). Industry 4.0 From the Systems Engineering Perspective: Alternative Holistic Framework Development. In R. Brunet-Thornton & F. Martinez (Eds.), *Analyzing the Impacts of Industry 4.0 in Modern Business Environments* (pp. 199–223). Hershey, PA: IGI Global. doi:10.4018/978-1-5225-3468-6.ch011

Buzady, Z. (2017). Resolving the Magic Cube of Effective Case Teaching: Benchmarking Case Teaching Practices in Emerging Markets – Insights from the Central European University Business School, Hungary. In D. Latusek (Ed.), *Case Studies as a Teaching Tool in Management Education* (pp. 79–103). Hershey, PA: IGI Global. doi:10.4018/978-1-5225-0770-3.ch005

Campatelli, G., Richter, A., & Stocker, A. (2016). Participative Knowledge Management to Empower Manufacturing Workers. *International Journal of Knowledge Management, 12*(4), 37–50. doi:10.4018/IJKM.2016100103

Căpusneanu, S., & Topor, D. I. (2018). Business Ethics and Cost Management in SMEs: Theories of Business Ethics and Cost Management Ethos. In I. Oncioiu (Ed.), *Ethics and Decision-Making for Sustainable Business Practices* (pp. 109–127). Hershey, PA: IGI Global. doi:10.4018/978-1-5225-3773-1.ch007

Carneiro, A. (2016). Maturity in Health Organization Information Systems: Metrics and Privacy Perspectives. *International Journal of Privacy and Health Information Management, 4*(2), 1–18. doi:10.4018/IJPHIM.2016070101

Chan, R. L., Mo, P. L., & Moon, K. K. (2018). Strategic and Tactical Measures in Managing Enterprise Risks: A Study of the Textile and Apparel Industry. In K. Strang, M. Korstanje, & N. Vajjhala (Eds.), *Research, Practices, and Innovations in Global Risk and Contingency Management* (pp. 1–19). Hershey, PA: IGI Global. doi:10.4018/978-1-5225-4754-9.ch001

Chandan, H. C. (2016). Motivations and Challenges of Female Entrepreneurship in Developed and Developing Economies. In N. Baporikar (Ed.), *Handbook of Research on Entrepreneurship in the Contemporary Knowledge-Based Global Economy* (pp. 260–286). Hershey, PA: IGI Global. doi:10.4018/978-1-4666-8798-1.ch012

Charlier, S. D., Burke-Smalley, L. A., & Fisher, S. L. (2018). Undergraduate Programs in the U.S: A Contextual and Content-Based Analysis. In J. Mendy (Ed.), *Teaching Human Resources and Organizational Behavior at the College Level* (pp. 26–57). Hershey, PA: IGI Global. doi:10.4018/978-1-5225-2820-3.ch002

Chaudhuri, S. (2016). Application of Web-Based Geographical Information System (GIS) in E-Business. In U. Panwar, R. Kumar, & N. Ray (Eds.), *Handbook of Research on Promotional Strategies and Consumer Influence in the Service Sector* (pp. 389–405). Hershey, PA: IGI Global. doi:10.4018/978-1-5225-0143-5.ch023

Choudhuri, P. S. (2016). An Empirical Study on the Quality of Services Offered by the Private Life Insurers in Burdwan. In U. Panwar, R. Kumar, & N. Ray (Eds.), *Handbook of Research on Promotional Strategies and Consumer Influence in the Service Sector* (pp. 31–55). Hershey, PA: IGI Global. doi:10.4018/978-1-5225-0143-5.ch002

Dahlberg, T., Kivijärvi, H., & Saarinen, T. (2017). IT Investment Consistency and Other Factors Influencing the Success of IT Performance. In S. De Haes & W. Van Grembergen (Eds.), *Strategic IT Governance and Alignment in Business Settings* (pp. 176–208). Hershey, PA: IGI Global. doi:10.4018/978-1-5225-0861-8.ch007

Damnjanović, A. M. (2017). Knowledge Management Optimization through IT and E-Business Utilization: A Qualitative Study on Serbian SMEs. In M. Vemić (Ed.), *Optimal Management Strategies in Small and Medium Enterprises* (pp. 249–267). Hershey, PA: IGI Global. doi:10.4018/978-1-5225-1949-2.ch012

Daneshpour, H. (2017). Integrating Sustainable Development into Project Portfolio Management through Application of Open Innovation. In M. Vemić (Ed.), *Optimal Management Strategies in Small and Medium Enterprises* (pp. 370–387). Hershey, PA: IGI Global. doi:10.4018/978-1-5225-1949-2.ch017

Daniel, A. D., & Reis de Castro, V. (2018). Entrepreneurship Education: How to Measure the Impact on Nascent Entrepreneurs. In A. Carrizo Moreira, J. Guilherme Leitão Dantas, & F. Manuel Valente (Eds.), *Nascent Entrepreneurship and Successful New Venture Creation* (pp. 85–110). Hershey, PA: IGI Global. doi:10.4018/978-1-5225-2936-1.ch004

David, F., van der Sijde, P., & van den Besselaar, P. (2016). Enterpreneurial Incentives, Obstacles, and Management in University-Business Co-Operation: The Case of Indonesia. In J. Saiz-Álvarez (Ed.), *Handbook of Research on Social Entrepreneurship and Solidarity Economics* (pp. 499–518). Hershey, PA: IGI Global. doi:10.4018/978-1-5225-0097-1.ch024

David, R., Swami, B. N., & Tangirala, S. (2018). Ethics Impact on Knowledge Management in Organizational Development: A Case Study. In N. Baporikar (Ed.), *Global Practices in Knowledge Management for Societal and Organizational Development* (pp. 19–45). Hershey, PA: IGI Global. doi:10.4018/978-1-5225-3009-1.ch002

Delias, P., & Lakiotaki, K. (2018). Discovering Process Horizontal Boundaries to Facilitate Process Comprehension. *International Journal of Operations Research and Information Systems*, *9*(2), 1–31. doi:10.4018/IJORIS.2018040101

Denholm, J., & Lee-Davies, L. (2018). Success Factors for Games in Business and Project Management. In *Enhancing Education and Training Initiatives Through Serious Games* (pp. 34–68). Hershey, PA: IGI Global. doi:10.4018/978-1-5225-3689-5.ch002

Deshpande, M. (2017). Best Practices in Management Institutions for Global Leadership: Policy Aspects. In N. Baporikar (Ed.), *Management Education for Global Leadership* (pp. 1–27). Hershey, PA: IGI Global. doi:10.4018/978-1-5225-1013-0.ch001

Deshpande, M. (2018). Policy Perspectives for SMEs Knowledge Management. In N. Baporikar (Ed.), *Knowledge Integration Strategies for Entrepreneurship and Sustainability* (pp. 23–46). Hershey, PA: IGI Global. doi:10.4018/978-1-5225-5115-7.ch002

Dezdar, S. (2017). ERP Implementation Projects in Asian Countries: A Comparative Study on Iran and China. *International Journal of Information Technology Project Management*, *8*(3), 52–68. doi:10.4018/IJITPM.2017070104

Domingos, D., Martinho, R., & Varajão, J. (2016). Controlled Flexibility in Healthcare Processes: A BPMN-Extension Approach. In M. Cruz-Cunha, I. Miranda, R. Martinho, & R. Rijo (Eds.), *Encyclopedia of E-Health and Telemedicine* (pp. 521–535). Hershey, PA: IGI Global. doi:10.4018/978-1-4666-9978-6.ch040

Domingos, D., Respício, A., & Martinho, R. (2017). Reliability of IoT-Aware BPMN Healthcare Processes. In C. Reis & M. Maximiano (Eds.), *Internet of Things and Advanced Application in Healthcare* (pp. 214–248). Hershey, PA: IGI Global. doi:10.4018/978-1-5225-1820-4.ch008

Dosumu, O., Hussain, J., & El-Gohary, H. (2017). An Exploratory Study of the Impact of Government Policies on the Development of Small and Medium Enterprises in Developing Countries: The Case of Nigeria. *International Journal of Customer Relationship Marketing and Management*, *8*(4), 51–62. doi:10.4018/IJCRMM.2017100104

Durst, S., Bruns, G., & Edvardsson, I. R. (2017). Retaining Knowledge in Smaller Building and Construction Firms. *International Journal of Knowledge and Systems Science*, *8*(3), 1–12. doi:10.4018/IJKSS.2017070101

Edvardsson, I. R., & Durst, S. (2017). Outsourcing, Knowledge, and Learning: A Critical Review. *International Journal of Knowledge-Based Organizations*, *7*(2), 13–26. doi:10.4018/IJKBO.2017040102

Edwards, J. S. (2018). Integrating Knowledge Management and Business Processes. In M. Khosrow-Pour, D.B.A. (Ed.), Encyclopedia of Information Science and Technology, Fourth Edition (pp. 5046-5055). Hershey, PA: IGI Global. doi:10.4018/978-1-5225-2255-3.ch437

Ejiogu, A. O. (2018). Economics of Farm Management. In *Agricultural Finance and Opportunities for Investment and Expansion* (pp. 56–72). Hershey, PA: IGI Global. doi:10.4018/978-1-5225-3059-6.ch003

Ekanem, I., & Abiade, G. E. (2018). Factors Influencing the Use of E-Commerce by Small Enterprises in Nigeria. *International Journal of ICT Research in Africa and the Middle East*, *7*(1), 37–53. doi:10.4018/IJICTRAME.2018010103

Ekanem, I., & Alrossais, L. A. (2017). Succession Challenges Facing Family Businesses in Saudi Arabia. In P. Zgheib (Ed.), *Entrepreneurship and Business Innovation in the Middle East* (pp. 122–146). Hershey, PA: IGI Global. doi:10.4018/978-1-5225-2066-5.ch007

El Faquih, L., & Fredj, M. (2017). Ontology-Based Framework for Quality in Configurable Process Models. *Journal of Electronic Commerce in Organizations*, *15*(2), 48–60. doi:10.4018/JECO.2017040104

El-Gohary, H., & El-Gohary, Z. (2016). An Attempt to Explore Electronic Marketing Adoption and Implementation Aspects in Developing Countries: The Case of Egypt. *International Journal of Customer Relationship Marketing and Management*, *7*(4), 1–26. doi:10.4018/IJCRMM.2016100101

Entico, G. J. (2016). Knowledge Management and the Medical Health Librarians: A Perception Study. In J. Yap, M. Perez, M. Ayson, & G. Entico (Eds.), *Special Library Administration, Standardization and Technological Integration* (pp. 52–77). Hershey, PA: IGI Global. doi:10.4018/978-1-4666-9542-9.ch003

Faisal, M. N., & Talib, F. (2017). Building Ambidextrous Supply Chains in SMEs: How to Tackle the Barriers? *International Journal of Information Systems and Supply Chain Management, 10*(4), 80–100. doi:10.4018/IJISSCM.2017100105

Fernandes, T. M., Gomes, J., & Romão, M. (2017). Investments in E-Government: A Benefit Management Case Study. *International Journal of Electronic Government Research, 13*(3), 1–17. doi:10.4018/IJEGR.2017070101

Fouda, F. A. (2016). A Suggested Curriculum in Career Education to Develop Business Secondary Schools Students' Career Knowledge Management Domains and Professional Thinking. *International Journal of Technology Diffusion, 7*(2), 42–62. doi:10.4018/IJTD.2016040103

Gallardo-Vázquez, D., & Pajuelo-Moreno, M. L. (2016). How Spanish Universities are Promoting Entrepreneurship through Your Own Lines of Teaching and Research? In L. Carvalho (Ed.), *Handbook of Research on Entrepreneurial Success and its Impact on Regional Development* (pp. 431–454). Hershey, PA: IGI Global. doi:10.4018/978-1-4666-9567-2.ch019

Gao, S. S., Oreal, S., & Zhang, J. (2018). Contemporary Financial Risk Management Perceptions and Practices of Small-Sized Chinese Businesses. In I. Management Association (Ed.), Global Business Expansion: Concepts, Methodologies, Tools, and Applications (pp. 917-931). Hershey, PA: IGI Global. doi:10.4018/978-1-5225-5481-3.ch041

Garg, R., & Berning, S. C. (2017). Indigenous Chinese Management Philosophies: Key Concepts and Relevance for Modern Chinese Firms. In B. Christiansen & G. Koc (Eds.), *Transcontinental Strategies for Industrial Development and Economic Growth* (pp. 43–57). Hershey, PA: IGI Global. doi:10.4018/978-1-5225-2160-0.ch003

Gencer, Y. G. (2017). Supply Chain Management in Retailing Business. In U. Akkucuk (Ed.), *Ethics and Sustainability in Global Supply Chain Management* (pp. 197–210). Hershey, PA: IGI Global. doi:10.4018/978-1-5225-2036-8.ch011

Giacosa, E. (2016). Innovation in Luxury Fashion Businesses as a Means for the Regional Development. In L. Carvalho (Ed.), *Handbook of Research on Entrepreneurial Success and its Impact on Regional Development* (pp. 206–222). Hershey, PA: IGI Global. doi:10.4018/978-1-4666-9567-2.ch010

Giacosa, E. (2018). The Increasing of the Regional Development Thanks to the Luxury Business Innovation. In L. Carvalho (Ed.), *Handbook of Research on Entrepreneurial Ecosystems and Social Dynamics in a Globalized World* (pp. 260–273). Hershey, PA: IGI Global. doi:10.4018/978-1-5225-3525-6.ch011

Gianni, M., & Gotzamani, K. (2016). Integrated Management Systems and Information Management Systems: Common Threads. In P. Papajorgji, F. Pinet, A. Guimarães, & J. Papathanasiou (Eds.), *Automated Enterprise Systems for Maximizing Business Performance* (pp. 195–214). Hershey, PA: IGI Global. doi:10.4018/978-1-4666-8841-4.ch011

Gianni, M., Gotzamani, K., & Linden, I. (2016). How a BI-wise Responsible Integrated Management System May Support Food Traceability. *International Journal of Decision Support System Technology*, *8*(2), 1–17. doi:10.4018/IJDSST.2016040101

Glykas, M., & George, J. (2017). Quality and Process Management Systems in the UAE Maritime Industry. *International Journal of Productivity Management and Assessment Technologies*, *5*(1), 20–39. doi:10.4018/IJPMAT.2017010102

Glykas, M., Valiris, G., Kokkinaki, A., & Koutsoukou, Z. (2018). Banking Business Process Management Implementation. *International Journal of Productivity Management and Assessment Technologies*, *6*(1), 50–69. doi:10.4018/IJPMAT.2018010104

Gomes, J., & Romão, M. (2017). The Balanced Scorecard: Keeping Updated and Aligned with Today's Business Trends. *International Journal of Productivity Management and Assessment Technologies*, *5*(2), 1–15. doi:10.4018/IJPMAT.2017070101

Gomes, J., & Romão, M. (2017). Aligning Information Systems and Technology with Benefit Management and Balanced Scorecard. In S. De Haes & W. Van Grembergen (Eds.), *Strategic IT Governance and Alignment in Business Settings* (pp. 112–131). Hershey, PA: IGI Global. doi:10.4018/978-1-5225-0861-8.ch005

Grefen, P., & Turetken, O. (2017). Advanced Business Process Management in Networked E-Business Scenarios. *International Journal of E-Business Research*, *13*(4), 70–104. doi:10.4018/IJEBR.2017100105

Haider, A., & Saetang, S. (2017). Strategic IT Alignment in Service Sector. In S. Rozenes & Y. Cohen (Eds.), *Handbook of Research on Strategic Alliances and Value Co-Creation in the Service Industry* (pp. 231–258). Hershey, PA: IGI Global. doi:10.4018/978-1-5225-2084-9.ch012

Haider, A., & Tang, S. S. (2016). Maximising Value Through IT and Business Alignment: A Case of IT Governance Institutionalisation at a Thai Bank. *International Journal of Technology Diffusion*, *7*(3), 33–58. doi:10.4018/IJTD.2016070104

Hajilari, A. B., Ghadaksaz, M., & Fasghandis, G. S. (2017). Assessing Organizational Readiness for Implementing ERP System Using Fuzzy Expert System Approach. *International Journal of Enterprise Information Systems*, *13*(1), 67–85. doi:10.4018/IJEIS.2017010105

Haldorai, A., Ramu, A., & Murugan, S. (2018). Social Aware Cognitive Radio Networks: Effectiveness of Social Networks as a Strategic Tool for Organizational Business Management. In H. Bansal, G. Shrivastava, G. Nguyen, & L. Stanciu (Eds.), *Social Network Analytics for Contemporary Business Organizations* (pp. 188–202). Hershey, PA: IGI Global. doi:10.4018/978-1-5225-5097-6.ch010

Hall, O. P. Jr. (2017). Social Media Driven Management Education. *International Journal of Knowledge-Based Organizations*, *7*(2), 43–59. doi:10.4018/IJKBO.2017040104

Hanifah, H., Halim, H. A., Ahmad, N. H., & Vafaei-Zadeh, A. (2017). Innovation Culture as a Mediator Between Specific Human Capital and Innovation Performance Among Bumiputera SMEs in Malaysia. In N. Ahmad, T. Ramayah, H. Halim, & S. Rahman (Eds.), *Handbook of Research on Small and Medium Enterprises in Developing Countries* (pp. 261–279). Hershey, PA: IGI Global. doi:10.4018/978-1-5225-2165-5.ch012

Hartlieb, S., & Silvius, G. (2017). Handling Uncertainty in Project Management and Business Development: Similarities and Differences. In Y. Raydugin (Ed.), *Handbook of Research on Leveraging Risk and Uncertainties for Effective Project Management* (pp. 337–362). Hershey, PA: IGI Global. doi:10.4018/978-1-5225-1790-0.ch016

Hass, K. B. (2017). Living on the Edge: Managing Project Complexity. In Y. Raydugin (Ed.), *Handbook of Research on Leveraging Risk and Uncertainties for Effective Project Management* (pp. 177–201). Hershey, PA: IGI Global. doi:10.4018/978-1-5225-1790-0.ch009

Hassan, A., & Privitera, D. S. (2016). Google AdSense as a Mobile Technology in Education. In J. Holland (Ed.), *Wearable Technology and Mobile Innovations for Next-Generation Education* (pp. 200–223). Hershey, PA: IGI Global. doi:10.4018/978-1-5225-0069-8.ch011

Hassan, A., & Rahimi, R. (2016). Consuming "Innovation" in Tourism: Augmented Reality as an Innovation Tool in Digital Tourism Marketing. In N. Pappas & I. Bregoli (Eds.), *Global Dynamics in Travel, Tourism, and Hospitality* (pp. 130–147). Hershey, PA: IGI Global. doi:10.4018/978-1-5225-0201-2.ch008

Hawking, P., & Carmine Sellitto, C. (2017). Developing an Effective Strategy for Organizational Business Intelligence. In M. Tavana (Ed.), *Enterprise Information Systems and the Digitalization of Business Functions* (pp. 222–237). Hershey, PA: IGI Global. doi:10.4018/978-1-5225-2382-6.ch010

Hawking, P., & Sellitto, C. (2017). A Fast-Moving Consumer Goods Company and Business Intelligence Strategy Development. *International Journal of Enterprise Information Systems, 13*(2), 22–33. doi:10.4018/IJEIS.2017040102

Hawking, P., & Sellitto, C. (2017). Business Intelligence Strategy: Two Case Studies. *International Journal of Business Intelligence Research, 8*(2), 17–30. doi:10.4018/IJBIR.2017070102

Haynes, J. D., Arockiasamy, S., Al Rashdi, M., & Al Rashdi, S. (2016). Business and E Business Strategies for Coopetition and Thematic Management as a Sustained Basis for Ethics and Social Responsibility in Emerging Markets. In M. Al-Shammari & H. Masri (Eds.), *Ethical and Social Perspectives on Global Business Interaction in Emerging Markets* (pp. 25–39). Hershey, PA: IGI Global. doi:10.4018/978-1-4666-9864-2.ch002

Hee, W. J., Jalleh, G., Lai, H., & Lin, C. (2017). E-Commerce and IT Projects: Evaluation and Management Issues in Australian and Taiwanese Hospitals. *International Journal of Public Health Management and Ethics, 2*(1), 69–90. doi:10.4018/IJPHME.2017010104

Hernandez, A. A. (2018). Exploring the Factors to Green IT Adoption of SMEs in the Philippines. *Journal of Cases on Information Technology, 20*(2), 49–66. doi:10.4018/JCIT.2018040104

Hernandez, A. A., & Ona, S. E. (2016). Green IT Adoption: Lessons from the Philippines Business Process Outsourcing Industry. *International Journal of Social Ecology and Sustainable Development, 7*(1), 1–34. doi:10.4018/IJSESD.2016010101

Hollman, A., Bickford, S., & Hollman, T. (2017). Cyber InSecurity: A Post-Mortem Attempt to Assess Cyber Problems from IT and Business Management Perspectives. *Journal of Cases on Information Technology, 19*(3), 42–70. doi:10.4018/JCIT.2017070104

Igbinakhase, I. (2017). Responsible and Sustainable Management Practices in Developing and Developed Business Environments. In Z. Fields (Ed.), *Collective Creativity for Responsible and Sustainable Business Practice* (pp. 180–207). Hershey, PA: IGI Global. doi:10.4018/978-1-5225-1823-5.ch010

Ilahi, L., Ghannouchi, S. A., & Martinho, R. (2016). A Business Process Management Approach to Home Healthcare Processes: On the Gap between Intention and Reality. In M. Cruz-Cunha, I. Miranda, R. Martinho, & R. Rijo (Eds.), *Encyclopedia of E-Health and Telemedicine* (pp. 439–457). Hershey, PA: IGI Global. doi:10.4018/978-1-4666-9978-6.ch035

Iwata, J. J., & Hoskins, R. G. (2017). Managing Indigenous Knowledge in Tanzania: A Business Perspective. In P. Jain & N. Mnjama (Eds.), *Managing Knowledge Resources and Records in Modern Organizations* (pp. 198–214). Hershey, PA: IGI Global. doi:10.4018/978-1-5225-1965-2.ch012

Jabeen, F., Ahmad, S. Z., & Alkaabi, S. (2016). The Internationalization Decision-Making of United Arab Emirates Family Businesses. In N. Zakaria, A. Abdul-Talib, & N. Osman (Eds.), *Handbook of Research on Impacts of International Business and Political Affairs on the Global Economy* (pp. 1–22). Hershey, PA: IGI Global. doi:10.4018/978-1-4666-9806-2.ch001

Jain, P. (2017). Ethical and Legal Issues in Knowledge Management Life-Cycle in Business. In P. Jain & N. Mnjama (Eds.), *Managing Knowledge Resources and Records in Modern Organizations* (pp. 82–101). Hershey, PA: IGI Global. doi:10.4018/978-1-5225-1965-2.ch006

Jamali, D., Abdallah, H., & Matar, F. (2016). Opportunities and Challenges for CSR Mainstreaming in Business Schools. *International Journal of Technology and Educational Marketing*, *6*(2), 1–29. doi:10.4018/IJTEM.2016070101

James, S., & Hauli, E. (2017). Holistic Management Education at Tanzanian Rural Development Planning Institute. In N. Baporikar (Ed.), *Management Education for Global Leadership* (pp. 112–136). Hershey, PA: IGI Global. doi:10.4018/978-1-5225-1013-0.ch006

Janošková, M., Csikósová, A., & Čulková, K. (2018). Measurement of Company Performance as Part of Its Strategic Management. In R. Leon (Ed.), *Managerial Strategies for Business Sustainability During Turbulent Times* (pp. 309–335). Hershey, PA: IGI Global. doi:10.4018/978-1-5225-2716-9.ch017

Jean-Vasile, A., & Alecu, A. (2017). Theoretical and Practical Approaches in Understanding the Influences of Cost-Productivity-Profit Trinomial in Contemporary Enterprises. In A. Jean Vasile & D. Nicolò (Eds.), *Sustainable Entrepreneurship and Investments in the Green Economy* (pp. 28–62). Hershey, PA: IGI Global. doi:10.4018/978-1-5225-2075-7.ch002

Jha, D. G. (2016). Preparing for Information Technology Driven Changes. In S. Tiwari & L. Nafees (Eds.), *Innovative Management Education Pedagogies for Preparing Next-Generation Leaders* (pp. 258–274). Hershey, PA: IGI Global. doi:10.4018/978-1-4666-9691-4.ch015

Joia, L. A., & Correia, J. C. (2018). CIO Competencies From the IT Professional Perspective: Insights From Brazil. *Journal of Global Information Management*, *26*(2), 74–103. doi:10.4018/JGIM.2018040104

Juma, A., & Mzera, N. (2017). Knowledge Management and Records Management and Competitive Advantage in Business. In P. Jain & N. Mnjama (Eds.), *Managing Knowledge Resources and Records in Modern Organizations* (pp. 15–28). Hershey, PA: IGI Global. doi:10.4018/978-1-5225-1965-2.ch002

K., I., & A, V. (2018). Monitoring and Auditing in the Cloud. In K. Munir (Ed.), *Cloud Computing Technologies for Green Enterprises* (pp. 318-350). Hershey, PA: IGI Global. doi:10.4018/978-1-5225-3038-1.ch013

Kabra, G., Ghosh, V., & Ramesh, A. (2018). Enterprise Integrated Business Process Management and Business Intelligence Framework for Business Process Sustainability. In A. Paul, D. Bhattacharyya, & S. Anand (Eds.), *Green Initiatives for Business Sustainability and Value Creation* (pp. 228–238). Hershey, PA: IGI Global. doi:10.4018/978-1-5225-2662-9.ch010

Kaoud, M. (2017). Investigation of Customer Knowledge Management: A Case Study Research. *International Journal of Service Science, Management, Engineering, and Technology*, *8*(2), 12–22. doi:10.4018/IJSSMET.2017040102

Kara, M. E., & Fırat, S. Ü. (2016). Sustainability, Risk, and Business Intelligence in Supply Chains. In M. Erdoğdu, T. Arun, & I. Ahmad (Eds.), *Handbook of Research on Green Economic Development Initiatives and Strategies* (pp. 501–538). Hershey, PA: IGI Global. doi:10.4018/978-1-5225-0440-5.ch022

Katuu, S. (2018). A Comparative Assessment of Enterprise Content Management Maturity Models. In N. Gwangwava & M. Mutingi (Eds.), *E-Manufacturing and E-Service Strategies in Contemporary Organizations* (pp. 93–118). Hershey, PA: IGI Global. doi:10.4018/978-1-5225-3628-4.ch005

Khan, M. A. (2016). MNEs Management Strategies in Developing Countries: Establishing the Context. In M. Khan (Ed.), *Multinational Enterprise Management Strategies in Developing Countries* (pp. 1–33). Hershey, PA: IGI Global. doi:10.4018/978-1-5225-0276-0.ch001

Khan, M. A. (2016). Operational Approaches in Organizational Structure: A Case for MNEs in Developing Countries. In M. Khan (Ed.), *Multinational Enterprise Management Strategies in Developing Countries* (pp. 129–151). Hershey, PA: IGI Global. doi:10.4018/978-1-5225-0276-0.ch007

Kinnunen, S., Ylä-Kujala, A., Marttonen-Arola, S., Kärri, T., & Baglee, D. (2018). Internet of Things in Asset Management: Insights from Industrial Professionals and Academia. *International Journal of Service Science, Management, Engineering, and Technology*, *9*(2), 104–119. doi:10.4018/IJSSMET.2018040105

Klein, A. Z., Sabino de Freitas, A., Machado, L., Freitas, J. C. Jr, Graziola, P. G. Jr, & Schlemmer, E. (2017). Virtual Worlds Applications for Management Education. In L. Tomei (Ed.), *Exploring the New Era of Technology-Infused Education* (pp. 279–299). Hershey, PA: IGI Global. doi:10.4018/978-1-5225-1709-2.ch017

Kożuch, B., & Jabłoński, A. (2017). Adopting the Concept of Business Models in Public Management. In M. Lewandowski & B. Kożuch (Eds.), *Public Sector Entrepreneurship and the Integration of Innovative Business Models* (pp. 10–46). Hershey, PA: IGI Global. doi:10.4018/978-1-5225-2215-7.ch002

Kumar, J., Adhikary, A., & Jha, A. (2017). Small Active Investors' Perceptions and Preferences Towards Tax Saving Mutual Fund Schemes in Eastern India: An Empirical Note. *International Journal of Asian Business and Information Management*, *8*(2), 35–45. doi:10.4018/IJABIM.2017040103

Lassoued, Y., Bouzguenda, L., & Mahmoud, T. (2016). Context-Aware Business Process Versions Management. *International Journal of e-Collaboration*, *12*(3), 7–33. doi:10.4018/IJeC.2016070102

Lavassani, K. M., & Movahedi, B. (2017). Applications Driven Information Systems: Beyond Networks toward Business Ecosystems. *International Journal of Innovation in the Digital Economy*, *8*(1), 61–75. doi:10.4018/IJIDE.2017010104

Lazzareschi, V. H., & Brito, M. S. (2017). Strategic Information Management: Proposal of Business Project Model. In G. Jamil, A. Soares, & C. Pessoa (Eds.), *Handbook of Research on Information Management for Effective Logistics and Supply Chains* (pp. 59–88). Hershey, PA: IGI Global. doi:10.4018/978-1-5225-0973-8.ch004

Lederer, M., Kurz, M., & Lazarov, P. (2017). Usage and Suitability of Methods for Strategic Business Process Initiatives: A Multi Case Study Research. *International Journal of Productivity Management and Assessment Technologies*, *5*(1), 40–51. doi:10.4018/IJPMAT.2017010103

Lee, I. (2017). A Social Enterprise Business Model and a Case Study of Pacific Community Ventures (PCV). In V. Potocan, M. Ünğan, & Z. Nedelko (Eds.), *Handbook of Research on Managerial Solutions in Non-Profit Organizations* (pp. 182–204). Hershey, PA: IGI Global. doi:10.4018/978-1-5225-0731-4.ch009

Lee, L. J., & Leu, J. (2016). Exploring the Effectiveness of IT Application and Value Method in the Innovation Performance of Enterprise. *International Journal of Enterprise Information Systems*, *12*(2), 47–65. doi:10.4018/IJEIS.2016040104

Lee, Y. (2016). Alignment Effect of Entrepreneurial Orientation and Marketing Orientation on Firm Performance. *International Journal of Customer Relationship Marketing and Management*, *7*(4), 58–69. doi:10.4018/IJCRMM.2016100104

Leon, L. A., Seal, K. C., Przasnyski, Z. H., & Wiedenman, I. (2017). Skills and Competencies Required for Jobs in Business Analytics: A Content Analysis of Job Advertisements Using Text Mining. *International Journal of Business Intelligence Research*, *8*(1), 1–25. doi:10.4018/IJBIR.2017010101

Leu, J., Lee, L. J., & Krischke, A. (2016). Value Engineering-Based Method for Implementing the ISO14001 System in the Green Supply Chains. *International Journal of Strategic Decision Sciences*, *7*(4), 1–20. doi:10.4018/IJSDS.2016100101

Levy, C. L., & Elias, N. I. (2017). SOHO Users' Perceptions of Reliability and Continuity of Cloud-Based Services. In M. Moore (Ed.), *Cybersecurity Breaches and Issues Surrounding Online Threat Protection* (pp. 248–287). Hershey, PA: IGI Global. doi:10.4018/978-1-5225-1941-6.ch011

Levy, M. (2018). Change Management Serving Knowledge Management and Organizational Development: Reflections and Review. In N. Baporikar (Ed.), *Global Practices in Knowledge Management for Societal and Organizational Development* (pp. 256–270). Hershey, PA: IGI Global. doi:10.4018/978-1-5225-3009-1.ch012

Lewandowski, M. (2017). Public Organizations and Business Model Innovation: The Role of Public Service Design. In M. Lewandowski & B. Kożuch (Eds.), *Public Sector Entrepreneurship and the Integration of Innovative Business Models* (pp. 47–72). Hershey, PA: IGI Global. doi:10.4018/978-1-5225-2215-7.ch003

Lhannaoui, H., Kabbaj, M. I., & Bakkoury, Z. (2017). A Survey of Risk-Aware Business Process Modelling. *International Journal of Risk and Contingency Management*, *6*(3), 14–26. doi:10.4018/IJRCM.2017070102

Li, J., Sun, W., Jiang, W., Yang, H., & Zhang, L. (2017). How the Nature of Exogenous Shocks and Crises Impact Company Performance?: The Effects of Industry Characteristics. *International Journal of Risk and Contingency Management*, *6*(4), 40–55. doi:10.4018/IJRCM.2017100103

Lu, C., & Liu, S. (2016). Cultural Tourism O2O Business Model Innovation-A Case Study of CTrip. *Journal of Electronic Commerce in Organizations*, *14*(2), 16–31. doi:10.4018/JECO.2016040102

Machen, B., Hosseini, M. R., Wood, A., & Bakhshi, J. (2016). An Investigation into using SAP-PS as a Multidimensional Project Control System (MPCS). *International Journal of Enterprise Information Systems*, *12*(2), 66–81. doi:10.4018/IJEIS.2016040105

Malega, P. (2017). Small and Medium Enterprises in the Slovak Republic: Status and Competitiveness of SMEs in the Global Markets and Possibilities of Optimization. In M. Vemić (Ed.), *Optimal Management Strategies in Small and Medium Enterprises* (pp. 102–124). Hershey, PA: IGI Global. doi:10.4018/978-1-5225-1949-2.ch006

Malewska, K. M. (2017). Intuition in Decision-Making on the Example of a Non-Profit Organization. In V. Potocan, M. Üngan, & Z. Nedelko (Eds.), *Handbook of Research on Managerial Solutions in Non-Profit Organizations* (pp. 378–399). Hershey, PA: IGI Global. doi:10.4018/978-1-5225-0731-4.ch018

Maroofi, F. (2017). Entrepreneurial Orientation and Organizational Learning Ability Analysis for Innovation and Firm Performance. In N. Baporikar (Ed.), *Innovation and Shifting Perspectives in Management Education* (pp. 144–165). Hershey, PA: IGI Global. doi:10.4018/978-1-5225-1019-2.ch007

Martins, P. V., & Zacarias, M. (2017). A Web-based Tool for Business Process Improvement. *International Journal of Web Portals*, *9*(2), 68–84. doi:10.4018/IJWP.2017070104

Matthies, B., & Coners, A. (2017). Exploring the Conceptual Nature of e-Business Projects. *Journal of Electronic Commerce in Organizations*, *15*(3), 33–63. doi:10.4018/JECO.2017070103

McKee, J. (2018). Architecture as a Tool to Solve Business Planning Problems. In M. Khosrow-Pour, D.B.A. (Ed.), Encyclopedia of Information Science and Technology, Fourth Edition (pp. 573-586). Hershey, PA: IGI Global. doi:10.4018/978-1-5225-2255-3.ch050

McMurray, A. J., Cross, J., & Caponecchia, C. (2018). The Risk Management Profession in Australia: Business Continuity Plan Practices. In N. Bajgoric (Ed.), *Always-On Enterprise Information Systems for Modern Organizations* (pp. 112–129). Hershey, PA: IGI Global. doi:10.4018/978-1-5225-3704-5.ch006

Meddah, I. H., & Belkadi, K. (2018). Mining Patterns Using Business Process Management. In R. Hamou (Ed.), *Handbook of Research on Biomimicry in Information Retrieval and Knowledge Management* (pp. 78–89). Hershey, PA: IGI Global. doi:10.4018/978-1-5225-3004-6.ch005

Mendes, L. (2017). TQM and Knowledge Management: An Integrated Approach Towards Tacit Knowledge Management. In D. Jaziri-Bouagina & G. Jamil (Eds.), *Handbook of Research on Tacit Knowledge Management for Organizational Success* (pp. 236–263). Hershey, PA: IGI Global. doi:10.4018/978-1-5225-2394-9.ch009

Mnjama, N. M. (2017). Preservation of Recorded Information in Public and Private Sector Organizations. In P. Jain & N. Mnjama (Eds.), *Managing Knowledge Resources and Records in Modern Organizations* (pp. 149–167). Hershey, PA: IGI Global. doi:10.4018/978-1-5225-1965-2.ch009

Mokoqama, M., & Fields, Z. (2017). Principles of Responsible Management Education (PRME): Call for Responsible Management Education. In Z. Fields (Ed.), *Collective Creativity for Responsible and Sustainable Business Practice* (pp. 229–241). Hershey, PA: IGI Global. doi:10.4018/978-1-5225-1823-5.ch012

Muniapan, B. (2017). Philosophy and Management: The Relevance of Vedanta in Management. In P. Ordóñez de Pablos (Ed.), *Managerial Strategies and Solutions for Business Success in Asia* (pp. 124–139). Hershey, PA: IGI Global. doi:10.4018/978-1-5225-1886-0.ch007

Muniapan, B., Gregory, M. L., & Ling, L. A. (2016). Marketing Education in Sarawak: Looking at It from the Employers' Viewpoint. In B. Smith & A. Porath (Eds.), *Global Perspectives on Contemporary Marketing Education* (pp. 112–130). Hershey, PA: IGI Global. doi:10.4018/978-1-4666-9784-3.ch008

Murad, S. E., & Dowaji, S. (2017). Using Value-Based Approach for Managing Cloud-Based Services. In A. Turuk, B. Sahoo, & S. Addya (Eds.), *Resource Management and Efficiency in Cloud Computing Environments* (pp. 33–60). Hershey, PA: IGI Global. doi:10.4018/978-1-5225-1721-4.ch002

Mutahar, A. M., Daud, N. M., Thurasamy, R., Isaac, O., & Abdulsalam, R. (2018). The Mediating of Perceived Usefulness and Perceived Ease of Use: The Case of Mobile Banking in Yemen. *International Journal of Technology Diffusion*, *9*(2), 21–40. doi:10.4018/IJTD.2018040102

Naidoo, V. (2017). E-Learning and Management Education at African Universities. In N. Baporikar (Ed.), *Management Education for Global Leadership* (pp. 181–201). Hershey, PA: IGI Global. doi:10.4018/978-1-5225-1013-0.ch009

Naidoo, V., & Igbinakhase, I. (2018). Opportunities and Challenges of Knowledge Retention in SMEs. In N. Baporikar (Ed.), *Knowledge Integration Strategies for Entrepreneurship and Sustainability* (pp. 70–94). Hershey, PA: IGI Global. doi:10.4018/978-1-5225-5115-7.ch004

Nayak, S., & Prabhu, N. (2017). Paradigm Shift in Management Education: Need for a Cross Functional Perspective. In N. Baporikar (Ed.), *Management Education for Global Leadership* (pp. 241–255). Hershey, PA: IGI Global. doi:10.4018/978-1-5225-1013-0.ch012

Ndede-Amadi, A. A. (2016). Student Interest in the IS Specialization as Predictor of the Success Potential of New Information Systems Programmes within the Schools of Business in Kenyan Public Universities. *International Journal of Information Systems and Social Change*, *7*(2), 63–79. doi:10.4018/IJISSC.2016040104

Nedelko, Z., & Potocan, V. (2016). Management Practices for Processes Optimization: Case of Slovenia. In G. Alor-Hernández, C. Sánchez-Ramírez, & J. García-Alcaraz (Eds.), *Handbook of Research on Managerial Strategies for Achieving Optimal Performance in Industrial Processes* (pp. 545–561). Hershey, PA: IGI Global. doi:10.4018/978-1-5225-0130-5.ch025

Nedelko, Z., & Potocan, V. (2017). Management Solutions in Non-Profit Organizations: Case of Slovenia. In V. Potocan, M. Üngan, & Z. Nedelko (Eds.), *Handbook of Research on Managerial Solutions in Non-Profit Organizations* (pp. 1–22). Hershey, PA: IGI Global. doi:10.4018/978-1-5225-0731-4.ch001

Nedelko, Z., & Potocan, V. (2017). Priority of Management Tools Utilization among Managers: International Comparison. In V. Wang (Ed.), *Encyclopedia of Strategic Leadership and Management* (pp. 1083–1094). Hershey, PA: IGI Global. doi:10.4018/978-1-5225-1049-9.ch075

Nedelko, Z., Raudeliūnienė, J., & Črešnar, R. (2018). Knowledge Dynamics in Supply Chain Management. In N. Baporikar (Ed.), *Knowledge Integration Strategies for Entrepreneurship and Sustainability* (pp. 150–166). Hershey, PA: IGI Global. doi:10.4018/978-1-5225-5115-7.ch008

Nguyen, H. T., & Hipsher, S. A. (2018). Innovation and Creativity Used by Private Sector Firms in a Resources-Constrained Environment. In S. Hipsher (Ed.), *Examining the Private Sector's Role in Wealth Creation and Poverty Reduction* (pp. 219–238). Hershey, PA: IGI Global. doi:10.4018/978-1-5225-3117-3.ch010

Nycz, M., & Pólkowski, Z. (2016). Business Intelligence as a Modern IT Supporting Management of Local Government Units in Poland. *International Journal of Knowledge and Systems Science*, *7*(4), 1–18. doi:10.4018/IJKSS.2016100101

Obaji, N. O., Senin, A. A., & Olugu, M. U. (2016). Supportive Government Policy as a Mechanism for Business Incubation Performance in Nigeria. *International Journal of Information Systems and Social Change*, *7*(4), 52–66. doi:10.4018/IJISSC.2016100103

Obicci, P. A. (2017). Risk Sharing in a Partnership. In *Risk Management Strategies in Public-Private Partnerships* (pp. 115–152). Hershey, PA: IGI Global. doi:10.4018/978-1-5225-2503-5.ch004

Obidallah, W. J., & Raahemi, B. (2017). Managing Changes in Service Oriented Virtual Organizations: A Structural and Procedural Framework to Facilitate the Process of Change. *Journal of Electronic Commerce in Organizations*, *15*(1), 59–83. doi:10.4018/JECO.2017010104

Ojasalo, J., & Ojasalo, K. (2016). Service Logic Business Model Canvas for Lean Development of SMEs and Start-Ups. In N. Baporikar (Ed.), *Handbook of Research on Entrepreneurship in the Contemporary Knowledge-Based Global Economy* (pp. 217–243). Hershey, PA: IGI Global. doi:10.4018/978-1-4666-8798-1.ch010

Ojo, O. (2017). Impact of Innovation on the Entrepreneurial Success in Selected Business Enterprises in South-West Nigeria. *International Journal of Innovation in the Digital Economy*, *8*(2), 29–38. doi:10.4018/IJIDE.2017040103

Okdinawati, L., Simatupang, T. M., & Sunitiyoso, Y. (2017). Multi-Agent Reinforcement Learning for Value Co-Creation of Collaborative Transportation Management (CTM). *International Journal of Information Systems and Supply Chain Management*, *10*(3), 84–95. doi:10.4018/IJISSCM.2017070105

Ortner, E., Mevius, M., Wiedmann, P., & Kurz, F. (2016). Design of Interactional Decision Support Applications for E-Participation in Smart Cities. *International Journal of Electronic Government Research*, *12*(2), 18–38. doi:10.4018/IJEGR.2016040102

Pal, K. (2018). Building High Quality Big Data-Based Applications in Supply Chains. In A. Kumar & S. Saurav (Eds.), *Supply Chain Management Strategies and Risk Assessment in Retail Environments* (pp. 1–24). Hershey, PA: IGI Global. doi:10.4018/978-1-5225-3056-5.ch001

Palos-Sanchez, P. R., & Correia, M. B. (2018). Perspectives of the Adoption of Cloud Computing in the Tourism Sector. In J. Rodrigues, C. Ramos, P. Cardoso, & C. Henriques (Eds.), *Handbook of Research on Technological Developments for Cultural Heritage and eTourism Applications* (pp. 377–400). Hershey, PA: IGI Global. doi:10.4018/978-1-5225-2927-9.ch018

Parry, V. K., & Lind, M. L. (2016). Alignment of Business Strategy and Information Technology Considering Information Technology Governance, Project Portfolio Control, and Risk Management. *International Journal of Information Technology Project Management*, *7*(4), 21–37. doi:10.4018/IJITPM.2016100102

Pashkova, N., Trujillo-Barrera, A., Apostolakis, G., Van Dijk, G., Drakos, P. D., & Baourakis, G. (2016). Business Management Models of Microfinance Institutions (MFIs) in Africa: A Study into Their Enabling Environments. *International Journal of Food and Beverage Manufacturing and Business Models*, *1*(2), 63–82. doi:10.4018/IJFBMBM.2016070105

Patiño, B. E. (2017). New Generation Management by Convergence and Individual Identity: A Systemic and Human-Oriented Approach. In N. Baporikar (Ed.), *Innovation and Shifting Perspectives in Management Education* (pp. 119–143). Hershey, PA: IGI Global. doi:10.4018/978-1-5225-1019-2.ch006

Pawliczek, A., & Rössler, M. (2017). Knowledge of Management Tools and Systems in SMEs: Knowledge Transfer in Management. In A. Bencsik (Ed.), *Knowledge Management Initiatives and Strategies in Small and Medium Enterprises* (pp. 180–203). Hershey, PA: IGI Global. doi:10.4018/978-1-5225-1642-2.ch009

Pejic-Bach, M., Omazic, M. A., Aleksic, A., & Zoroja, J. (2018). Knowledge-Based Decision Making: A Multi-Case Analysis. In R. Leon (Ed.), *Managerial Strategies for Business Sustainability During Turbulent Times* (pp. 160–184). Hershey, PA: IGI Global. doi:10.4018/978-1-5225-2716-9.ch009

Perano, M., Hysa, X., & Calabrese, M. (2018). Strategic Planning, Cultural Context, and Business Continuity Management: Business Cases in the City of Shkoder. In A. Presenza & L. Sheehan (Eds.), *Geopolitics and Strategic Management in the Global Economy* (pp. 57–77). Hershey, PA: IGI Global. doi:10.4018/978-1-5225-2673-5.ch004

Pereira, R., Mira da Silva, M., & Lapão, L. V. (2017). IT Governance Maturity Patterns in Portuguese Healthcare. In S. De Haes & W. Van Grembergen (Eds.), *Strategic IT Governance and Alignment in Business Settings* (pp. 24–52). Hershey, PA: IGI Global. doi:10.4018/978-1-5225-0861-8.ch002

Perez-Uribe, R., & Ocampo-Guzman, D. (2016). Conflict within Colombian Family Owned SMEs: An Explosive Blend between Feelings and Business. In J. Saiz-Álvarez (Ed.), *Handbook of Research on Social Entrepreneurship and Solidarity Economics* (pp. 329–354). Hershey, PA: IGI Global. doi:10.4018/978-1-5225-0097-1.ch017

Pérez-Uribe, R. I., Torres, D. A., Jurado, S. P., & Prada, D. M. (2018). Cloud Tools for the Development of Project Management in SMEs. In R. Perez-Uribe, C. Salcedo-Perez, & D. Ocampo-Guzman (Eds.), *Handbook of Research on Intrapreneurship and Organizational Sustainability in SMEs* (pp. 95–120). Hershey, PA: IGI Global. doi:10.4018/978-1-5225-3543-0.ch005

Petrisor, I., & Cozmiuc, D. (2017). Global Supply Chain Management Organization at Siemens in the Advent of Industry 4.0. In L. Saglietto & C. Cezanne (Eds.), *Global Intermediation and Logistics Service Providers* (pp. 123–142). Hershey, PA: IGI Global. doi:10.4018/978-1-5225-2133-4.ch007

Pierce, J. M., Velliaris, D. M., & Edwards, J. (2017). A Living Case Study: A Journey Not a Destination. In N. Silton (Ed.), *Exploring the Benefits of Creativity in Education, Media, and the Arts* (pp. 158–178). Hershey, PA: IGI Global. doi:10.4018/978-1-5225-0504-4.ch008

Radosavljevic, M., & Andjelkovic, A. (2017). Multi-Criteria Decision Making Approach for Choosing Business Process for the Improvement: Upgrading of the Six Sigma Methodology. In J. Stanković, P. Delias, S. Marinković, & S. Rochhia (Eds.), *Tools and Techniques for Economic Decision Analysis* (pp. 225–247). Hershey, PA: IGI Global. doi:10.4018/978-1-5225-0959-2.ch011

Radovic, V. M. (2017). Corporate Sustainability and Responsibility and Disaster Risk Reduction: A Serbian Overview. In M. Camilleri (Ed.), *CSR 2.0 and the New Era of Corporate Citizenship* (pp. 147–164). Hershey, PA: IGI Global. doi:10.4018/978-1-5225-1842-6.ch008

Raghunath, K. M., Devi, S. L., & Patro, C. S. (2018). Impact of Risk Assessment Models on Risk Factors: A Holistic Outlook. In K. Strang, M. Korstanje, & N. Vajjhala (Eds.), *Research, Practices, and Innovations in Global Risk and Contingency Management* (pp. 134–153). Hershey, PA: IGI Global. doi:10.4018/978-1-5225-4754-9.ch008

Raman, A., & Goyal, D. P. (2017). Extending IMPLEMENT Framework for Enterprise Information Systems Implementation to Information System Innovation. In M. Tavana (Ed.), *Enterprise Information Systems and the Digitalization of Business Functions* (pp. 137–177). Hershey, PA: IGI Global. doi:10.4018/978-1-5225-2382-6.ch007

Rao, Y., & Zhang, Y. (2017). The Construction and Development of Academic Library Digital Special Subject Databases. In L. Ruan, Q. Zhu, & Y. Ye (Eds.), *Academic Library Development and Administration in China* (pp. 163–183). Hershey, PA: IGI Global. doi:10.4018/978-1-5225-0550-1.ch010

Ravasan, A. Z., Mohammadi, M. M., & Hamidi, H. (2018). An Investigation Into the Critical Success Factors of Implementing Information Technology Service Management Frameworks. In K. Jakobs (Ed.), *Corporate and Global Standardization Initiatives in Contemporary Society* (pp. 200–218). Hershey, PA: IGI Global. doi:10.4018/978-1-5225-5320-5.ch009

Renna, P., Izzo, C., & Romaniello, T. (2016). The Business Process Management Systems to Support Continuous Improvements. In W. Nuninger & J. Châtelet (Eds.), *Handbook of Research on Quality Assurance and Value Management in Higher Education* (pp. 237–256). Hershey, PA: IGI Global. doi:10.4018/978-1-5225-0024-7.ch009

Rezaie, S., Mirabedini, S. J., & Abtahi, A. (2018). Designing a Model for Implementation of Business Intelligence in the Banking Industry. *International Journal of Enterprise Information Systems, 14*(1), 77–103. doi:10.4018/IJEIS.2018010105

Riccò, R. (2016). Diversity Management: Bringing Equality, Equity, and Inclusion in the Workplace. In J. Prescott (Ed.), *Handbook of Research on Race, Gender, and the Fight for Equality* (pp. 335–359). Hershey, PA: IGI Global. doi:10.4018/978-1-5225-0047-6.ch015

Romano, L., Grimaldi, R., & Colasuonno, F. S. (2017). Demand Management as a Success Factor in Project Portfolio Management. In L. Romano (Ed.), *Project Portfolio Management Strategies for Effective Organizational Operations* (pp. 202–219). Hershey, PA: IGI Global. doi:10.4018/978-1-5225-2151-8.ch008

Rostek, K. B. (2016). Risk Management: Role and Importance in Business Organization. In D. Jakóbczak (Ed.), *Analyzing Risk through Probabilistic Modeling in Operations Research* (pp. 149–178). Hershey, PA: IGI Global. doi:10.4018/978-1-4666-9458-3.ch007

Rouhani, S., & Savoji, S. R. (2016). A Success Assessment Model for BI Tools Implementation: An Empirical Study of Banking Industry. *International Journal of Business Intelligence Research, 7*(1), 25–44. doi:10.4018/IJBIR.2016010103

Ruan, Z. (2016). A Corpus-Based Functional Analysis of Complex Nominal Groups in Written Business Discourse: The Case of "Business". *International Journal of Computer-Assisted Language Learning and Teaching*, *6*(2), 74–90. doi:10.4018/IJCALLT.2016040105

Ruhi, U. (2018). Towards an Interdisciplinary Socio-Technical Definition of Virtual Communities. In M. Khosrow-Pour, D.B.A. (Ed.), Encyclopedia of Information Science and Technology, Fourth Edition (pp. 4278-4295). Hershey, PA: IGI Global. doi:10.4018/978-1-5225-2255-3.ch371

Ryan, J., Doster, B., Daily, S., & Lewis, C. (2016). A Case Study Perspective for Balanced Perioperative Workflow Achievement through Data-Driven Process Improvement. *International Journal of Healthcare Information Systems and Informatics*, *11*(3), 19–41. doi:10.4018/IJHISI.2016070102

Safari, M. R., & Jiang, Q. (2018). The Theory and Practice of IT Governance Maturity and Strategies Alignment: Evidence From Banking Industry. *Journal of Global Information Management*, *26*(2), 127–146. doi:10.4018/JGIM.2018040106

Sahoo, J., Pati, B., & Mohanty, B. (2017). Knowledge Management as an Academic Discipline: An Assessment. In B. Gunjal (Ed.), *Managing Knowledge and Scholarly Assets in Academic Libraries* (pp. 99–126). Hershey, PA: IGI Global. doi:10.4018/978-1-5225-1741-2.ch005

Saini, D. (2017). Relevance of Teaching Values and Ethics in Management Education. In N. Baporikar (Ed.), *Management Education for Global Leadership* (pp. 90–111). Hershey, PA: IGI Global. doi:10.4018/978-1-5225-1013-0.ch005

Sambhanthan, A. (2017). Assessing and Benchmarking Sustainability in Organisations: An Integrated Conceptual Model. *International Journal of Systems and Service-Oriented Engineering*, *7*(4), 22–43. doi:10.4018/IJSSOE.2017100102

Sambhanthan, A., & Potdar, V. (2017). A Study of the Parameters Impacting Sustainability in Information Technology Organizations. *International Journal of Knowledge-Based Organizations*, *7*(3), 27–39. doi:10.4018/IJKBO.2017070103

Sánchez-Fernández, M. D., & Manríquez, M. R. (2018). The Entrepreneurial Spirit Based on Social Values: The Digital Generation. In P. Isaias & L. Carvalho (Eds.), *User Innovation and the Entrepreneurship Phenomenon in the Digital Economy* (pp. 173–193). Hershey, PA: IGI Global. doi:10.4018/978-1-5225-2826-5.ch009

Sanchez-Ruiz, L., & Blanco, B. (2017). Process Management for SMEs: Barriers, Enablers, and Benefits. In M. Vemić (Ed.), *Optimal Management Strategies in Small and Medium Enterprises* (pp. 293–319). Hershey, PA: IGI Global. doi:10.4018/978-1-5225-1949-2.ch014

Sanz, L. F., Gómez-Pérez, J., & Castillo-Martinez, A. (2018). Analysis of the European ICT Competence Frameworks. In V. Ahuja & S. Rathore (Eds.), *Multidisciplinary Perspectives on Human Capital and Information Technology Professionals* (pp. 225–245). Hershey, PA: IGI Global. doi:10.4018/978-1-5225-5297-0.ch012

Sarvepalli, A., & Godin, J. (2017). Business Process Management in the Classroom. *Journal of Cases on Information Technology*, *19*(2), 17–28. doi:10.4018/JCIT.2017040102

Satpathy, B., & Muniapan, B. (2016). Ancient Wisdom for Transformational Leadership and Its Insights from the Bhagavad-Gita. In U. Aung & P. Ordoñez de Pablos (Eds.), *Managerial Strategies and Practice in the Asian Business Sector* (pp. 1–10). Hershey, PA: IGI Global. doi:10.4018/978-1-4666-9758-4.ch001

Saygili, E. E., Ozturkoglu, Y., & Kocakulah, M. C. (2017). End Users' Perceptions of Critical Success Factors in ERP Applications. *International Journal of Enterprise Information Systems*, *13*(4), 58–75. doi:10.4018/IJEIS.2017100104

Saygili, E. E., & Saygili, A. T. (2017). Contemporary Issues in Enterprise Information Systems: A Critical Review of CSFs in ERP Implementations. In M. Tavana (Ed.), *Enterprise Information Systems and the Digitalization of Business Functions* (pp. 120–136). Hershey, PA: IGI Global. doi:10.4018/978-1-5225-2382-6.ch006

Seidenstricker, S., & Antonino, A. (2018). Business Model Innovation-Oriented Technology Management for Emergent Technologies. In M. Khosrow-Pour, D.B.A. (Ed.), Encyclopedia of Information Science and Technology, Fourth Edition (pp. 4560-4569). Hershey, PA: IGI Global. doi:10.4018/978-1-5225-2255-3.ch396

Senaratne, S., & Gunarathne, A. D. (2017). Excellence Perspective for Management Education from a Global Accountants' Hub in Asia. In N. Baporikar (Ed.), *Management Education for Global Leadership* (pp. 158–180). Hershey, PA: IGI Global. doi:10.4018/978-1-5225-1013-0.ch008

Sensuse, D. I., & Cahyaningsih, E. (2018). Knowledge Management Models: A Summative Review. *International Journal of Information Systems in the Service Sector*, *10*(1), 71–100. doi:10.4018/IJISSS.2018010105

Sensuse, D. I., Wibowo, W. C., & Cahyaningsih, E. (2016). Indonesian Government Knowledge Management Model: A Theoretical Model. *Information Resources Management Journal*, *29*(1), 91–108. doi:10.4018/irmj.2016010106

Seth, M., Goyal, D., & Kiran, R. (2017). Diminution of Impediments in Implementation of Supply Chain Management Information System for Enhancing its Effectiveness in Indian Automobile Industry. *Journal of Global Information Management*, *25*(3), 1–20. doi:10.4018/JGIM.2017070101

Seyal, A. H., & Rahman, M. N. (2017). Investigating Impact of Inter-Organizational Factors in Measuring ERP Systems Success: Bruneian Perspectives. In M. Tavana (Ed.), *Enterprise Information Systems and the Digitalization of Business Functions* (pp. 178–204). Hershey, PA: IGI Global. doi:10.4018/978-1-5225-2382-6.ch008

Shaikh, A. A., & Karjaluoto, H. (2016). On Some Misconceptions Concerning Digital Banking and Alternative Delivery Channels. *International Journal of E-Business Research*, *12*(3), 1–16. doi:10.4018/IJEBR.2016070101

Shams, S. M. (2016). Stakeholder Relationship Management in Online Business and Competitive Value Propositions: Evidence from the Sports Industry. *International Journal of Online Marketing*, *6*(2), 1–17. doi:10.4018/IJOM.2016040101

Shamsuzzoha, A. (2016). Management of Risk and Resilience within Collaborative Business Network. In R. Addo-Tenkorang, J. Kantola, P. Helo, & A. Shamsuzzoha (Eds.), *Supply Chain Strategies and the Engineer-to-Order Approach* (pp. 143–159). Hershey, PA: IGI Global. doi:10.4018/978-1-5225-0021-6.ch008

Shaqrah, A. A. (2018). Analyzing Business Intelligence Systems Based on 7s Model of McKinsey. *International Journal of Business Intelligence Research*, *9*(1), 53–63. doi:10.4018/IJBIR.2018010104

Sharma, A. J. (2017). Enhancing Sustainability through Experiential Learning in Management Education. In N. Baporikar (Ed.), *Management Education for Global Leadership* (pp. 256–274). Hershey, PA: IGI Global. doi:10.4018/978-1-5225-1013-0.ch013

Shetty, K. P. (2017). Responsible Global Leadership: Ethical Challenges in Management Education. In N. Baporikar (Ed.), *Innovation and Shifting Perspectives in Management Education* (pp. 194–223). Hershey, PA: IGI Global. doi:10.4018/978-1-5225-1019-2.ch009

Sinthupundaja, J., & Kohda, Y. (2017). Effects of Corporate Social Responsibility and Creating Shared Value on Sustainability. *International Journal of Sustainable Entrepreneurship and Corporate Social Responsibility*, *2*(1), 27–38. doi:10.4018/IJSECSR.2017010103

Škarica, I., & Hrgović, A. V. (2018). Implementation of Total Quality Management Principles in Public Health Institutes in the Republic of Croatia. *International Journal of Productivity Management and Assessment Technologies*, *6*(1), 1–16. doi:10.4018/IJPMAT.2018010101

Smuts, H., Kotzé, P., Van der Merwe, A., & Loock, M. (2017). Framework for Managing Shared Knowledge in an Information Systems Outsourcing Context. *International Journal of Knowledge Management*, *13*(4), 1–30. doi:10.4018/IJKM.2017100101

Soares, E. R., & Zaidan, F. H. (2016). Information Architecture and Business Modeling in Modern Organizations of Information Technology: Professional Career Plan in Organizations IT. In G. Jamil, J. Poças Rascão, F. Ribeiro, & A. Malheiro da Silva (Eds.), *Handbook of Research on Information Architecture and Management in Modern Organizations* (pp. 439–457). Hershey, PA: IGI Global. doi:10.4018/978-1-4666-8637-3.ch020

Sousa, M. J., Cruz, R., Dias, I., & Caracol, C. (2017). Information Management Systems in the Supply Chain. In G. Jamil, A. Soares, & C. Pessoa (Eds.), *Handbook of Research on Information Management for Effective Logistics and Supply Chains* (pp. 469–485). Hershey, PA: IGI Global. doi:10.4018/978-1-5225-0973-8.ch025

Spremic, M., Turulja, L., & Bajgoric, N. (2018). Two Approaches in Assessing Business Continuity Management Attitudes in the Organizational Context. In N. Bajgoric (Ed.), *Always-On Enterprise Information Systems for Modern Organizations* (pp. 159–183). Hershey, PA: IGI Global. doi:10.4018/978-1-5225-3704-5.ch008

Steenkamp, A. L. (2018). Some Insights in Computer Science and Information Technology. In *Examining the Changing Role of Supervision in Doctoral Research Projects: Emerging Research and Opportunities* (pp. 113–133). Hershey, PA: IGI Global. doi:10.4018/978-1-5225-2610-0.ch005

Studdard, N., Dawson, M., Burton, S. L., Jackson, N., Leonard, B., Quisenberry, W., & Rahim, E. (2016). Nurturing Social Entrepreneurship and Building Social Entrepreneurial Self-Efficacy: Focusing on Primary and Secondary Schooling to Develop Future Social Entrepreneurs. In Z. Fields (Ed.), *Incorporating Business Models and Strategies into Social Entrepreneurship* (pp. 154–175). Hershey, PA: IGI Global. doi:10.4018/978-1-4666-8748-6.ch010

Sun, Z. (2016). A Framework for Developing Management Intelligent Systems. *International Journal of Systems and Service-Oriented Engineering*, *6*(1), 37–53. doi:10.4018/IJSSOE.2016010103

Swami, B., & Mphele, G. T. (2016). Problems Preventing Growth of Small Entrepreneurs: A Case Study of a Few Small Entrepreneurs in Botswana Sub-Urban Areas. In N. Baporikar (Ed.), *Handbook of Research on Entrepreneurship in the Contemporary Knowledge-Based Global Economy* (pp. 479–508). Hershey, PA: IGI Global. doi:10.4018/978-1-4666-8798-1.ch020

Tabach, A., & Croteau, A. (2017). Configurations of Information Technology Governance Practices and Business Unit Performance. *International Journal of IT/Business Alignment and Governance*, *8*(2), 1–27. doi:10.4018/IJITBAG.2017070101

Talaue, G. M., & Iqbal, T. (2017). Assessment of e-Business Mode of Selected Private Universities in the Philippines and Pakistan. *International Journal of Online Marketing*, *7*(4), 63–77. doi:10.4018/IJOM.2017100105

Tam, G. C. (2017). Project Manager Sustainability Competence. In *Managerial Strategies and Green Solutions for Project Sustainability* (pp. 178–207). Hershey, PA: IGI Global. doi:10.4018/978-1-5225-2371-0.ch008

Tambo, T. (2018). Fashion Retail Innovation: About Context, Antecedents, and Outcome in Technological Change Projects. In I. Management Association (Ed.), Fashion and Textiles: Breakthroughs in Research and Practice (pp. 233-260). Hershey, PA: IGI Global. doi:10.4018/978-1-5225-3432-7.ch010

Tambo, T., & Mikkelsen, O. E. (2016). Fashion Supply Chain Optimization: Linking Make-to-Order Purchasing and B2B E-Commerce. In S. Joshi & R. Joshi (Eds.), *Designing and Implementing Global Supply Chain Management* (pp. 1–21). Hershey, PA: IGI Global. doi:10.4018/978-1-4666-9720-1.ch001

Tandon, K. (2016). Innovative Andragogy: The Paradigm Shift to Heutagogy. In S. Tiwari & L. Nafees (Eds.), *Innovative Management Education Pedagogies for Preparing Next-Generation Leaders* (pp. 238–257). Hershey, PA: IGI Global. doi:10.4018/978-1-4666-9691-4.ch014

Tantau, A. D., & Frăţilă, L. C. (2018). Information and Management System for Renewable Energy Business. In *Entrepreneurship and Business Development in the Renewable Energy Sector* (pp. 200–244). Hershey, PA: IGI Global. doi:10.4018/978-1-5225-3625-3.ch006

Teixeira, N., Pardal, P. N., & Rafael, B. G. (2018). Internationalization, Financial Performance, and Organizational Challenges: A Success Case in Portugal. In L. Carvalho (Ed.), *Handbook of Research on Entrepreneurial Ecosystems and Social Dynamics in a Globalized World* (pp. 379–423). Hershey, PA: IGI Global. doi:10.4018/978-1-5225-3525-6.ch017

Trad, A., & Kalpić, D. (2016). The E-Business Transformation Framework for E-Commerce Architecture-Modeling Projects. In I. Lee (Ed.), *Encyclopedia of E-Commerce Development, Implementation, and Management* (pp. 733–753). Hershey, PA: IGI Global. doi:10.4018/978-1-4666-9787-4.ch052

Trad, A., & Kalpić, D. (2016). The E-Business Transformation Framework for E-Commerce Control and Monitoring Pattern. In I. Lee (Ed.), *Encyclopedia of E-Commerce Development, Implementation, and Management* (pp. 754–777). Hershey, PA: IGI Global. doi:10.4018/978-1-4666-9787-4.ch053

Trad, A., & Kalpić, D. (2018). The Business Transformation Framework, Agile Project and Change Management. In M. Khosrow-Pour, D.B.A. (Ed.), Encyclopedia of Information Science and Technology, Fourth Edition (pp. 620-635). Hershey, PA: IGI Global. doi:10.4018/978-1-5225-2255-3.ch054

Trad, A., & Kalpić, D. (2018). The Business Transformation and Enterprise Architecture Framework: The Financial Engineering E-Risk Management and E-Law Integration. In B. Sergi, F. Fidanoski, M. Ziolo, & V. Naumovski (Eds.), *Regaining Global Stability After the Financial Crisis* (pp. 46–65). Hershey, PA: IGI Global. doi:10.4018/978-1-5225-4026-7.ch003

Turulja, L., & Bajgoric, N. (2018). Business Continuity and Information Systems: A Systematic Literature Review. In N. Bajgoric (Ed.), *Always-On Enterprise Information Systems for Modern Organizations* (pp. 60–87). Hershey, PA: IGI Global. doi:10.4018/978-1-5225-3704-5.ch004

van Wessel, R. M., de Vries, H. J., & Ribbers, P. M. (2016). Business Benefits through Company IT Standardization. In K. Jakobs (Ed.), *Effective Standardization Management in Corporate Settings* (pp. 34–53). Hershey, PA: IGI Global. doi:10.4018/978-1-4666-9737-9.ch003

Vargas-Hernández, J. G. (2017). Professional Integrity in Business Management Education. In N. Baporikar (Ed.), *Management Education for Global Leadership* (pp. 70–89). Hershey, PA: IGI Global. doi:10.4018/978-1-5225-1013-0.ch004

Vasista, T. G., & AlAbdullatif, A. M. (2017). Role of Electronic Customer Relationship Management in Demand Chain Management: A Predictive Analytic Approach. *International Journal of Information Systems and Supply Chain Management*, *10*(1), 53–67. doi:10.4018/IJISSCM.2017010104

Vergidis, K. (2016). Rediscovering Business Processes: Definitions, Patterns, and Modelling Approaches. In P. Papajorgji, F. Pinet, A. Guimarães, & J. Papathanasiou (Eds.), *Automated Enterprise Systems for Maximizing Business Performance* (pp. 97–122). Hershey, PA: IGI Global. doi:10.4018/978-1-4666-8841-4.ch007

Vieru, D., & Bourdeau, S. (2017). Survival in the Digital Era: A Digital Competence-Based Multi-Case Study in the Canadian SME Clothing Industry. *International Journal of Social and Organizational Dynamics in IT*, *6*(1), 17–34. doi:10.4018/IJSODIT.2017010102

Vijayan, G., & Kamarulzaman, N. H. (2017). An Introduction to Sustainable Supply Chain Management and Business Implications. In M. Khan, M. Hussain, & M. Ajmal (Eds.), *Green Supply Chain Management for Sustainable Business Practice* (pp. 27–50). Hershey, PA: IGI Global. doi:10.4018/978-1-5225-0635-5.ch002

Vlachvei, A., & Notta, O. (2017). Firm Competitiveness: Theories, Evidence, and Measurement. In A. Vlachvei, O. Notta, K. Karantininis, & N. Tsounis (Eds.), *Factors Affecting Firm Competitiveness and Performance in the Modern Business World* (pp. 1–42). Hershey, PA: IGI Global. doi:10.4018/978-1-5225-0843-4.ch001

von Rosing, M., Fullington, N., & Walker, J. (2016). Using the Business Ontology and Enterprise Standards to Transform Three Leading Organizations. *International Journal of Conceptual Structures and Smart Applications*, *4*(1), 71–99. doi:10.4018/IJCSSA.2016010104

von Rosing, M., & von Scheel, H. (2016). Using the Business Ontology to Develop Enterprise Standards. *International Journal of Conceptual Structures and Smart Applications*, *4*(1), 48–70. doi:10.4018/IJCSSA.2016010103

Walczak, S. (2016). Artificial Neural Networks and other AI Applications for Business Management Decision Support. *International Journal of Sociotechnology and Knowledge Development*, *8*(4), 1–20. doi:10.4018/IJSKD.2016100101

Wamba, S. F., Akter, S., Kang, H., Bhattacharya, M., & Upal, M. (2016). The Primer of Social Media Analytics. *Journal of Organizational and End User Computing*, *28*(2), 1–12. doi:10.4018/JOEUC.2016040101

Wang, C., Schofield, M., Li, X., & Ou, X. (2017). Do Chinese Students in Public and Private Higher Education Institutes Perform at Different Level in One of the Leadership Skills: Critical Thinking?: An Exploratory Comparison. In V. Wang (Ed.), *Encyclopedia of Strategic Leadership and Management* (pp. 160–181). Hershey, PA: IGI Global. doi:10.4018/978-1-5225-1049-9.ch013

Wang, F., Raisinghani, M. S., Mora, M., & Wang, X. (2016). Strategic E-Business Management through a Balanced Scored Card Approach. In I. Lee (Ed.), *Encyclopedia of E-Commerce Development, Implementation, and Management* (pp. 361–386). Hershey, PA: IGI Global. doi:10.4018/978-1-4666-9787-4.ch027

Wang, J. (2017). Multi-Agent based Production Management Decision System Modelling for the Textile Enterprise. *Journal of Global Information Management, 25*(4), 1–15. doi:10.4018/JGIM.2017100101

Wiedemann, A., & Gewald, H. (2017). Examining Cross-Domain Alignment: The Correlation of Business Strategy, IT Management, and IT Business Value. *International Journal of IT/Business Alignment and Governance, 8*(1), 17–31. doi:10.4018/IJITBAG.2017010102

Wolf, R., & Thiel, M. (2018). Advancing Global Business Ethics in China: Reducing Poverty Through Human and Social Welfare. In S. Hipsher (Ed.), *Examining the Private Sector's Role in Wealth Creation and Poverty Reduction* (pp. 67–84). Hershey, PA: IGI Global. doi:10.4018/978-1-5225-3117-3.ch004

Wu, J., Ding, F., Xu, M., Mo, Z., & Jin, A. (2016). Investigating the Determinants of Decision-Making on Adoption of Public Cloud Computing in E-government. *Journal of Global Information Management, 24*(3), 71–89. doi:10.4018/JGIM.2016070104

Xu, L., & de Vrieze, P. (2016). Building Situational Applications for Virtual Enterprises. In I. Lee (Ed.), *Encyclopedia of E-Commerce Development, Implementation, and Management* (pp. 715–724). Hershey, PA: IGI Global. doi:10.4018/978-1-4666-9787-4.ch050

Yablonsky, S. (2018). Innovation Platforms: Data and Analytics Platforms. In *Multi-Sided Platforms (MSPs) and Sharing Strategies in the Digital Economy: Emerging Research and Opportunities* (pp. 72–95). Hershey, PA: IGI Global. doi:10.4018/978-1-5225-5457-8.ch003

Yusoff, A., Ahmad, N. H., & Halim, H. A. (2017). Agropreneurship among Gen Y in Malaysia: The Role of Academic Institutions. In N. Ahmad, T. Ramayah, H. Halim, & S. Rahman (Eds.), *Handbook of Research on Small and Medium Enterprises in Developing Countries* (pp. 23–47). Hershey, PA: IGI Global. doi:10.4018/978-1-5225-2165-5.ch002

Zanin, F., Comuzzi, E., & Costantini, A. (2018). The Effect of Business Strategy and Stock Market Listing on the Use of Risk Assessment Tools. In *Management Control Systems in Complex Settings: Emerging Research and Opportunities* (pp. 145–168). Hershey, PA: IGI Global. doi:10.4018/978-1-5225-3987-2.ch007

Zgheib, P. W. (2017). Corporate Innovation and Intrapreneurship in the Middle East. In P. Zgheib (Ed.), *Entrepreneurship and Business Innovation in the Middle East* (pp. 37–56). Hershey, PA: IGI Global. doi:10.4018/978-1-5225-2066-5.ch003

About the Contributors

Tarryn Kille is a Senior Lecturer at the University of Southern Queensland (USQ) lecturing in aviation management, human factors and commercial UAV operations. Tarryn commenced her aviation career as a professional pilot and flight instructor. Whilst working in the industry, she completed an undergraduate degree in aviation at Curtin University and a Masters of Business Administration, majoring in Aviation Management. Prior to joining USQ, Tarryn was appointed as Program Director and Senior Lecturer at Griffith University. Tarryn served as Manager, Policy and Business Development, with the Australian Civil Aviation Safety Authority and held membership on the Operations Panel with the International Civil Aviation Organization.

Paul Bates has held the positions of Chair of the Qantas Group Flight Training Research board and a consultant to Qantas Group Flight Training. He has been a member of the Industry Board of Aviation Australia. He has previously been a member of the International Advisory Board of the Singapore Association of Aviation Focused Enterprises, the Queensland Government's Aviation Reference Group and President of the Australasian University Aviation Association. He is currently a member of the Education and Research Committee and The Council of Associates of the International Air Cargo Association. Dr. Bates is the Head of Aviation at the University of Southern Queensland and previous Chairs the outreach committee of ICAO's Next Generation of Aviation Professionals Task Force as well as a member of the IPTC Outreach workstream.

Seung Yong (Paul) Lee is a senior lecturer at University of Southern Queensland, Australia. Before becoming a full-time academic, he was a flight instructor and an assistant flight examiner for almost 10 years. He is currently in the final stage of completing his PhD. His research interest is in the field of human factors and flight training in general aviation.

* * *

Berk Anbaroglu received his PhD from the Civil, Environmental and Geomatics Engineering Department of University College London in 2013. His main research interests include spatio-temporal data mining and citizen science. Specifically, he is interested in developing novel methods to enhance state-of-the-practice solutions on health care and intelligent transportation systems. He is actively involved in international organisations such as ISPRS and COST EU. He has also been delivering courses focusing on the various aspects of geoinformatics at the Department of Geomatics Engineering at Hacettepe University, Turkey since 2015. The courses he delivers include programming Python, data structures and algorithms, database management systems, geographical information systems and programming in geographical information systems.

Dmitriy Bondarev, simulator instructor of Air Navigation Systems Department at the Institute of Aeronavigation of the National Aviation University. Education: Air Traffic Service (NAU), postgraduate education PhD. Occupational skills covered: Air traffic services, Air Traffic Control, Procedures and Phraseology of Air Traffic Control, Aviation English, Meteorology, Human Factors in aviation, Aircraft, Safety of Flights, Decision Making and Radio Navigation facilities. Required Actions.

Aleksandar Erceg, Ph.D., is an Assistant Professor at the Faculty of Economics in Osijek, J.J. Strossmayer University of Osijek. He previously worked as deputy purchasing manager at two local production companies in Osijek (Saponia d.d. and Kandit d.o.o.). During his 20 years of experience in Saponia and Kandit, his tasks included organization of purchasing departments and supply chain management, cooperation with suppliers, customs authorities and logistics companies. He was part of the project team for the implementation of the ISO 9001 and 14001 standards in Saponia, and the IFS standard in Kandit. He has published more than 30 scientific articles in peer-reviewed journals. His research focuses on entrepreneurship, franchising, operations management, and lean thinking. As part of the team of the Franchise Centre Osijek, he has conducted more than 30 seminars on franchising throughout Croatia and advises entrepreneurs at various stages of implementation of franchising in their business. He is teaching several courses, including Operations Management, Franchising Entrepreneurship, and Corporate Entrepreneurship at various levels of study programmes. In addition to regular education, he is committed to continuous learning in the field of entrepreneurship, operations management, lean thinking, and franchising, by participating in domestic and international educational programmes. He earned his Ph.D. in Management from Faculty of Economics in Osijek in 2012. He is a member of the Croatian Association of Purchasing (HUND).

Isaac Levi Henderson, MAv, is an Assistant Lecturer teaching in the areas of human factors, aeronautics and air transport management. He is working towards a PhD at Massey University. Isaac also provides consultancy services to organisations wishing to be certificated as commercial unmanned aircraft operators. His research interests are in the areas of tourism and air transport management, with a particular focus on consumer perspectives. His current research focusses on defining the brand construct, airline brand choice and the creation of airport brands. Isaac is also interested in technological innovation within the aviation industry, in particular, unmanned aircraft and hybrid airships.

Robyn Kamira is Founder and CEO Incredible Skies + Medical Drones Aotearoa + Paua Interface Ltd: Te Rarawa (tribal affiliation), BSocSci (computer science, Māori), PGDip (Info systems), Board Member and Full member of Institute of IT Professionals, Hi-Tech Awards Judge 2017-2019, emerging technologies, IT strategies + project management.

Zafer Kilic is the experienced head of International Relations/Business with a demonstrated history of working in higher education. Skilled in Microsoft Word, Educational Technology, English and Non-Governmental Organizations. Strong community and social service professional with a Bachelor's degree in focused in International trade and logistics from Hasan Kaloyncu University. Currently finishing MA thesis on SWOT Analyses of Transportation in Turkey.

David Murray Kille is a casual lecturer in the Aviation discipline of the School of Commerce at the University of Southern Queensland. Murray currently lectures in Safety Management, Human Factors, International Aviation Systems and Safety Culture in Aviation. Murray has over thirty years of experience in the aviation industry. He has held senior management and training roles in a range of sub-sectors including international airlines, corporate and general aviation. As Chief Pilot and Head of Check and Training, Murray has developed a keen appreciation of the development and implementation of safety management systems and the impact of human factors on safety.

Doo Hwan Kim is originally from Seoul, South Korea. He is now an worldwide famous professor in the field air and space law. His foreign 59 Articles written by English, Chinese and Japanese language in the field of air and space law etc. has been published by the famous law Journal of USA, the U.K. Canada, Germany, The Netherlands, China, Japan and India etc. as well as domestic 119 Articles publishd by the Korean famous Law Journal in the field of air and space policy etc. His foreign and domestic articles are total 178 articles. He has published three books in the field of "air & space law and policy" by the Hague, The Netherlands and the Korean Publication Co. from 2,000 to 2011. He had got J.D. & LL.M degree from Seoul National University on 1953-1959. He had got J.S.D, Degree from Graduate School of Kyung Hee University in Seoul, Korea on 1980-1984. He was a Professor & Dean (6 years) of Law College, Soongsil University, Seoul, Korea on 1981-1999. He was a Visiting Scholar, Law School, University of California at Los Angeles (UCLA), Washington College of Law, The American University in USA and Institute of Air and Space Law, McGill University (Montreal, Canada) on 1990-91. He was a Visiting Prof.of Korea Aerospace University on 2001-2009 and Chuo Gakuin University in Japan on 2000-2013 as well as Honorary Prof. of Gujarat National Law University in India on Nov. 2004~2010. He received a "National Gold Magnolia Medal" from President of the Korean Government on December 1994 as well as an "Outstanding Contribution Award" from Director of Institute of Space Law under Beijing Institute of Technology in China on October, 2016. His Biography has published to the Asia/Pacific-Who's-Who (India) from 1993 to present and Who's Who in the World (USA) from 1999 to present. He was a President of the Korean Association of Air and Space Law 1993-1999. He was a legal adviser of Prime Minister office on1980-1983 and Ministry of Transportation on 1980-85 as well as Ministry of Justice of the Korean Government on 1985-1989. He is now a Honorary President, Korea Society of Air & Space Law and Policy, Seoul. Prof. Kim is teaching now "air and space law" Law School of the Beijing Institute of Technology and Nanjing University of Aeronautics and Astronautics in China as a Visiting Professor.

Kristina Marintseva, 2019 - Current Coventry University Lecturer in Aviation Management, Faculty of Engineering, Environment & Computing 2017 – 2018 National Aviation University Acting Dean of the Faculty of Transport Technologies Encouraging high teaching standards, methods of evaluating teaching effectiveness in the Faculty of Transport Technology. Collaborating with other University Faculties, as well as establishing and maintaining contacts with other institutions and academic bodies. Responsible to the Rector of the University for Faculty administration and performs Faculty Dean duties in accordance with the policies and procedures of the National Aviation University. 2015 - 2019 National Aviation University Professor of the Air Transportation Management Department Scientific research in air transportation management: analysis, mathematical modelling (2000 – Current). Lecturer in Technologies of Air Transportation; Aviation Management and Tourism. 2002 – 2015 National Aviation University Associate Professor of the Air Transportation Management Department Lecturer in General Theory of Systems and System Analysis, Research Methods, Air Transportation of Passenger and Cargo. Supervising of the postgraduate educational projects in Air Transportation Management (2002 – 2016). Supervising of syllabi development for the Air Transportation Management Department. 2000 - 2002 National Aviation University PhD Study, Lecturer / Part-time Employment Scientific research in the Modeling of Aircraft Leasing. Lecturer in Cargo Air Transportation.

Savern Reweti, PhD, is a Senior Lecturer in aviation management at Massey University's School of Aviation. During his previous service in the Royal New Zealand Air Force he developed a PC-based IFR/VFR procedural simulator and an air traffic control module for the ab-initio pilot training programme. This led to further research into the development of low cost PC-based aviation training devices (PCATD's). Additional projects included the development of customised PCATD systems for the Massey University School of Aviation and the Auckland Rescue Helicopter Trust. His current research focuses on the development of low-cost twin-engine flight simulation devices, low-cost force-feedback flight controls and the development of iPad apps for automating the calculation of take-off & landing chart data for pilot training.

Tetiana Shmelova, PhD, DSc., Professor of Department of Air Navigation Systems in National Aviation University (Ukraine). Thesis in the field of Scientific and Methodological Basis of Decision-Making in the Air Navigation System (2012). Areas of Scientific Interests: Mathematical Models of Decision-Making by Human operator, especially in Emergency Situation; Decision Support System; Research of Air Navigation system as Socio-technical system; Aviation Sociometry and Socionics; Management and Marketing in aviation; Information technology and Informatics of Decision-Making; Problems of Human factors in aviation. Author of more than 250 scientific articles and guides (about 200 articles is in Ukrainian and Russian, about 45 articles in English, 14 methodical manuals, 12 copyright certificates for computer programs), 14 monographs in fields of aviation, economics, mathematics, theory of system. Teaching courses: Theory of decision-making, Mathematical Programming, Effectiveness of Air Traffic Management.

Peter Tatham joined the (UK) Royal Navy as a logistics officer in 1970, and served in a variety of appointments during a career of some 35 years during which he rose to the rank of Commodore. Following his retirement, he joined Cranfield University where he completed his doctoral thesis which investigated the role of shared values within UK military supply networks. This work received the 2010 Emerald/EFMD award for the year's best logistics/supply network management-related PhD. Peter joined Griffith University in July 2010, and was appointed as a Professor in July 2015. He teaches and researches in the fields of commercial and humanitarian logistics, and is a member of the Editorial Boards Journal of Humanitarian Logistics and Supply Chain Management, the International Journal of Physical Distribution and Logistics Management and the Journal of Defense Analytics and Logistics.

Index

T

U

V

W

www.ingramcontent.com/pod-product-compliance
Lightning Source LLC
LaVergne TN
LVHW080341011026
849161LV00009B/2142